Photonics Meeting 2019

2nd Annual Conference and Workshop

Journal of Physics: Conference Series Volume 1371

Terengganu, Malaysia
8 – 9 July 2019

Editors:

Mohd Fauzi Ab Rahman
Cheng Xiau San
Anas Abdul Latiff
Sulaiman Wadi Harun

ISBN: 978-1-7138-1128-2
ISSN: 1742-6588

Printed from e-media with permission by:

Curran Associates, Inc.
57 Morehouse Lane
Red Hook, NY 12571

Some format issues inherent in the e-media version may also appear in this print version.

Institute of Physics
Dirac House, Temple Back
Bristol BS1 6BE UK

Phone: 44 1 17 929 7481
Fax: 44 1 17 920 0979

techtracking@iop.org

Additional copies of this publication are available from:

Curran Associates, Inc.
57 Morehouse Lane
Red Hook, NY 12571 USA
Phone: 845-758-0400
Fax: 845-758-2633
Email: curran@proceedings.com
Web: www.proceedings.com

TABLE OF CONTENTS

Journal of Physics: Conference Series **1371** (2019) 011001 doi:10.1088/1742-6596/1371/1/011001

Preface

Photonics Meeting 2019 (PM19) is an annual event organized by the Optical Society of Malaysia (OSM) in celebrating UNESCO International Day of Light. It was held on 8th and 9th July 2019 at Universiti Teknologi MARA, Dungun, Terengganu, Malaysia. This two-day event consists of a conference and a workshop that receives nearly 40 participants from the photonics and optics field. The primary goal of PM19 is to promoting research and development activities, and to exchange scientific knowledge among the participants. A compilation of articles that represent the significant results of the conference has been selected to publish in this volume.

Editors
MOHD FAUZI AB RAHMAN
CHENG XIAU SAN
ANAS ABDUL LATIFF
SULAIMAN WADI HARUN

PM19 activities

Journal of Physics: Conference Series **1371** (2019) 011001 doi:10.1088/1742-6596/1371/1/011001

Keynote speaker

Participant Part 1

Journal of Physics: Conference Series **1371** (2019) 011001 doi:10.1088/1742-6596/1371/1/011001

Participant Part 2

Workshop Session

Journal of Physics: Conference Series **1371** (2019) 011001 doi:10.1088/1742-6596/1371/1/011001

Organizing Commitee

List of Committee

Advisor
Sulaiman Wadi Harun, Malaysia (University of Malaya)

Conference Chair
Mohd Hafiz Bin Jali (University of Malaya)

Secretary
Nur Farhanah Zulkipli (University of Malaya)
Ahmad Haziq Aiman Rosol (Optical Society of Malaysia)

Bursary
Maslinda Mat Sharif (University of Malaya)
Hazli Rafis Abdul Rahim (Universiti Teknikal Malaysia Melaka)

Registration
Md Ashadi Md Johari (University of Malaya)
Aminah Ahmad (University of Malaya)
Siti Halma Johari (Universiti Teknikal Malaysia Melaka)

Logistic & Local Arrangment
Baktiar Musa (Universiti Teknologi MARA, Dungun)
Zulzilawati Jusoh (Universiti Teknologi MARA, Dungun)

Journal of Physics: Conference Series **1371** (2019) 011001 doi:10.1088/1742-6596/1371/1/011001

Marketing & Publicity
Ahmad Razif Muhammad (Optical Society of Malaysia)
Muhammad Taufiq Ahmad (Optical Society of Malaysia)

Publication & Technical Program
Mohd Fauzi Ab Rahman (University of Malaya)
Cheng Xiau San (KDU University College)
Anas Abdul Latiff (Universiti Teknikal Malaysia Melaka)

Program Coordinator & Protocol
Mohamad Faizal Baharom (University of Malaya)
Ummu Umairah Mohd Ali (Universiti Teknologi MARA, Dungun)
Farina Saffa Mohamad Samsamnun (University of Malaya)

Media & Communication
Haziezol Helmi Yusof (University of Malaya)
Muhammad Farid Mohd Rosdi (University of Malaya)
Mohd Adzimnuddin (Universiti Teknikal Malaysia Melaka)

Peer review statement

All papers published in this volume of *Journal of Physics: Conference Series* have been peer reviewed through processes administered by the proceedings Editors. Reviews were conducted by expert referees to the professional and scientific standards expected of a proceedings journal published by IOP Publishing.

Journal of Physics: Conference Series **1371** (2019) 012001 doi:10.1088/1742-6596/1371/1/012001

Microsecond pulse erbium-doped fiber laser using WS_2 deposited on D-shaped fiber fabricated by polishing wheel technique

A A A Jafry[1], N Kasim[1], Y Munajat[1], R A M Yusoff[1], S W Harun[2]

[1] Department of Physics, Faculty of Science, Universiti Teknologi Malaysia, 81310 Skudai, Johor, Malaysia.
[2] Department of Electrical Engineering, University of Malaya, 50603 Kuala Lumpur, Malaysia

afiqarif.fst@gmail.com; k.nabilah@utm.my

Abstract. We demonstrate a stable Q-switching pulse train output centered at 1561 nm by implementing D-shaped fiber deposited tungsten disulphide (WS_2) as a saturable absorber in erbium-doped fiber laser. D-shaped fiber was fabricated using a polishing wheel technique with two stages polishing. Proposed Q-switched EDFL generates a maximum repetition rate of 47.28 kHz at a pulse duration of 5.4 μs. Under pump power of 104.6 mW, the laser consumed pulse energy of 99.4 nJ with an output power of 4.7 mW. This WS_2-D-shaped SA is excellence for the production of Q-switched laser source.

1. Introduction

A laser operating at microsecond to nanosecond pulse duration (Q-switched) is essential for various applications in industry. Conversion of continuous-wave into energetic short pulses is dependable on a nonlinear optical element called saturable absorber (SA). Passively Q-switched is first demonstrated using semiconductor saturable absorber mirrors (SESAMs). However, SESAMs has narrow tuning range due to the limited wavelength spectrum and complex fabrication method [1]. Thus, an all-fiber laser is introduced with the implementation of various materials as a SA such as carbon nanotubes (CNTs), graphene, topological insulators (TIs), and transition metal dichalcogenides (TMDs) including several others [2-6]. The all-fiber laser has shown excellence lasing performance due to compact structure, high-quality optical pulses and adjust-free configuration [7].

Among all materials, TMDs exhibit excellent saturable absorption due to the sizable band gap structure from bulk to 2-dimensional (2D) [8]. For instance, bulk-tungsten disulphide (WS_2) has an indirect band gap (1.3 eV) which can be engineered to form 2D-WS_2 with direct band gap (2.1 eV) [9]. Recently, it is proven that nano-sheets WS_2 exhibit high second order susceptibility and saturable absorption [10]. In addition, superior properties of monolayer WS_2 including high carrier mobility and strong orbit coupling [11] make it a suitable saturable absorber for the generation of Q-switched pulse laser.

There are two ways to implement SA into all-fiber laser cavity; sandwiching thin film structure in between fiber-ferrule and implementing SA material on a special fiber such as D-shaped fiber, microfiber and photonic-crystal fiber. The first technique has few disadvantages including limited nonlinear interaction length between material and light [12] as well as inducing mechanical and thermal damage to the end face of fiber connector [13]. Therefore, the mechanism of evanescent field interaction between light and material is introduced using D-shaped fiber. D-shaped fiber has a long nonlinear interaction length and high optical damage threshold which makes it superior for the generation of Q-switched [14]. Passively Q-switched fiber laser is introduced with the implementation of WS_2 on D-shaped fiber as a saturable absorber for erbium-doped fiber laser.

2. Preparation of D-shaped fiber deposited WS_2

ASE source was launched into single mode optical fiber (CORNING SMF-28) with core/cladding diameter of 9/125 μm while optical power meter (THORLABS) was connected at the other end to observe the insertion loss of the polished fiber. Single mode optical fiber was clamped using fiber holder at both ends to ensure less vibration during polishing. Rotating wheel with 5V DC motor was set up with 800-grid sandpaper and 1500-grid fine sandpaper pasted around the wheel. D-shaped fiber was polished using 800-grid sandpaper until desired insertion loss was obtained on optical power meter. The D-shaped fiber was then polished by 1500-grid fine sandpaper for 15 minutes. Two stages polishing were carried out to ensure that the polishing surface is smooth thus allowing good propagation of light inside the D-shaped fiber. D-shaped fiber has insertion loss of 3 dB, length of 1.8 mm and diameter of 64 μm. Experimental setup for D-shaped fiber fabrication technique is shown in Figure 1. WS_2 used was bought from Graphene Supermarket. A 3 μL WS_2 solution was dropped on D-shaped fiber polished surface using micropipette (EPPENDORF) with 10 mW laser diode pump launched into the laser cavity.

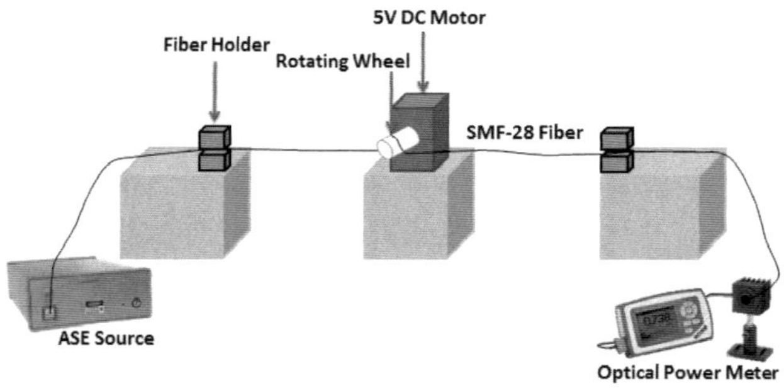

Figure 1. Experimental setup for D-shaped fiber polishing technique

3. Erbium Doped Fiber Laser Ring Cavity

Passively Q-switched erbium-doped fiber laser was generated using the setup as shown in Figure 2. A 980 nm laser diode was launched into 980/1550 nm wavelength division multiplexer (WDM). Pumped light was then propagated into 2.8 m erbium-doped fiber with an absorption coefficient of 23 dB/m, core diameter of 4 μm and numerical aperture of 0.16. The light was then directed into isolator to ensure unidirectional operation in the laser cavity. D-shaped fiber deposited WS_2 was placed inside the ring cavity and the light was then propagated to 90/10 coupler, where 90% of the light was allowed to propagate back inside the ring cavity. 10% of light was used for analysis purposes. Optical spectrum analyzer (YOKOGAWA, AQ6370D) with a resolution of 0.02 nm and optical power meter (THORLABS) was used to characterize output signal. A 350 MHz digital oscilloscope (GWINSTEK, GDS-3352) and 9 kHz-7.8 GHz RF spectrum analyser (ANRITSU, MS2683A) connected via a 1.2-GHz InGaAs photodetector were used for temporal characteristics analysis.

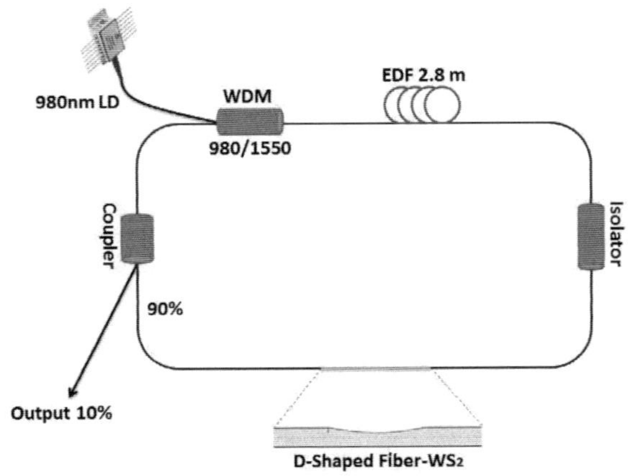

Figure 2. Experimental setup for Q-switched erbium-doped fiber laser using WS_2-D-shaped fiber as saturable absorber.

4. Result and Discussion

The laser cavity starts to generate Q-switched at pump power of 63.4 mW. Figure 3(a) shows continuous-wave spectrum centered at 1567 nm. Center wavelength shifted to shorter spectrum, 1561 nm when D-shaped fiber deposited WS_2 was incorporated into the laser cavity. This indicates lasing operation of developed laser cavity. Oscilloscope trace of Q-switched at pump power of 88 mW was shown in Figure 3(b) with inset of single pulse at the same pump power was also plotted. It is observed that the pulse generated has a full-width half maximum (FWHM) of 6.4 µs. RF spectrum shown in Figure 3(c) exhibit signal-to-noise ratio (SNR) of 52.06 dB which indicates stability of the pulses generated.

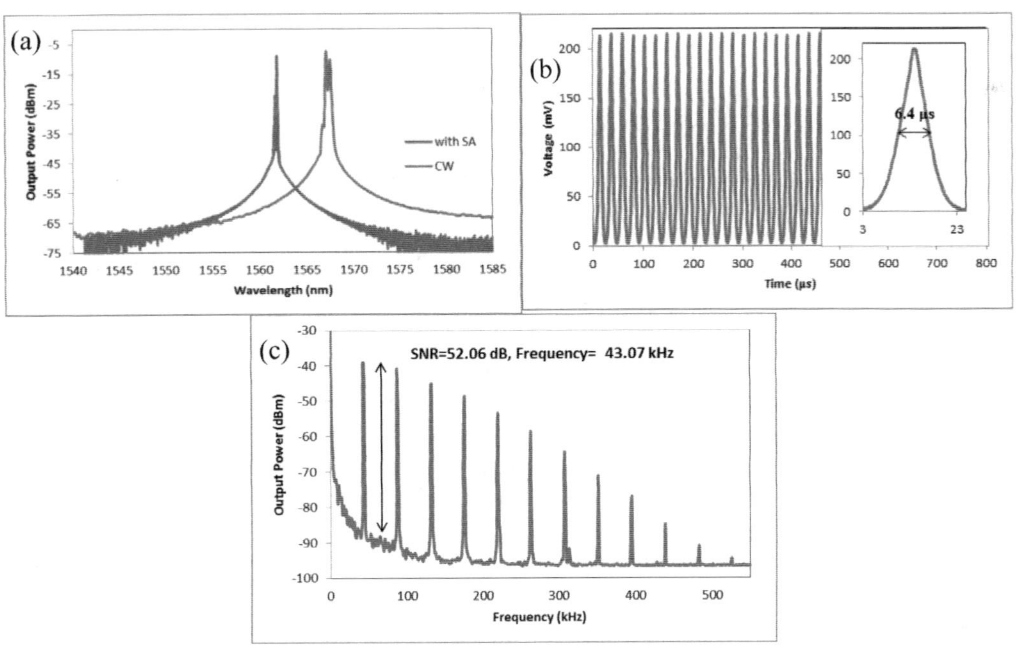

Figure 3. Spectral and temporal performances of the Q-switched laser at 88 mW pump power. (a) Output spectrum. (b) Typical oscilloscope trace with two pulses envelops (c) RF spectrum.

Self-started Q-switching visible at the input pump power of 63.4 mW and disappeared above pump power of 104.6 mW. As the pump power increased from 63.4 mW to 104.6 mW, the repetition rate of the pulses increases from 39.22 kHz to 47.28 kHz whereas the pulse width decrease from 8.1 µs to 5.4 µs. The relationship between pulse width and repetition rate with pump power was shown in Figure 4(a). Figure 4(b) shows measured output power which was 3.5 mW to 4.7 mW plotted against pump power. Whereas, calculated pulse energy from 88.73 nJ to 99.40 nJ was also shown in Figure 4(b).

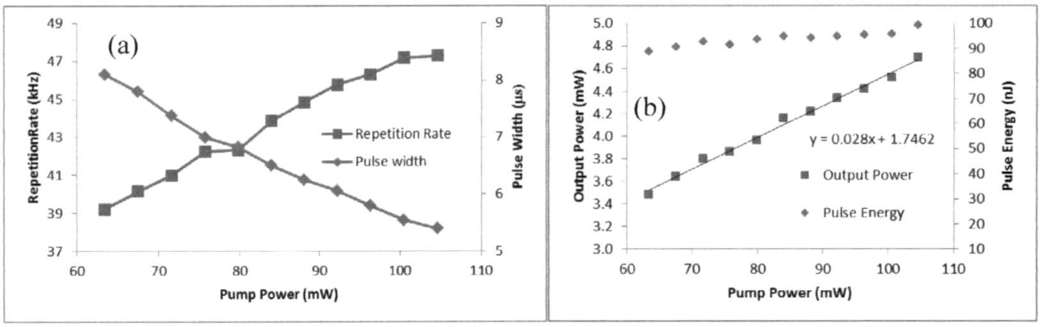

Figure 4. Q-switching performance (a) Repetition rate and pulse width against pump power (b) Output power and pulse energy against pump power.

5. Conclusion

We present Q-switched erbium-doped fiber laser with WS_2 deposited on D-shaped fiber incorporated into laser cavity as a saturable absorber. D-shaped fiber was prepared by polishing wheel technique with improved two stages polishing to ensure efficient light-matter interaction inside D-shaped fiber. With an appropriate amount of WS_2 deposited onto D-shaped fiber, stable Q-switched was initiated within 63.4 mW to 104.6 mW. The generated pulses exhibit the highest repetition rate of 47.28 kHz and the shortest pulse width of 5.4 µs. SNR measured was 52.06 dB indicating the stability of the laser. The maximum pulse energy generated was 99.4 nJ corresponds to pump power of 104.6 mW.

Acknowledgements

This research is fully supported by FRGS grant, R.J130000.7826.4F928. The authors fully acknowledged the Ministry of Education (MOE) Malaysia and Universiti Teknologi Malaysia for the approved fund which makes this important research viable and effective.

References

[1] Sun, Z., et al., *Graphene mode-locked ultrafast laser.* ACS nano, 2010. **4**(2): p. 803-810.

[2] Rusdi, M., et al. *Nickel Oxide as a Q-switcher for Short Pulsed Thulium Doped Fiber Laser Generation.* in *Journal of Physics: Conference Series*. 2019. IOP Publishing.

[3] Baharom, M., et al., *Lutetium oxide film as a passive saturable absorber for generating Q-switched fiber laser at 1570 nm wavelength.* Optical Fiber Technology, 2019. **50**: p. 82-86.

[4] Ab Rahman, M.F., et al., *Ultrashort Pulse Soliton Fiber Laser Generation With Integration of Antimony Film Saturable Absorber.* Journal of Lightwave Technology, 2018. **36**(16): p. 3522-3527.

[5] Reddy, P.H., et al., *Titanium dioxide doped fiber as a new saturable absorber for generating mode-locked erbium doped fiber laser.* Optik, 2018. **158**: p. 1327-1333.

[6] Rahman, M., et al., *An 8 cm long holmium-doped fiber saturable absorber for Q-switched fiber laser generation at 2-µm region.* Optical Fiber Technology, 2018. **43**: p. 67-71.

[7] Zhao, J., et al., *Three operation regimes with an L-band ultrafast fiber laser passively mode-locked by graphene oxide saturable absorber.* JOSA B, 2014. **31**(4): p. 716-722.

[8] Jafry, A., et al., *Q-switched Erbium-Doped Fiber Laser Using MoS 2 Deposited Side-polished*

D-shape Fiber. Nonlinear Optics, Quantum Optics: Concepts in Modern Optics, 2018. **49**.

[9] Khazaeinezhad, R., et al., *Femtosecond soliton pulse generation using evanescent field interaction through Tungsten disulfide (WS 2) film.* Journal of Lightwave Technology, 2015. **33**(17): p. 3550-3557.

[10] Kassani, S.H., et al., *All-fiber Er-doped Q-switched laser based on tungsten disulfide saturable absorber.* Optical Materials Express, 2015. **5**(2): p. 373-379.

[11] Ahmad, H., et al., *Passively Q-switched erbium-doped fiber laser at C-band region based on WS 2 saturable absorber.* Applied optics, 2016. **55**(5): p. 1001-1005.

[12] Steinberg, D., et al., *Graphene oxide and reduced graphene oxide as saturable absorbers onto D-shaped fibers for sub 200-fs EDFL mode-locking.* Optical Materials Express, 2018.

[13] Lee, E.J., et al., *Active control of all-fibre graphene devices with electrical gating.* Nature communications, 2015. **6**: p. 6851.

[14] Park, N.H., et al., *Monolayer graphene saturable absorbers with strongly enhanced evanescent-field interaction for ultrafast fiber laser mode-locking.* Optics express, 2015. **23**(15): p. 19806-19812.

Journal of Physics: Conference Series **1371** (2019) 012002 doi:10.1088/1742-6596/1371/1/012002

Development of portable solar storage device

M S Roslan[1], F Adan[1], M A Mohammad[1], N Norazman[1] , S Haizam[1] and Z Haider[2].

[1]Center for Diploma Studies, Universiti Tun Hussein Onn Malaysia, Hab Pendidikan Tinggi Pagoh, 84600 Panchor, Muar, Johor, Malaysia.
[2]Physics and Chemistry Department, Faculty of Science, Universiti Tun Hussein Onn Malaysia, Hab Pendidikan Tinggi Pagoh, 84600 Panchor, Muar, Johor, Malaysia.

sufi@uthm.edu.my

Abstract. The demanding for energy in Malaysia to use for all-purpose of small device charging has been developed. The purpose of this project is to develop portable solar storage (PSS) device with all the components of an off-grid solar station compact into a small portable handheld casing. The PSS is aimed for outdoor activity passionate such as hikers, campers and climbers that need a portable power charger that can charge their electronic gadgets while they carry on with their activities. The input current and voltage of the solar panel had been recorded over a span of two weeks. The generation of electricity is harvested from solar irradiation on 12 V and 18 V solar panel and stored upon 98 W energy storage beneath the solar panel. The power stored put on the test and capable to recharge up to 10,000 mAh capacity for one-time charging. As a comparison between 12V and 18 V solar panel, the result shows the average charging rate of the 12 V and 18 V solar panels could reach as high as 16.086 W/h and 13.35 W/h respectively which considered had reached 12.75% efficiency.

1. Introduction

As we all know, the demand for the usage of solar energy in Malaysia is not as high as the demand overseas even though some of us are aware of the advantages of solar energy. The sunlight is the main and biggest source of all source of energy [1]. Solar energy is free energy for the entire life. In order to consume and use it wisely, there are certain devices based on the solar panel have been invented for such purposes. The major benefit of Malaysia's climate is the exposure to sunlight increases the ability of usage of the renewable resources of energy such as the Sun. Malaysia is known as a country which located near the equator alongside with Thailand, Indonesia and Singapore. Peninsular Malaysia with the coordinates of 4°0'N 102°30'E is in the humid province. The characteristics of Malaysia's climates may include unvarying temperature, high humidity and abundant rainfall. The solar radiation received in January ranges from 6.5kWh/m^2 and decreases lower to 6.0kWh/m^2 in August [2]. The amount of exposure to sunlight in Malaysia qualifies the usage of solar panel as the source of electricity.

Humans depend on gadgets in daily lives same as how gadgets need batteries to stay alive and functioning well [3]. Even though we are off the grid, we have to admit we are living in a power-based world such cell phones, laptops and others devices especially for those who love to do outdoor activities [4]. According to Jake Walnut, 2017, while on outdoor adventures, the greatest way to stay plugged in is by portable power supply [5]. Portable power supply may include alternators and generators. Generators are originally used for backup power however, they are noisy, produce fumes and very unfriendly environmental [6]. The limitation of noise set by the national parks is to 60 decibels at fifty feet [7]. In contrast, the usage of solar panels or solar power for campers is better since it is a silent power supply and most importantly environmental-friendly where there is no release of toxic fumes [6]. Moreover, the US Department of Energy's National Renewable Energy Laboratories stated that the cost of utility-scale solar has fallen 30%. This falling cost is the reason why the utilization of solar increases by companies as the hardware can be bought at low prices [8]. Thus, we would like to develop a portable solar storage device which can be used as a power supply that depends on solar energy. This will prevent the pollutions made by the generators.

2. Methodology

The procedure to conduct this project is first, in the morning on the first day according to Figure 1.(a), a solar panel of 18V was connected to the solar charge controller (SCC). The initial voltage and resistance of the power supplied by the solar panel to the SCC were recorded by using a multimeter.

Every hour starts from 7.00 am till 12.00 pm, the reading of the voltage and resistance were recorded until the PSS battery capacity is fully charged which have been displayed by the SCC. The power of energy captured by the solar was calculated with the formula of P=IV and I=V/R. The experiment conducted during the day will be continued at night after all the data required have been collected

(a) (b)

Figure 1.(a) The PSS under the sunlight and (b) the PSS connected to a load

During the night, the experiment is set up where the SCC is being connected to the PSS battery and a load of 1400 mAh as shown in Figure 1.(b) The load is at 0% battery performance before being used in PSS for being able to compare its performance. The reading of voltage and resistance were recorded every hour from 7.00 pm till 12.00 am to calculate the power discharged to the load. The increments of the percentage of the load during charging also being recorded simultaneously with the reading of voltage and resistance.

The 18 V solar panel is used 3 times for loads of 1400 mAh, 4400 mAh and 10,000 mAh. The experiments were repeated for 3 days and 3 nights for the 18 V solar panel. All the procedures were repeated by charging the 18V solar panel to 12V solar panel. The 12V solar panel was also being used 3 times for load of 1400 mAh, 4400 mAh and 10,000 mAh. It was also repeated for 3 days and 3 nights for data collection.

3. Result and Discussion

The data collected on the charging and discharge of voltage on the Portable Solar Storage (PSS) has been done at Student's Residential College, University Tun Hussein Onn, Pagoh, Johor over the span of two weeks from 10th until 27th October 2018. The data was collected on the average from the 7 am until 3 pm depending on the weather and the time it takes for the power storage to reach maximum capacity. Two different voltages of solar panels used which were 12 V and 18 V where the data collection starts with the later. Furthermore, the desired altitude and latitude of Malaysia which is 19 m and 101.5°E respectively and the irradiation in the month of October according to the Meteorology Data (refer appendices) is 135.5 kWh/m^2 which makes it possible to harvest energy during the 6 hours span of the day. Also, contribute to the energy harvesting is the Ambient Temperature of the month of October which is 27.48°C. Both solar panels were setup up at a 20° angle for the best tilt.

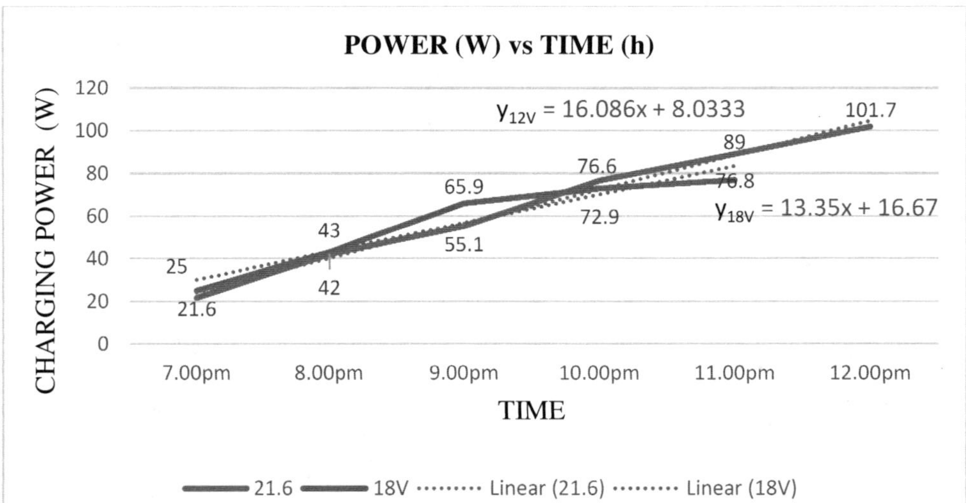

Figure 2. Charging power (W) against the time of different panels.

On average of both days, the 18 V solar panel managed to produce a power of 25.0 W at 7.00 am where the sun was about to radiate. Next, for the 12 V solar panel, it took 5 hours to completely charge the battery storage. It took a higher amount of power to charge the battery storage compared to using 18 V solar panel due to the small area of the 12 V solar panel. The charging rate of the 18 V solar panel is 13.35 W/h (177 Wh/m^2) whereas for the 12 V solar panel is 16.086 W/h. (765 Wh/m^2). This result showing 12 V panel provides great energy harvesting as compared with 18 V one.

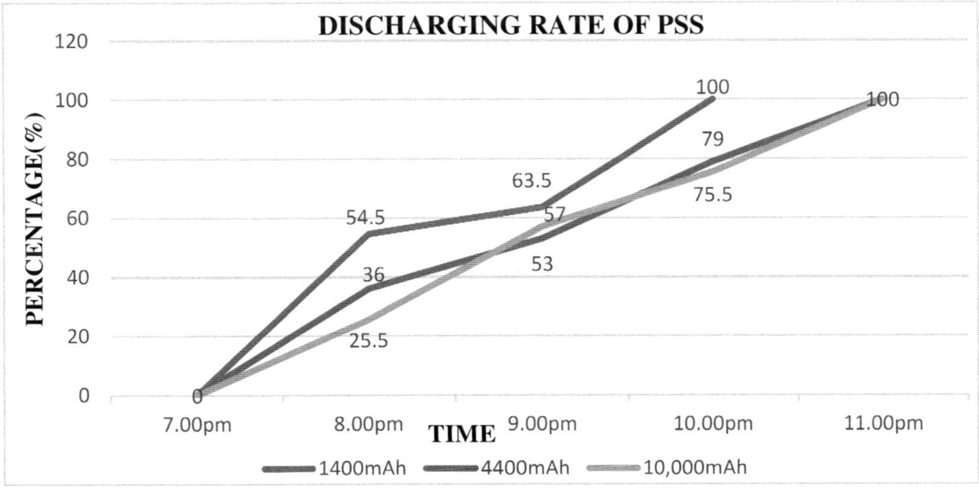

Figure 3. Percentage (%) of different load applied charging by PSS.

The data of PSS was collected at night starting from 7.00 pm with a 1400 mAh load capacity of a phone at 0% power percentage. Different from the 1400 mAh, the 4400 mAh and 10,000 mAh load capacity took 4 hours to charge until it reaches full capacity. The 4400 mAh load has a constant increase in percentage which is between the ranges of 17% to 26% until it reaches full percentage by 11.00 pm. The 10,000 mAh is able to charge at a constant pace as the 4400 mAh where by 9.00 pm it reaches 57% and 2 hours later it reaches 100%. The fast pace of charging was due to the power supplied by the solar panel when charging the PSS. The device had been charged earlier in the morning until afternoon by the 12V solar panel. After two attempts, the percentage of load for the 4400 mAh and 10,000 mAh load could not be completed due to low voltage released by the solar charge controller. This shows that the PSS is not suitable in being charged by a 12 V solar panel and the SCC cannot retrieve as the same amount of power as when the device is charged with the 18 V solar panel

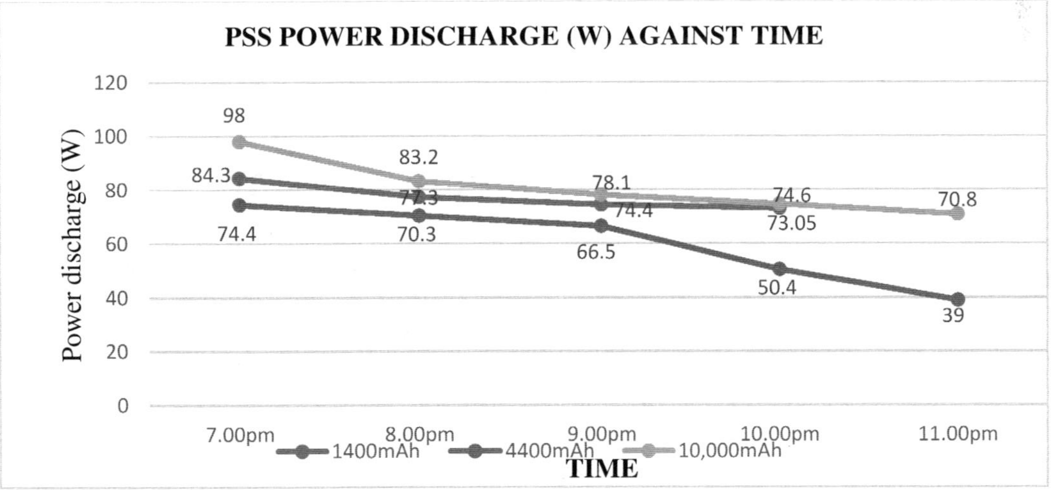

Figure 4. Graph of PSS power discharge (W) against the time of different load capacity.

This was done at night after the PSS was charged earlier in the morning until the afternoon with a 12 V and 18 V solar panel. The power discharge from the PSS causes a decrease in the voltage of storage capacity. It took 11.27 W to charge the load capacity of 1400 mAh meaning that the PSS has the ability to charge 1400 mAh load capacity twice in a day although not fully charged when the second time the load is charged when using 18 V solar panel. While discharging from the load of 1400 mAh, it needed 72.7 W to fully charge due to less power given by the 12 V solar panel means there were restrictions when doing that experiment. Attempts on discharging the 4400 mAh and 10,000 mAh load cannot be completed due to some technical errors in the solar charge controller.

4. Conclusions

Peninsular Malaysia which is situated near the equator has a great advantage of retrieving a 6 hours' sunlight in a day. This benefits the Portable Solar Storage (PSS) device in being able to generate energy and to supply power to the load. The PSS has an average charging rate of 16.086 W/h and 13.35 W/h of power rate generated by the 12 V and 18 V solar panels respectively. The power is enough to supply power to the 1400 mAh, 4400 mAh and 10,000 mAh load capacity.

Load with capacity 1400 mAh load such as a Nokia phone that can only be used for calling and messaging needed less power during charging to full load capacity. Compared to the 4400mAh load capacity such as the Smartphone and iPhone that uses many features such as WhatsApp and Instagram consume higher power during charging to full load capacity. For 10,000 mAh the power bank uses high power to reach full load capacity but took almost the same time to charge the same as the 4400 mAh. In this study, we had shown that the capability of the solar panel to harvest solar energy with efficiency of 12.75%.

References

[1] S. Ashok, "Solar Energy," 2018. [Online]. Available: https://www.britainnica.com/print/article/552905. .
[2] A. Shavavalipour, M. H. Hakemzadeh, K. Sopian, S. M. Haris and S. H. Zaidi, "New Formulation of the Estimation of Montly Average Daily Solar Irradiation for the Tropics: A case study for Peninsular Malaysia," Hindawi, pp. 1-2, 2013
[3] J. Capistrano, "Are We too Dependent on Technology?," XEN Life, 23 April 2016. [Online]. Available: https://www.google.com/amp/s/xenlife.com.au/are-we-being-too-dependent-on-technology/amp/.
[4] "Portable Rechargeable Power Supply - Perfect for Camping," Generator Power Source, [Online]. Available: https://generatorpowersource.com/portable-rechargeable-power-supply-perfect-for-camping/
[5] J. Walnut, "Ultimate Guide to the Best Portable Power for Camping," Beyond The Tent, 2017. [Online]. Available: https://www.beyondthetent.com/guide-to-portable-power-for-camping/.
[6] P. Quilty, "Solar VS Generators: The Pros & Cons," Advantures Groups Holdings, 23 September 2016. [Online]. Available: https://www.caravanworld.com.au/1609/solar-vs-generators-the-pros-and-cons.
[7] D. Gum, "Solar Panels or Generators for Campers and RVs - What's best?," Camper Report, 2017. [Online]. Available: https://camperreport.com/solar-panels-or-generators-for-campers-and-rvs-whats-best/.
[8] B. Jones, "Solar Power is Becoming Accessible, Dropping More Than a Quarter in Just One Year," Futurism, 20 September 2017. [Online]. Available: https://futurism.com/by-2100-our-carbon-usage-may-be-enough-to-trigger-a-10000-year-long-ecological-disaster.
[9] K. Ranabhat, L. Patrikeev, A. Antal'evna Revina, K. Andrianov, V. Lapshinsky and E. Sofronova, "An Introduction to Solar Cell Technology," pp. 482-488, 2016.
[10] STIKopedia, "Power Generation, Solar PV Panels," 2018. [Online]. Available: www.solarstik.com/stikopedia/pv-panels/.
[11] M. A. Green, K. Emery, Y. Hishikawa, W. Warta and E. D. Dunlop, Solar cell efficiency tables (version 48), John Wiley & Sons, Ltd, 2016.

Journal of Physics: Conference Series **1371** (2019) 012003 doi:10.1088/1742-6596/1371/1/012003

Photochemical reduction of carbonyl group of polyimide by 450 nm diode laser

M S Roslan[1,2], H Kamarudin[2], N Sharip[2], A F M Zin[2], A Muhamad[2], A Agam[3], N Nayan[1], and U Tariq[4]

[1]Microelectronic and Nanotechnology-Shamsuddin Research Centre (MiNT-SRC), Universiti Tun Hussein Onn Malaysia, 86400 Parit Raja, Batu Pahat Johor, Malaysia
[2]Center for Diploma Studies, Universiti Tun Hussein Onn Malaysia, Muar 84600, Malaysia.
[3]Physics Department, Faculty of Applied Science and Technology, Universiti Tun Hussein Onn Malaysia, Muar 84600, Malaysia
[4]Physics Department, Faculty of Science, Universiti Teknologi Malaysia, 81310 Skudai, Johor Bahru, Malaysia

sufi@uthm.edu.my

Abstract. In this article, we report directly patterned graphene nanoribbons by laser irradiation. The transformation of polymer into graphene nanostructure by breaking oxygen bonding and followed by graphitization of carbon atoms has been employed by using the photochemical technique. The scalable and fast growth of graphene nanoribbon was carried out by irradiating 450 nm diode laser directly onto polyimide polymer. Patterned graphene was characterized by Raman spectroscopy, Fourier transform infrared spectroscopy (FTIR) and Field emission scanning electron microscopy (FESEM). The ability of polyimide to absorb laser light in the near-ultraviolet region contribute to the tailoring of graphene nanoribbon which reduces oxygen bonding as well as paves way for the development of optoelectronics applications.

1. Introduction

Graphene is an emerging two-dimensional carbon allotrope that has attracted tremendous research interest. The unprecedented feature of graphene including being lightweight and highly flexible with high charge carrier mobility of 200000 $cm^2 V^{-1} s^{-1}$ [1] and Young's modulus of ~1.0 TPa [2] has lead to various potential future applications in photonics [3], optoelectronics [4] microelectronics [5], biosensors [6] and energy [7, 8]. Graphene has been produced in many different ways including mechanically exfoliation [9] chemical and thermal reduction [10], and chemical vapour deposition [11]. Laser-induced synthesis of graphene is being carried out using CO_2 laser [12], Nd:Yag laser [13], and KrF laser [14] but the synthesis using diode laser has not been reported yet to the best of our knowledge. Laser irradiation technique applied in the fabrication of integrated graphene microstructures has attracted increasing attention due to its advantages of single-step capability, rapid material processing speed, and large scan area. In some cases, the processing of graphene can be easily achieved in free-space intense laser heating. Numerous laser-assisted methods including pulsed and continuous lasers have been developed to produce graphene. These methods involve laser-induced unzipping of carbon nanotube [15], pulsed laser deposition [16], laser-induced chemical vapor deposition [17] and laser-

Journal of Physics: Conference Series **1371** (2019) 012003 doi:10.1088/1742-6596/1371/1/012003

induced epitaxial growth [18]. Until now, various integrated, flexible and wearable graphene-based architectures have been developed through direct growth and patterning, with laser irradiation technique for wide applications. In this study, we are investigating the fast synthesis of graphene by the photochemical reaction of polymer chain induced by laser diode irradiation highly potential for sensors, energy conversion, and optoelectronics applications.

2. Methodology

Polyimide was characterized by repeating the imide structure as a linear or heterocyclic unit. Polyimide is considered a strong synthetic polymer with astounding heat and chemical resistant. Local heating caused by laser irradiation brought local chemical/physical reactional changes in the polyimide structures. Laser irradiations broke oxygen-containing bonds of polyimide and new chemical bonds with the molecules/compounds around the local ambient environment are formed.

The transformation from polyimide to graphene via laser irradiation strategy usually follows two basic processes (i) first one is light absorption and photochemical removal of oxygen from the polyimide and (ii) structural reorganization of the newly formed, reduced carbon lattice into the planar, hexagonal, sp^2-conjugated graphene structure. The experimental setup for direct patterning graphene nanoribbon is shown in figure 1. Power supply connected with diode laser was used to control the output power of the laser. The laser head was equipped with an adjustable laser focus. The focus of the laser was set at 8 cm. The polyimide tape was attached on a glass substrate. Intense laser power up to 2500 mW was applied through polyimide tape. The polyimide tape absorbed the laser light and broke the oxygen bond. The remains C atoms then rearranged themselves to form graphene nanoribbon replacing polyimide.

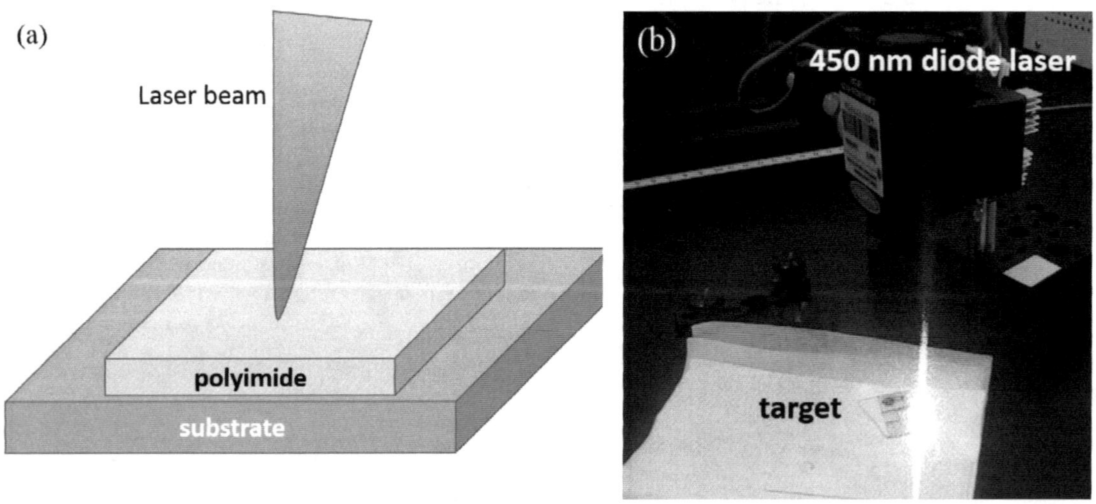

Figure 1. (a) Schematic diagram for laser irradiation and (b) Experimental setup for direct patterning of graphene by 450 nm diode laser.

3. Result and Discussion

Patterned graphene nanoribbon as grown by laser irradiation left a black mark on the glass substrate and are easily distinguishable by naked eyesight (figure 2). Intense laser heating left graphene with C-C bonding discernible with polyimide tape. As the polyimide tape was peeled off from substrate mechanically, the graphene remained on top of the substrate.

Journal of Physics: Conference Series **1371** (2019) 012003 doi:10.1088/1742-6596/1371/1/012003

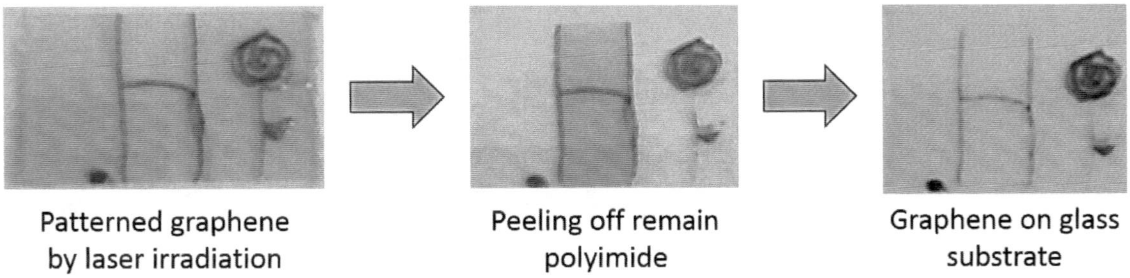

Figure 2. Process of graphene patterning on a glass substrate.

3.1. Electron microscopy

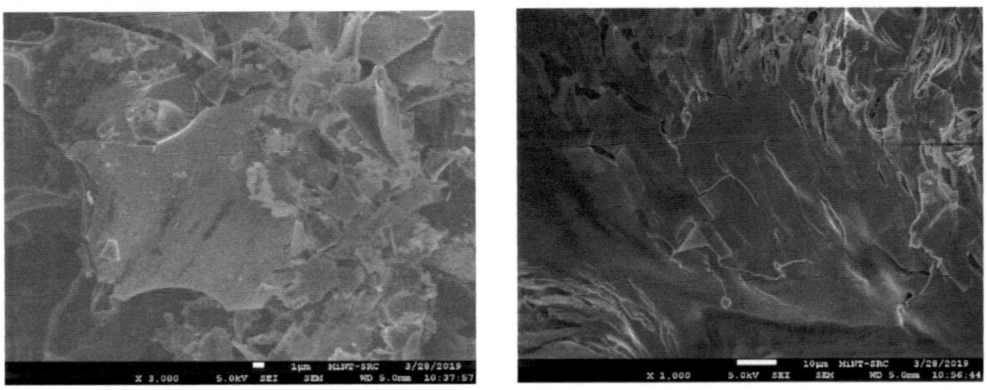

(a) (b)

Figure 3. FESEM images of carbon layer with different magnifications (a) ×3000 (b) ×1000

The details of the carbon atom that formed after laser irradiation has been captured by field emission scanning electron microscope (FESEM). The image of carbon atom under FESEM as shown in figure 3.

3.2. Raman spectroscopy

Raman spectroscopy used a laser to excite molecular atom of carbon. The scattering effect by lattice vibration gives the Raman signal. It's prior to emitting signal at 1580 cm^{-1} that indicate G band and signal at 1350 cm^{-1} of D band released by lattice vibration of carbon. The result of Raman spectroscopy for the carbon upon laser irradiation on polyimide are shown in figure 4.

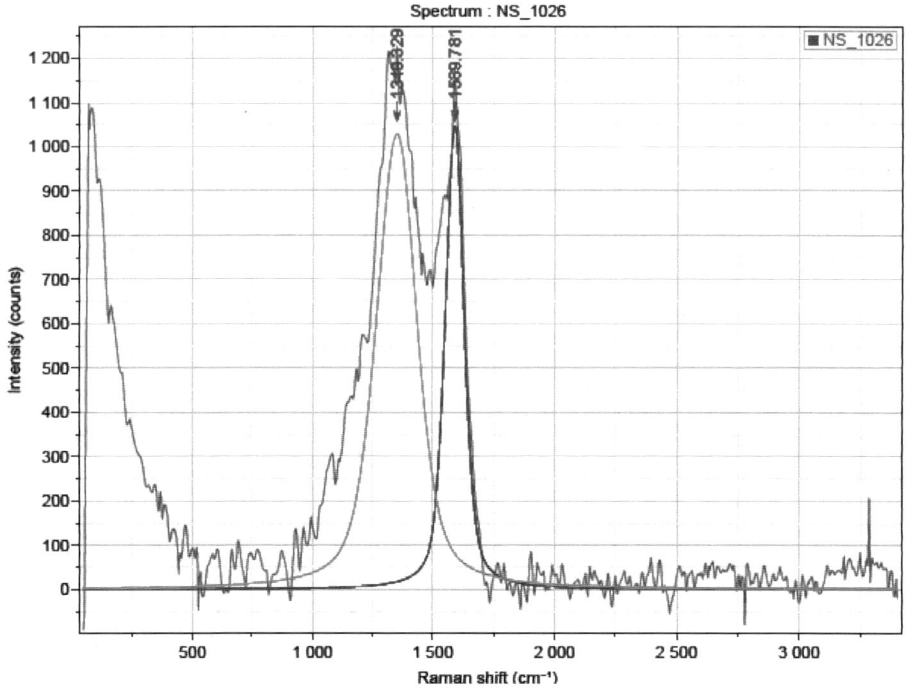

Figure 4. Raman spectrum of the irradiation area.

3.3. Fourier Transform Infrared Spectroscopy (FTIR)

The Fourier Transform Infrared (FTIR) spectroscopy results are shown in figure 5. C=O bond lies at wavenumber between 1720 cm^{-1} – 1708 cm^{-1} [19]. For the C=O bond at 1775.20 cm^{-1}, the peak intensity after the laser irradiation was also slightly reduced. The intensity of the peak corresponding to the C=O bond reduced after the laser irradiation on polyimide polymer as shown in figure 5. For C-O bond lies around wavenumber 1225.84 cm^{-1} as shown in figure 5(a), this peak totally disappeared after laser irradiation on polyimide polymer as depicted in figure 5(b). This means that the laser irradiation on polyimide broke the C-O bond on the polyimide polymer.

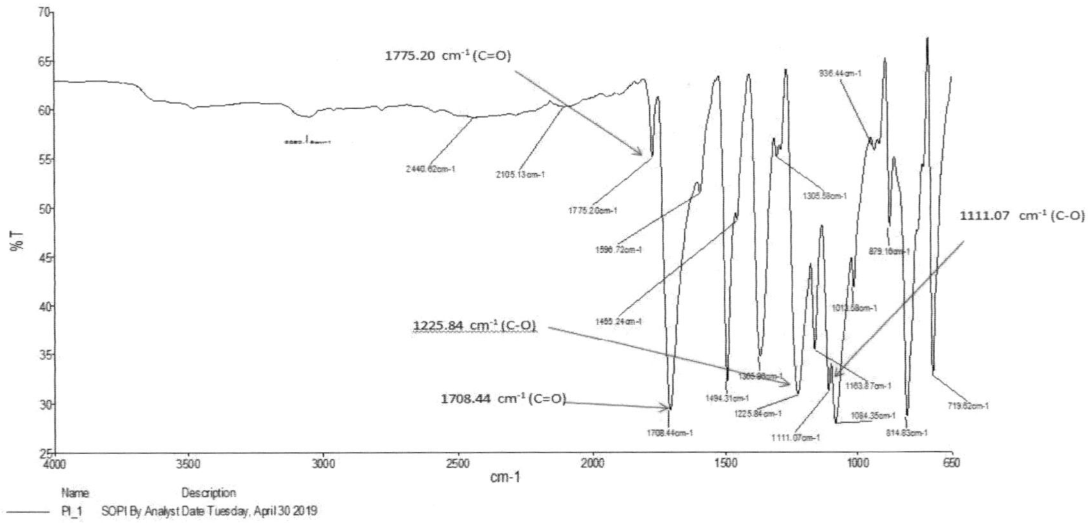

(a)

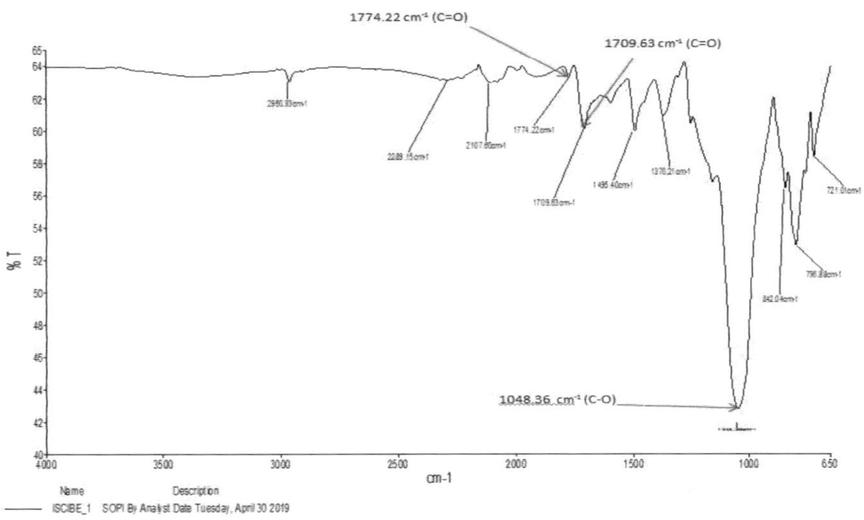

(b)

Figure 5. FTIR spectra of polyimide tape (a) before irradiation and (b) after laser irradiation.

4. Conclusion

Graphene nanoribbons have been successfully directed patterned on a substrate by using 450 nm diode laser with a power of 2500 mW. This method can be considered a fast and economical way to produce graphene. We have also studied the breaking of the chain of carbon atoms in the polymer by laser irradiation to produce graphene verified by FTIR spectroscopy results. Images of carbon atom layer by FESEM spectroscopy test support the formation of layered carbon and Raman spectra confirm the presence of sp^2 orbital of honeycomb carbon formation upon laser irradiation.

Acknowledgement

Specials thanks to Research Management Center (RMC) Universiti Tun Hussein Onn Malaysia for funding the project under Grant Tier-1 Vot U929 and Vot H112.

References

[1] E. V Castro *et al.*, "Limits on Charge Carrier Mobility in Suspended Graphene due to Flexural Phonons," *Phys. Rev. Lett.*, vol. 105, no. 26, p. 266601, Dec. 2010.
[2] C. Lee, X. Wei, J. W. Kysar, and J. Hone, "Measurement of the Elastic Properties and Intrinsic Strength of Monolayer Graphene," *Science (80-.).*, vol. 321, no. 5887, p. 385 LP-388, Jul. 2008.
[3] Q. Bao *et al.*, "Atomic-layer graphene as a saturable absorber for ultrafast pulsed lasers," *Adv. Funct. Mater.*, vol. 19, no. 19, pp. 3077–3083, 2009.
[4] Q. Bao and K. P. Loh, "Graphene Photonics, Plasmonics, and Broadband Optoelectronic Devices," *ACS Nano*, vol. 6, no. 5, pp. 3677–3694, May 2012.
[5] F. Schwierz, "Graphene transistors," *Nat. Nanotechnol.*, vol. 5, no. 7, p. 487, 2010.
[6] T. Kuila, S. Bose, P. Khanra, A. K. Mishra, N. H. Kim, and J. H. Lee, "Recent advances in graphene-based biosensors," *Biosens. Bioelectron.*, vol. 26, no. 12, pp. 4637–4648, 2011.
[7] X. Wang, L. Zhi, and K. Müllen, "Transparent, conductive graphene electrodes for dye-sensitized solar cells," *Nano Lett.*, vol. 8, no. 1, pp. 323–327, 2008.
[8] Y. Fu, G. Wang, T. Mei, J. Li, J. Wang, and X. Wang, "Accessible graphene aerogel for efficiently harvesting solar energy," *ACS Sustain. Chem. Eng.*, vol. 5, no. 6, pp. 4665–4671, 2017.
[9] A. K. Geim and K. S. Novoselov, "The rise of graphene," in *Nanoscience and Technology: A Collection of Reviews from Nature Journals*, World Scientific, 2010, pp. 11–19.
[10] M. S. A. Bhuyan, M. N. Uddin, M. M. Islam, F. A. Bipasha, and S. S. Hossain, "Synthesis of graphene," *Int. Nano Lett.*, vol. 6, no. 2, pp. 65–83, 2016.
[11] X. Li, L. Colombo, and R. S. Ruoff, "Synthesis of graphene films on copper foils by chemical vapor deposition," *Adv. Mater.*, vol. 28, no. 29, pp. 6247–6252, 2016.
[12] J. Lin *et al.*, "Laser-induced porous graphene films from commercial polymers," *Nat. Commun.*, vol. 5, p. 5714, Dec. 2014.
[13] M. Qian *et al.*, "Formation of graphene sheets through laser exfoliation of highly ordered pyrolytic graphite," *Appl. Phys. Lett.*, vol. 98, no. 17, p. 173108, Apr. 2011.
[14] A. T. T. Koh, Y. M. Foong, and D. H. C. Chua, "Cooling rate and energy dependence of pulsed laser fabricated graphene on nickel at reduced temperature," *Appl. Phys. Lett.*, vol. 97, no. 11, p. 114102, Sep. 2010.
[15] P. Kumar, L. S. Panchakarla, and C. N. R. Rao, "Laser-induced unzipping of carbon nanotubes to yield graphene nanoribbons," *Nanoscale*, vol. 3, no. 5, pp. 2127–2129, 2011.
[16] K. Wang, G. Tai, K. H. Wong, S. P. Lau, and W. Guo, "Ni induced few-layer graphene growth at low temperature by pulsed laser deposition," *AIP Adv.*, vol. 1, no. 2, p. 22141, Jun. 2011.
[17] J. B. Park *et al.*, "Fast growth of graphene patterns by laser direct writing," *Appl. Phys. Lett.*, vol. 98, no. 12, p. 123109, Mar. 2011.
[18] Y. Zhao, Q. Han, Z. Cheng, L. Jiang, and L. Qu, "Integrated graphene systems by laser irradiation for advanced devices," *Nano Today*, vol. 12, pp. 14–30, 2017.
[19] Elzein, Tamara, et al. "FTIR study of polycaprolactone chain organization at interfaces." Journal of colloid and interface science 237.2 (2014): 381-387.

Journal of Physics: Conference Series **1371** (2019) 012004 doi:10.1088/1742-6596/1371/1/012004

Q-switched ytterbium-doped fiber laser using graphene oxide as passive saturable absorber

R A M Yusoff[1], A A A Jafry[1], N Kasim[1], Y Munajat[1], S W Harun[2], N A H Halim[1]

[1] Department of Physics, Faculty of Science, Universiti Teknologi Malaysia, 81310 Skudai, Johor, Malaysia.
[2] Department of Electrical Engineering, University of Malaya, 50603 Kuala Lumpur, Malaysia

r.addayusoff92@gmail.com; k.nabilah@utm.my

Abstract. A Q-switched ytterbium-doped fiber laser (YDFL) was demonstrated using Graphene Oxide (GO) as a saturable absorber (SA). Without SA, the ring cavity operates in a continuous wave laser at 1038 nm which is shifted to 1030 nm with the implementation of SA. The laser has a threshold pump power of 175 mW, a maximum repetition rate of 141 kHz and the shortest pulse width of 1.94 μs. The highest pulse energy of 5.65 nJ is achieved at the pump power of 175mW. A stable Q-switched ytterbium doped fibre laser was successfully achieved in this experiment as the pulses measured SNR of 56.52 dB.

1. Introduction

Pulsed laser has received tremendous focus among researchers within the field of laser because it plays crucial roles in wide selection of applications including optical communication, optical sensing, materials technology, life sciences, precision mechanics, national defense science and etc [1-3]. There are two techniques to realize pulsed laser which are active and passive. The latter is more favourable due to few advantages including ease of preparation, high flexibility and high efficiency. The generation of pulsed laser via passive means is highly dependent on the implementation of material as a saturable absorber inside a laser cavity. Recent years, many materials were implemented inside laser cavity to initiate pulsed laser such as semiconductor saturable absorber mirrors (SESAMs) [4] and carbon nanotubes (CNTs) [5].

For SESAMs, pulsing method require additional devices such as mirror, lenses and U-bench units which introduce extra losses to the cavity [6]. Therefore, researcher starts to use carbon-based materials to generate pulsed laser. CNTs, which seems very promising in terms of electron mobility exhibit ultra-fast recovery time (< 1 ps) and wide absorption spectrum (1-2 μm) [7]. However, tunability of wavelength for CNTs are dependable on it shapes and sizes, thus provide complexity for pulse generation. For instance, structural, morphological and alignment control needed to be establish for the purpose of band gap alteration for CNTs [8]. Thus, 2-dimensional carbon allotrope (graphene) is introduced due to excellent electronic and optical properties. In terms of saturable absorption ability, graphene provide significant improvement compare to CNTs. Hexagonal honeycomb lattice of graphene exhibit wide-band operation due to its ultrafast recovery time (~200 fs) and point band gap structure [8]. In fact, graphene oxide (GO), precursor of graphene offers easier sample preparation than it parents. This is attributed by hydrophilic properties of graphene oxide which makes it easier to dissolve in water [9]. On the other hand, most work in pulsed laser focused on the generation of Q-

Journal of Physics: Conference Series **1371** (2019) 012004 doi:10.1088/1742-6596/1371/1/012004

switched at 1.55 μm region [10] but only few works produced Q-switched at 1 μm regime. In this research, a stable Q-switched pulses is generated in ytterbium-doped fiber laser using graphene oxide as a saturable absorber.

2. Experimental setup

Figure 1 shows the experimental setup of the Ytterbium-doped fiber laser with Graphene oxide as a saturable absorber. The ring cavity consists of a 2 m long Ytterbium doped fiber (YDF) as the active medium, a 980/1050 nm wavelength-division multiplexer (WDM), an optical isolator, a GO based SA, and a 90:10 output coupler. The GO as a SA is sandwiched between two FC/PC fiber connectors and inserted into the fiber cavity. The insertion loss of the SA is measured to be around 2dB at 1030nm. It is pumped by the 980 nm laser diode via a 980/1050 nm WDM. Isolator is located between coupler and gain medium to allow light travel in a one direction. The output of the laser is tapped out of the cavity through a 90:10 output coupler, which allows 10% of the output to be extracted for analysis.

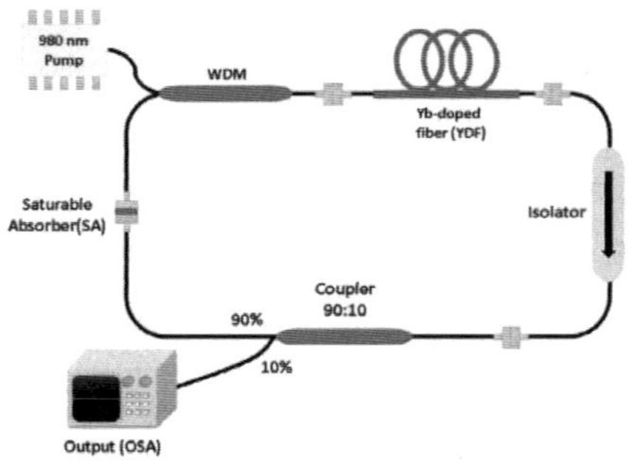

Figure 1. The experimental setup of the Ytterbium fiber laser with Graphene oxide as a passive saturable absorber.

3. Results and Discussion

In the experiment, the YDFL started to generate pulse at threshold pump power of 154 mW. Figure 2 shows the comparison of the output spectrum of the proposed YDFL with and without using GO as a SA at the pump power of 175 mW. As seen, the operating wavelength of the laser shifts to the left from 1038nm to 1030 nm with the incorporation of the SA. The output spectrum with SA was measured at 1038nm. Without the SA, the ring cavity operates in a continuous wave method and output spectrum at 1030nm.

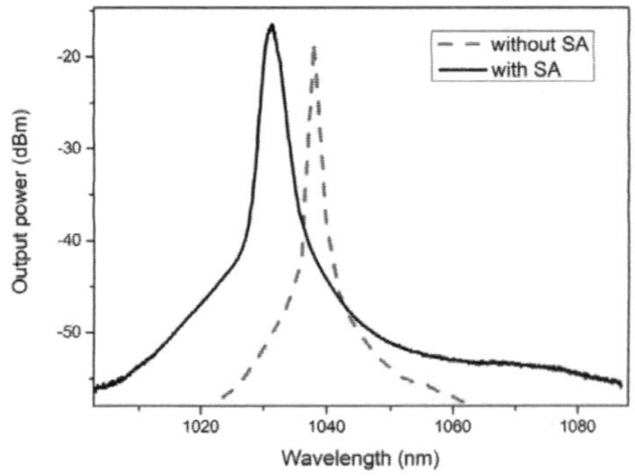

Figure 2. Output spectrum of the YDFL with and without GO as a SA at pump power of 175 mW

The stable pulse trains with the different of repetition rate were observed when the pump sources were progressively increased from 154 mW to 184 mW. The YDFL started to produce a Q-switched pulse at threshold pump power of 154 mW. Figures 3, 4 and 5 shows (a) typical oscilloscope trace of the Q-switched pulse train (b) a single envelop of the Q-switched at three different pump power which is 154 mW, 171 mW and 184 mW. The pulse trains have the pulse width of 3.72 μs, 2.44 μs and 1.94 μs, which is corresponds to the repetition rate of 88 kHz, 115kHz and 141 kHz, respectively.

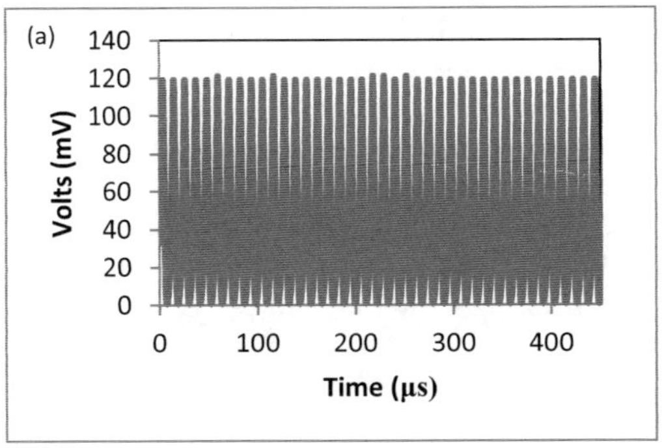

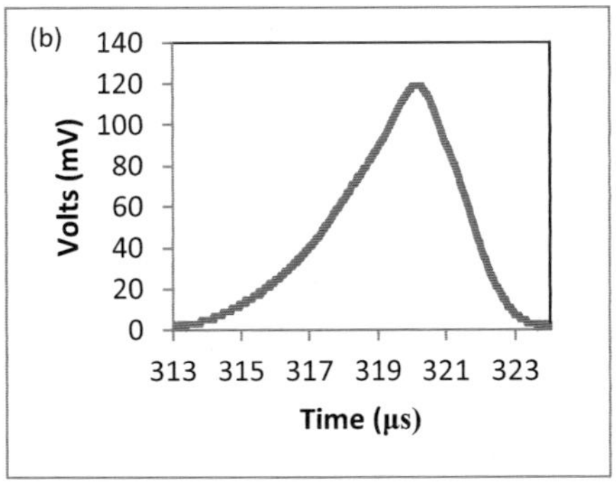

Figure 3. (a) Oscilloscope pulse train and (b) a single envelop of the Q-switched at power pump which is 154 mW.

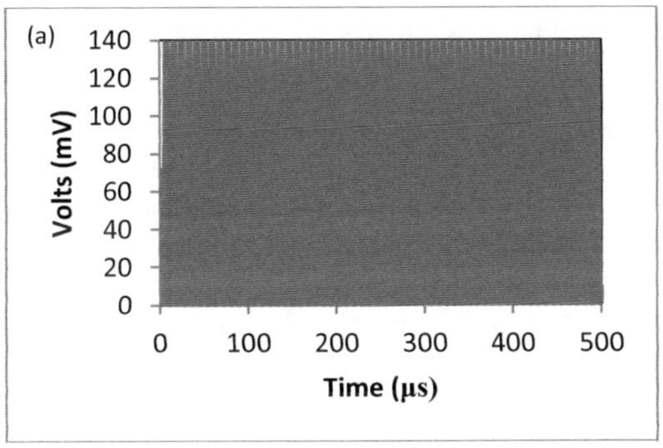

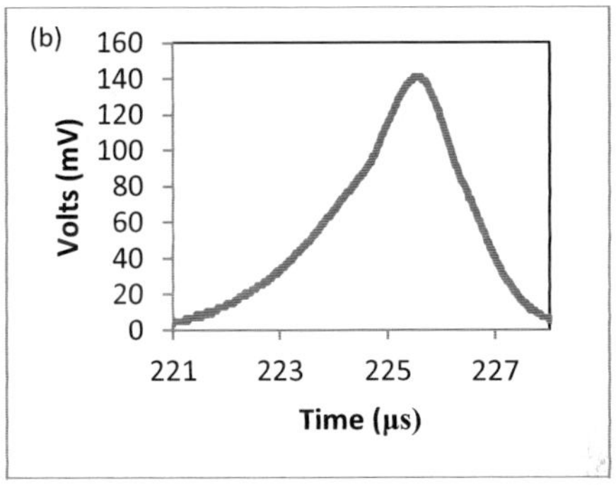

Figure 4. (a) Oscilloscope pulse train and (b) a single envelop of the Q-switched at power pump which is 171 mW.

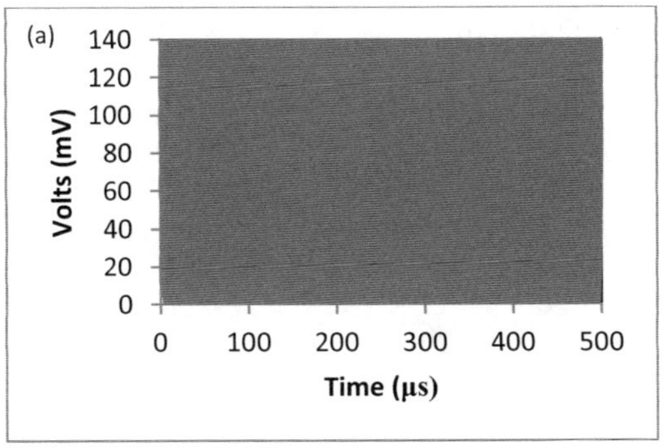

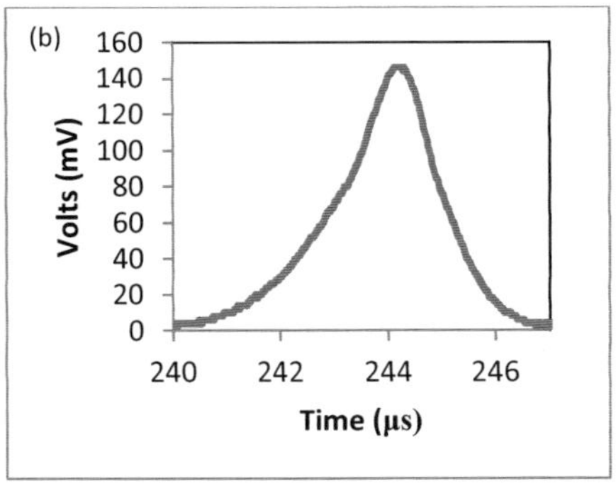

Figure 5. (a) Oscilloscope pulse train and (b) a single envelop of the Q-switched at power pump which is 184 mW.

Figure 6 shows the relationship between the repetition rate and the pulse width versus the pumping sources. The repetition rate of the proposed YDFL can be changed from 88 kHz to 141 kHz when the pump was gradually increased from 154 mW to 184 mW. From that, the pulse width reduces from 3.72 μs to 1.94 μs as the pump increased.

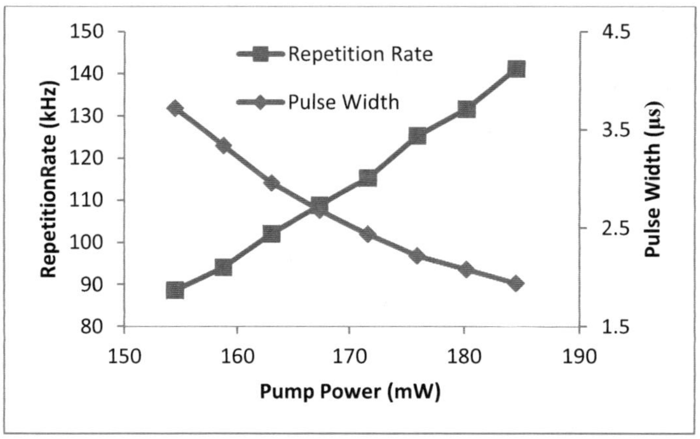

Figure 6. Repetition rate and pulse width of Q-switched YDFL within 154 mW to 184 mW

Figure 7 shows the repetition rate and the average output power as the function of the pump power. As seen, it shows both of repetition rate and the average output power was increased as the pumping sources increased. The highest pulse energy obtained was 5.65 nJ at the power pump of 184 mW. From that, the constructed laser is working under passively Q-switched. The slope efficiency of 1.3% is obtained.

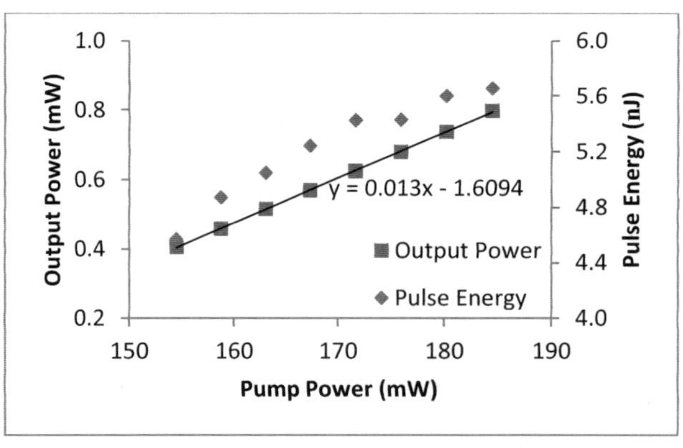

Figure 7. Repetition rate and output power as a function of 980nm of pump power.

The radio frequency (RF) spectrum analyser at pump power of 171 mW was measured. Figure 8 shows the RF spectrum of the Q-switched laser output of the pumping sources. It shows a stable repetition rate of 115.2 kHz and peak to background ratio of about 56.52 dB. This indicates stable pulses are produced by our laser cavity.

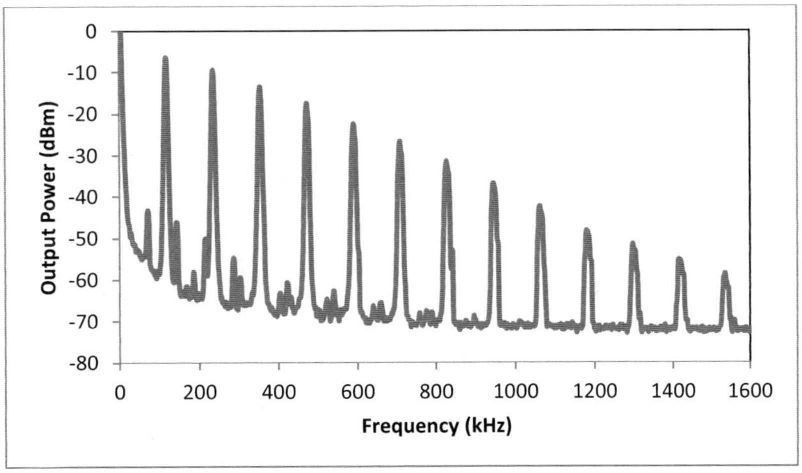

Figure 8. RF spectrum at pump power of 171 mW (SNR: 56.52 dB)

4. Conclusions

A YDFL is proposed and demonstrated using graphene oxide as a passive saturable absorber (SA). GO film is sandwiched between two FC/PC fiber connector to analyze the performance of Ytterbium doped fibre laser based on Graphene Oxide as a saturable absorber. Without SA, the ring cavity operates in a continuous wave regime with output spectrum at 1038 nm. By inserting SA, the YDFL operates at 1030nm. Stable Q-switched starts at the threshold pump power of 154 mW producing pulse with the repetition rate of 88 kHz and the pulse width of 3.72 μs. At maximum pump power of 184 mW, we achieved the maximum pulse energy of 5.65 nJ and the shortest pulse duration of 1.94 μs. A passively Q-switched ytterbium doped fibre laser was successfully achieved in this experiment.

5. Acknowledgements

This research is fully supported by FRGS grant, R.J130000.7826.4F928. The authors fully acknowledged Ministry of Education (MOE) Malaysia and Universiti Teknologi Malaysia for the approved fund which makes this important research viable and effective.

References

[1] Graf S, Staupendahl G, Kramer A, Muller FA. High precision materials processing using a novel Q-switched CO2 laser. *Optics and Lasers in Engineering*. **66:**152-7.
[2] Shah S, Alster TS. Laser Treatment of Dark Skin An Updated Review. *American Journal of Clinical Dermatology*. **11(6)**:389-97.
[3] Walsh BM, Lee HR, Barnes NP. Mid infrared lasers for remote sensing applications. *Journal of Luminescence*. **169**:400-5.
[4] Keller U. Recent developments in compact ultrafast lasers. *Nature*. **424**:831.
[5] Kasim N, A-Masoodi AHH, Ahmad F, Munajat Y, Ahmad H, Harun SW. Q-switched ytterbium doped fiber laser using multi-walled carbon nanotubes saturable absorber. *Chinese Optics Letters*. **12(3)**:4.
[6] Ismail MA, Harun SW, Zulkepely NR, Nor RM, Ahmad F, Ahmad H. Nanosecond soliton pulse generation by mode-locked erbium-doped fiber laser using single-walled carbon-nanotube-based saturable absorber. *Applied Optics*. **51(36)**:8621-4.
[7] Jung M, Koo J, Chang YM, Debnath P, Song YW, Lee JH. An all fiberized, 1.89-mu m Q-switched laser employing carbon nanotube evanescent field interaction. *Laser Physics Letters*. **9(9)**:669-73.
[8] Song YW, Jang SY, Han WS, Bae MK. Graphene mode-lockers for fiber lasers functioned with

evanescent field interaction. *Applied Physics Letters*. **96(5)**:3.

[9] Zhao JQ, Wang YG, Ruan SC, Yan PG, Zhang H, Tsang YH, Yang JH, Huang GX. Three operation regimes with an L-band ultrafast fiber laser passively mode-locked by graphene oxide saturable absorber. *Journal of the Optical Society of America B-Optical Physics*. **31(4)**:716-22.

[10] Jafry AAA, Arof H, Kasim N, Munajat Y, Apsari R, Harun SW. Q-switched Erbium-doped Fiber Laser Using MoS_2 Deposited Side-polished D-shape Fiber. *Nonlinear Optics Quantum Optics-Concepts in Modern Optics*. **49(3-4)**:285-293.

Optimization of ZnO nanorods growth duration for humidity sensing application

Mohd Hafiz Jali[1,2], Hazli Rafis Abdul Rahim[a], Haziezol Helmi Mohd Yusof[1,3], Md Ashadi Md Johari[1,4], Siddharth Thokchom [5], Sulaiman Wadi Harun[1]

[1]Department of Electrical Engineering, Faculty of Engineering, University of Malaya, 50603 Kuala Lumpur, Malaysia
[2]Faculty of Electrical Engineering, Universiti Teknikal Malaysia Melaka, 76100 Melaka, Malaysia
[3]Faculty of Electronic and Computer Engineering, Universiti Teknikal Malaysia Melaka, 76100 Melaka, Malaysia
[4]Faculty of Electrical and Electronic Engineering Technology, Universiti Teknikal Malaysia Melaka, 76100 Melaka, Malaysia
5School of Technology, Assam Don Bosco University, Guwahati, Assam,781017, India

mohd.hafiz@utem.edu.my

Abstract. This paper described an optimization approach of ZnO nanorods growth duration based on humidity sensing scheme. The proposed structure comprises of silica microfiber integrated with Zinc Oxide (ZnO) nanorods coated glass surface. The silica microfiber was tapered into constant waist diameter of 10 μm using flame brushing technique. The glass surface was grown with ZnO nanorods using hydrothermal method for 6 hours, 9 hours, 12 hours, 15 hours and 18 hours growth time. ZnO nanorods growth time would affect morphological structures such as density which change the light absorption into the nanorods. Samples were exposed to the difference humidity level ranging from 35%RH to 85%RH to observe the change of ZnO refractive index on the glass surface resulting different light attenuation in the silica microfiber. It utilizes the unique features of the scattering and surface absorption capability of the microfiber and ZnO nanomaterials coated glass surface to alter the output light intensity. The reported results may contribute to the optimal ZnO nanorods growth time for humidity sensing application.

1. Introduction

Zinc oxide (ZnO) nanomaterial has wide band gap (3.37 eV), high exciton binding energy (~60 meV at room temperature), direct charge transport along with ZnO arrays, higher surface-to-volume ratio and chemically reactive surface [1]. It became desirable material for various applications such as biosensors [2], resonators [3] and medical devices [4]. Guided light couple into ZnO nanorod waveguides due to higher refractive index (RI) as compared to silica fiber [5]. Moreover, high concentrations of oxygen vacancies create active sites for water molecules adsorption for humidity sensor [1]. Even though numbers of synthesis techniques for ZnO nanorods coating process such as chemical vapour depositions [6], physical vapour depositions [7] and vapour transport [8] has been implemented, hydrothermal synthesis method has been selected because it is inexpensive and avoid the use of complex vacuum environment and high temperature.

Another major component for the proposed sensor is microfiber. It demonstrates unique properties such as large evanescent field, strong optical confinement and configurability which make them promising element for physical sensing [9]. Evanescent field propagates outside the microfiber make it a very sensitive medium to a change in the surrounding refractive index (RI) [10]. The surrounding material's refractive index would increase as the power fraction propagating in the evanescent field. Thus, the evanescent wave coupled with other waveguides such as metal, and semiconductor forming near-field interaction with its surrounding. Thus, a high fractional evanescent field allows great responses towards physical sensing [11-13].

There are numerous advantages of an optical device as compared to electronic counterpart such as can be employed at a higher pressure and temperature level as well as in flammable environments [14, 15]. The sensing works when the sensitive material interacts with the evanescent field surrounding it and changes the transmitted output power [15]. Normally the sensitive material is directly coated which make the handling process of the become complicated and weaken the performance of the microfiber [16, 17]. The microfiber has been integrated with ZnO nanorods coated glass as a potential structure for humidity sensing. It eases the interaction between ZnO coating layer and the evanescent wave from the microfiber with the easy fabrication process, low cost and simple structure. In this paper, the optimization approach of the ZnO nanorods growth duration for the proposed sensor has been explained in this work. The demand for an optical sensor in many fields has attracted researchers to explore numerous sensing structure for physical sensing [18].

2. Sensing mechanism

Figure 1 illustrated the sensing mechanism of the proposed sensor. When the proposed sensor is exposed to the humidity, the water molecule chemisorbed onto the ZnO nanorods surface due to weak hydrogen bonding and the air medium is replaced by humidity. This cause the refractive index of the ZnO becomes higher as compared to the microfiber. Thus, the interaction between the evanescent field and nanorods surface occur which highly reliance on the refractive index and the extinction coefficient [19]. The evanescent wave absorbance depends on the concentrations of the analyte if other parameters remain constant [20]. The effective index between the surrounding medium and ZnO nanorods would increase as the humidity level increase. More light coupling into the ZnO nanorods waveguides due to the increment of the forward scattering coefficient which reduce the light transmission through the sensing region [21]

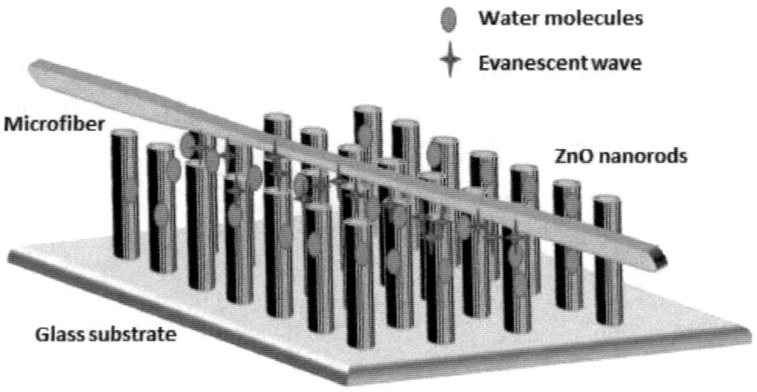

Figure 1. Proposed humidity sensor structure.

3. Material, fabrication and characterization

3.1. Glass substrate and microfiber preparation

Microfiber and ZnO nanorods coated glass has been prepared according to our previous work [22]. ZnO nanorods have been grown onto glass substrates (Heathrow Scientific LLC, USA) using hydrothermal synthesis technique for 6 hours, 9 hours,12 hours, 15 hours and 18 hours. Single mode fiber (Corning SMF-28, USA) with 125 µm diameter was tapered into waist diameters of 10 µm with a tapered length of 2 cm using flame brushing technique as shown in figure 2. It is realized by controlling several elements such as motor speed, fiber stretching length and flame movement. The length and diameter of the microfibers were verified and measured using a microscope (Medilux-12) with 20X magnification.

Figure 2. Microfiber with diameter around 10 µm.

3.2. Characterization & experiment

In order to view the physical structures of ZnO nanorods growth and determine the chemical content of the samples, Field emission Scanning electron microscopy (FESEM) and Energy dispersive X-ray (EDX) was performed. Figure 3 depicted the experiment setups. The microfiber was laid on the ZnO nanorods coated glass surface and placed inside a sealed chamber (22 x 12 x 12 cm). Amplified Spontaneous Emission (ASE) from an erbium doped fiber amplifier (EDFA) was used as a light source and connected to an optical spectrum analyser (OSA) (Anritsu: MS9710C) for output power measurement. The output spectrum was measured in the bandwidth between 1500 to 1600 nm measured in dBm. The humidity level was increased from 35%RH to 85%RH at almost constant room temperature, 27°C. The probe of %RH meter (Hygrometer RS 1365, Sensitivity: 1%) was placed close to the samples to monitor the actual humidity level around the sample's surface. The readings were recorded several times to verify the stability and repeatability of the experiment results.

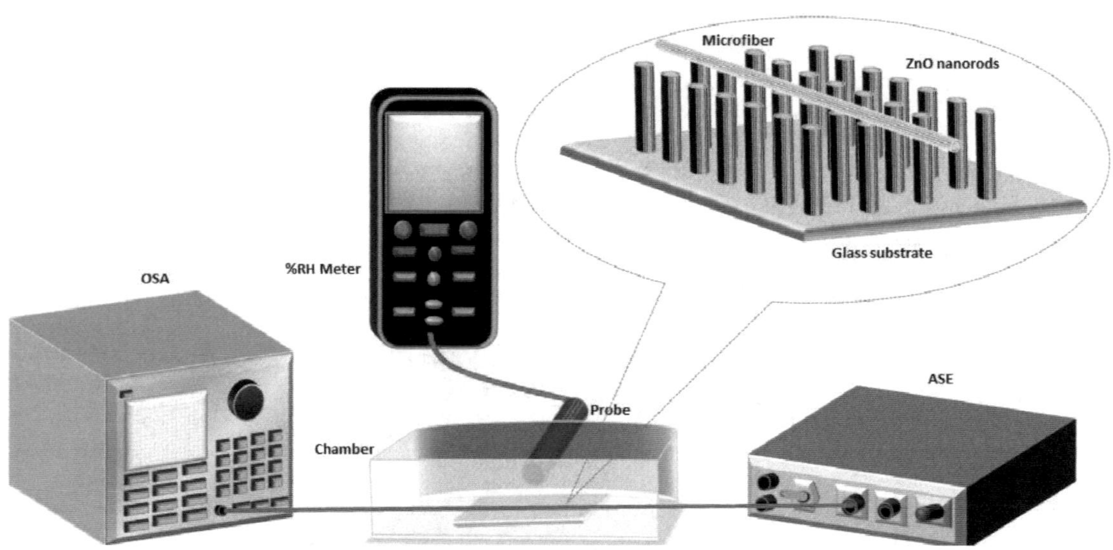

Figure 3. Experimental setup to observe the effect of humidity level.

4. Result and discussion

Figure 4 shows the EDX elemental analysis of the ZnO nanorods samples consist of only zinc and oxygen**.** Figure 5 shows the FESEM images of the nanorods density on the glass surface at different growth times at 20.00 kX magnifications. The average nanorods density decrease monotonically with the increment of growth time because the diameter of the nanorods increases as shown in the graph in figure 6. It would affect the forward and backward scattering into the nanorods which lead to variation light transmission behaviour inside the microfiber. Thus, the output light intensity would change with respect to difference nanorods structure.

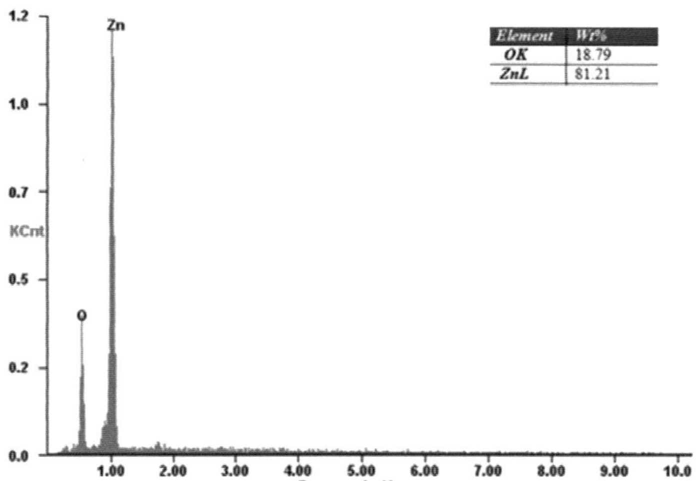

Element	Wt%
OK	18.79
ZnL	81.21

Figure 4. EDX elemental analysis shows the samples only consist of zinc and oxygen

(a) 6 hours (b) 9 hours (c) 12 hours (d) 15 hours (e) 18 hours

Figure 5. FESEM images of the ZnO nanorods density at different growth duration.

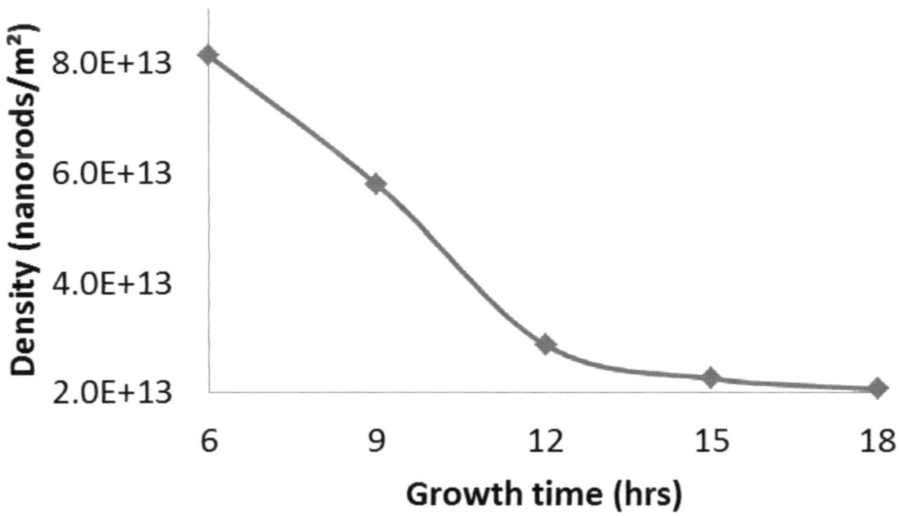

Figure 6. Average nanorods density at different growth hours duration.

Figure 7 shows the output power difference when the proposed sensor was exposed at maximum humidity and minimum humidity level. Output power fluctuated according to the difference nanorods morphological structure during the exposure to different humidity level [23]. The output power difference of the 12 hours sample has the highest value because larger light absorption into the nanorods. This is due to the sample has the best trade-off between the diameter and density which allow more light to absorb into the ZnO nanorods sample. Figure 8 shows the trendline of every growth time samples. It shows that the output power of 12 growth time sample reduced tremendously as compared to other samples. The linearity of 9 hours, 12 hours and 15 hours growth time samples have more than 95% which show an acceptable result for sensing application as depicted in figure 9. The 12 hours growth time sample has the highest sensitivity with 0.0316 dBm/%RH compared to the other samples as shown in figure 10. Thus, the 12 hours growth time was an optimal sample for humidity sensing application because it exhibits the highest light absorption inside ZnO nanorods which was the utmost criteria for the proposed sensing mechanism.

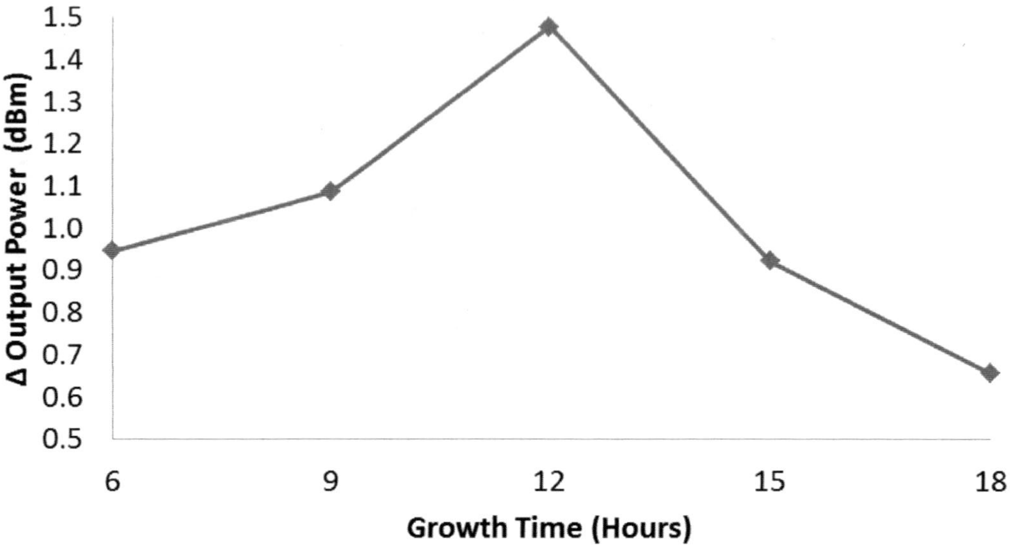

Figure 7: Output different from every growth time samples.

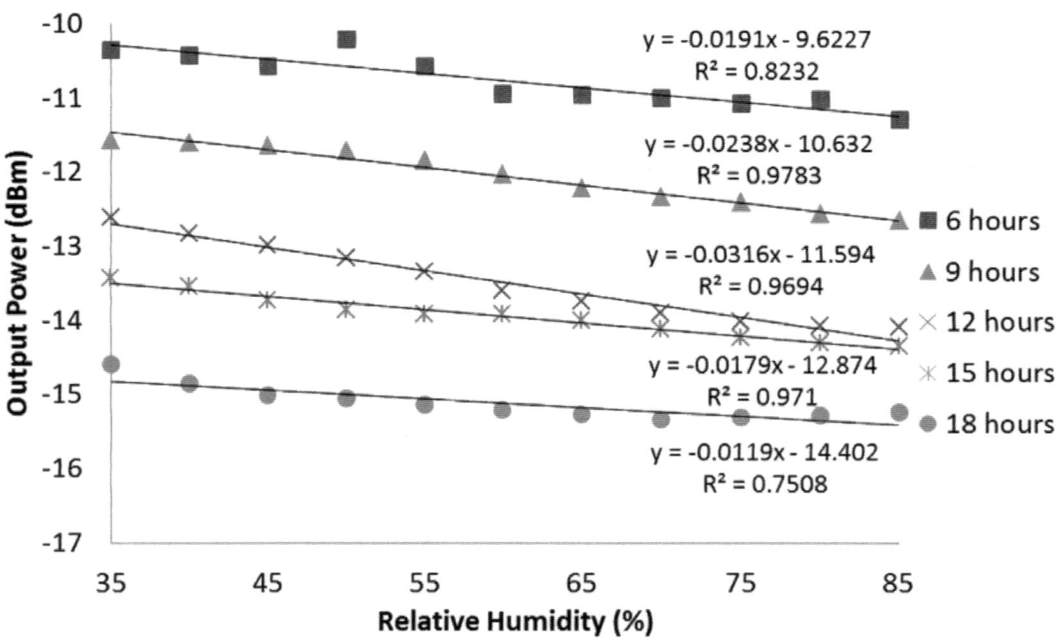

Figure 8. Trendline of every growth time samples at when %RH increases

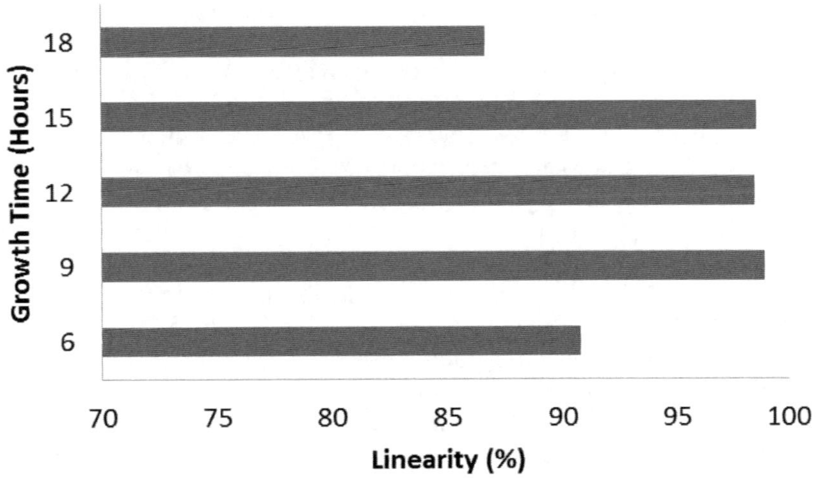

Figure 9: Linearity of every growth time samples.

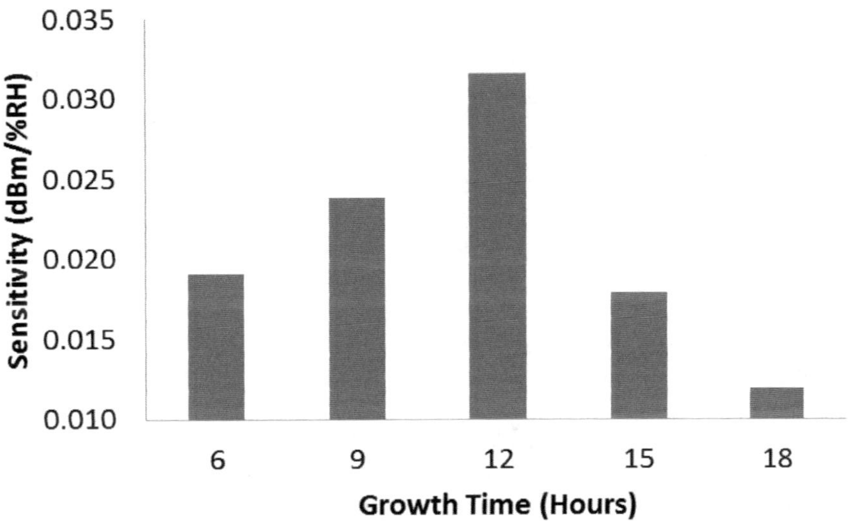

Figure 10. The sensitivity of every growth time samples

5. Conclusion

This report demonstrated the optimization technique of the ZnO nanorods growth time for the proposed sensor. This study is essential to understand the relationship between difference nanorods physical structure and the light scattering behaviour inside it. The best growth time sample was found to be 12 hours growth time sample. The output power reduced tremendously when humidity level increase due to light leakage from the microfiber which absorbs into the nanorods. The linearity and sensitivity also show the highest result as compared to the other samples. The outcomes of this work provide an optimal ZnO nanorods growth time samples for humidity sensing application.

Acknowledgements

The authors would like to acknowledge the University of Malaya, Universiti Teknikal Malaysia Melaka and Ministry of Education, Malaysia for their financial support. (PJP/2018/FKEKK(4B)/S01616).

References

[1] Ismail A S, Mamat M H and Mahmood M R 2017 *Nanostructured Materials-Fabrication to Applications*: InTech)
[2] Chiavaioli F, Baldini F, Tombelli S, Trono C and Giannetti A 2017 Biosensing with optical fiber gratings *Nanophotonics* **6** 663-79
[3] Irawati N, Rahman H, Ahmad H and Harun S 2017 A PMMA microfiber loop resonator based humidity sensor with ZnO nanorods coating *Measurement* **99** 128-33
[4] Quy C T, Hung C M, Van Duy N, Hoa N D, Jiao M and Nguyen H 2017 Ethanol-sensing characteristics of nanostructured ZnO: nanorods, nanowires, and porous nanoparticles *Journal of Electronic Materials* **46** 3406-11
[5] Voss T, Svacha G T, Mazur E, Müller S, Ronning C, Konjhodzic D and Marlow F 2007 High-order waveguide modes in ZnO nanowires *Nano letters* **7** 3675-80
[6] Wu J J and Liu S C 2002 Low-temperature growth of well-aligned ZnO nanorods by chemical vapor deposition *Advanced materials* **14** 215-8
[7] Kong Y, Yu D, Zhang B, Fang W and Feng S 2001 Ultraviolet-emitting ZnO nanowires synthesized by a physical vapor deposition approach *Applied Physics Letters* **78** 407-9
[8] Huang M H, Wu Y, Feick H, Tran N, Weber E and Yang P 2001 Catalytic growth of zinc oxide nanowires by vapor transport *Advanced Materials* **13** 113-6
[9] Jali M H, Rahim H R A, Johari M A M, Hamid S S, Yusof H H M, Thokchom S, Wang P and Harun S W 2019 Optical characterization of different waist diameter on microfiber loop resonator humidity sensor *Sensors and Actuators A: Physical* **285** 200-9
[10] Lim K, Harun S, Arof H and Ahmad H 2012 *Selected Topics on Optical Fiber Technology*: InTech)
[11] Lou J, Wang Y and Tong L 2014 Microfiber optical sensors: A review *Sensors* **14** 5823-44
[12] Chen G Y, Ding M, Newson T and Brambilla G 2013 A review of microfiber and nanofiber based optical sensors *The Open Optics Journal* **7**
[13] Wu X and Tong L 2013 Optical microfibers and nanofibers *Nanophotonics* **2** 407-28
[14] Lokman A, Arof H, Harun S W, Harith Z, Rafaie H A and Nor R M 2016 Optical fiber relative humidity sensor based on inline Mach–Zehnder interferometer with ZnO nanowires coating *IEEE Sensors Journal* **16** 312-6
[15] Ascorbe J, Corres J, Arregui F and Matias I 2017 Recent developments in fiber optics humidity sensors *Sensors* **17** 893
[16] Bhatia V, Campbell D K, Sherr D, D'Alberto T, Zabaronick N, Ten Eyck G A, Murphy K A and Claus R O 1997 Temperature-insensitive and strain-insensitive long-period grating sensors for smart structures *Optical Engineering* **36** 1872-7
[17] Yao Q, Meng H, Wang W, Xue H, Xiong R, Huang B, Tan C and Huang X 2014 Simultaneous measurement of refractive index and temperature based on a core-offset Mach–Zehnder interferometer combined with a fiber Bragg grating *Sensors and Actuators A: Physical* **209** 73-7
[18] Misra S K, Pandey N K, Shakya V and Roy A 2015 Application of undoped and Al 2 O 3-doped ZnO nanomaterials as solid-state humidity sensor and its characterization studies *IEEE Sensors Journal* **15** 3582-9
[19] Azad S, Sadeghi E, Parvizi R, Mazaheri A and Yousefi M 2017 Sensitivity optimization of ZnO clad-modified optical fiber humidity sensor by means of tuning the optical fiber waist diameter *Optics & Laser Technology* **90** 96-101
[20] Punjabi N, Satija J and Mukherji S 2015 *Sensing Technology: Current Status and Future Trends III*: Springer) pp 25-45
[21] Yusof H H M, Harun S W, Dimyati K, Bora T, Mohammed W S and Dutta J 2018 Optical dynamic range maximization for humidity sensing by controlling growth of zinc oxide nanorods *Photonics and Nanostructures-Fundamentals and Applications* **30** 57-64
[22] Jali M H, Rahim H R A, Ashadi M J M, Thokchom S and Harun S W 2018 Applied microfiber evanescent wave on ZnO nanorods coated glass surface towards temperature sensing *Sensors and Actuators A: Physical* **277** 103-11

[23] Azad S, Sadeghi E, Parvizi R and Mazaheri A 2017 Fast response relative humidity clad-modified multimode optical fiber sensor with hydrothermally dimension controlled ZnO nanorods *Materials Science in Semiconductor Processing* **66** 200-6

Journal of Physics: Conference Series **1371** (2019) 012006 doi:10.1088/1742-6596/1371/1/012006

Microbottle resonator for temperature sensing

Malathy Batumalay[3], Md Ashadi Md Johari[1,2], Muhammad Imran Mustafa Abdul Khudus[3], Mohd Hafiz Bin Jali[1,4], Abdullah Al Noman[1], Sulaiman Wadi Harun[1]

[1] Department of Electrical Engineering, Faculty of Engineering, University of Malaya, 50603 Kuala Lumpur, Malaysia
[2] Fakulti Teknologi Kejuruteraan Elektrik dan Elektronik, Universiti Teknikal Malaysia Melaka, 76100 Melaka, Malaysia
[3] Fakulti IT, INTI International University, 71800 Nilai, Negeri Sembilan, Malaysia
[4] Department of Physics, Faculty of Science, University of Malaya, 50603 Kuala Lumpur, Malaysia
[5] Fakulti Kejuruteraan Elektrik, Universiti Teknikal Malaysia Melaka, 76100 Melaka, Malaysia

malathy.batumalay@newinti.edu.my

Abstract. The whispering gallery mode on the surface of the microbottle optical resonator (MBR) was studied in this research paper as a relative temperature sensor. The MBR is formed by a technique known as "soften-and-compress" allowed SMF-28 silica fiber to be formed in bottle structure with a bottle diameter $D_b = 190\mu m$, stem diameter of $D_s = 125\mu m$ and bottle length of $L_b = 182\mu m$. The Q-factor of the MBR is defined by excited with bare microfiber with 2 μm diameter and managed to have >10^5. The range of temperature between 40^0C to 100^0C is then employed to the MBR as a temperature sensor for analysis purpose. The performance of the MBR is promising with sensitivity 0.0149 dB/^{0}C with linearity 94% and P-value >10^{-5} which is defined as a good sensor. The sensitivity value from the wavelength shift is 1.3 pm/^{0}C. The repeatability and stability of the MBR can be employed as a temperature sensor.

1. Introduction

Recently, whispering gallery modes (WGMs) optical microresonator (OMRs) have captured significant interest due to their potential applications as optical sensors, lasers and related plasmonic devices [1-6]. OMRs confine light in spatially and temporally, allowing the production of high optical intensities, large-quality factor and extending photon lifetimes in the resonator structure and have been realized in a variety of different geometries such as microdisks, micropillars, microrings and photonic-crystal cavities [7-17]. WGMs OMRs have also been intensively studied, exhibiting the advantages of a number such high-quality factor, low intrinsic loss and simple fabrication methods [2, 18-20].

The micro bottle resonator (MBR) is a sub of OMR which is made from the silica fiber by changing the fiber structure into a bump, which looks like a bottle. For the operation of MBR, the WGMs is used to circulate around the MBR surface continuously perpendicular from the bottle axis. The electromagnetic field was located around the MBR surface which is useful for sensing application [21, 22]. This concept is similar to the electromagnetic field on surface plasmon-polaritons (SPPs) [23-29]. Though the MBR has been suggested to be ideal for silica-based OMR sensors as they have very high-quality factor and free spectral range (FSR) [2, 30].

Journal of Physics: Conference Series **1371** (2019) 012006 doi:10.1088/1742-6596/1371/1/012006

In this paper, we proposed to use the MBR for temperature sensing. The MBR structured from SMF-28 silica fiber by "soften-and-compress" technique which then coupled with the taper microfiber with 2μm wrist diameter to performed sensing [31-33]. As the diameter of the taper microfiber is maintained at 2um, it experienced small wavelength shifting with the shifting factor < 55.25nm/um. Hence, the decreased of the external refractive index influenced the wavelength shift and contributed to the high negative thermo-optic coefficient. As per the theoretical analysis, the MBR in the setup will increase the sensitivity of the temperature sensor. The vacuum oven is used to provide a closed environment with a range of temperature between 40^0C to 100^0C. The purposed MBR is observed to have high sensitivity, stability with consistence repeatability.

2. Characterization of MBR

A standard silica fiber SMF28 is used to form MBR using technique known as "softened-and-compress", which by then used by WGMs in regulating spectral characterization [34]. The splicing machine (Furukawa Electric Fitel S178A) is used to hold and compressed on both sides of SMF28 in inward direction after being heated with an electrical arc. The number of arcs applied will determine the size of the bottle in the centre of heated fiber [30]. **Figure 1** shows physical structure characteristic of MBR with three parameters, as known as the bottle diameter D_b, the stem diameter D_s and the neck-to-neck length L_b. In this work, the size of MBR was set at 190 μm for D_b diameter which is similar to other work previously [30]. The MBR is employed with microfiber with 2 μm of waist diameter fabricated using flame brushing method [35].

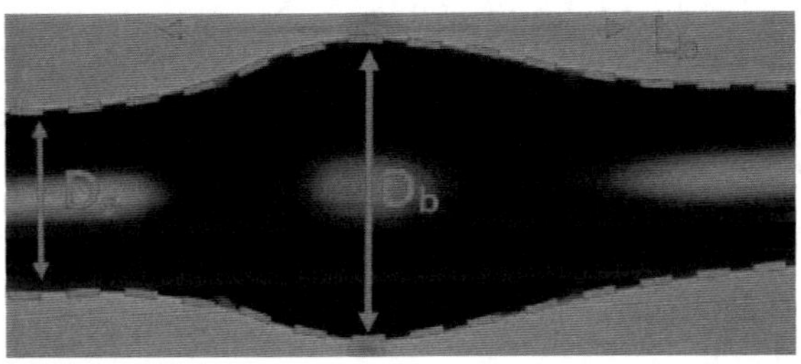

Figure 1. Sample fabricated MBR of L_b = 182 μm, D_b = 190 μm and D_s = 125 μm

The MBR which employed with 2 μm tapered microfiber, is firstly characterized by launched wavelength range between 1551 nm to 1560nm from a tuneable laser source (ANDO AQ4321D) into it. The tuneable laser source used in this experiment is able to produce a laser with a wavelength range between 1520 nm until 1620 nm, which is compatible with the proposed experiment. The wavelength interval of 0.001 nm was tuned from laser source collected by an optical power meter (THORLABS S145C) as output transmitted power. **Figure 2** shows the transmission spectra of MBR, where numbers of sharp resonation were defined clearly and used as a quality factor [36]. The MBR and the tapered microfiber manage to have the insertion loss for more than -5dBm and its may accede by controlling the position of microfiber and MBR as mention in previous research [37]. The Q-factor for the MBR is defined by calculation and also by Lorentzian fitting where both are found to have 1.555 x 10^5 which is similar than previous work [30]. The calculation of Q-factor is defined by $\Delta\lambda/\lambda$ where λ is the resonant wavelength and the Lorentzian fitting is from the Origin software application. The insertion loss of the entire MBR and the microfiber diameter contributes to the value of Q-factor.

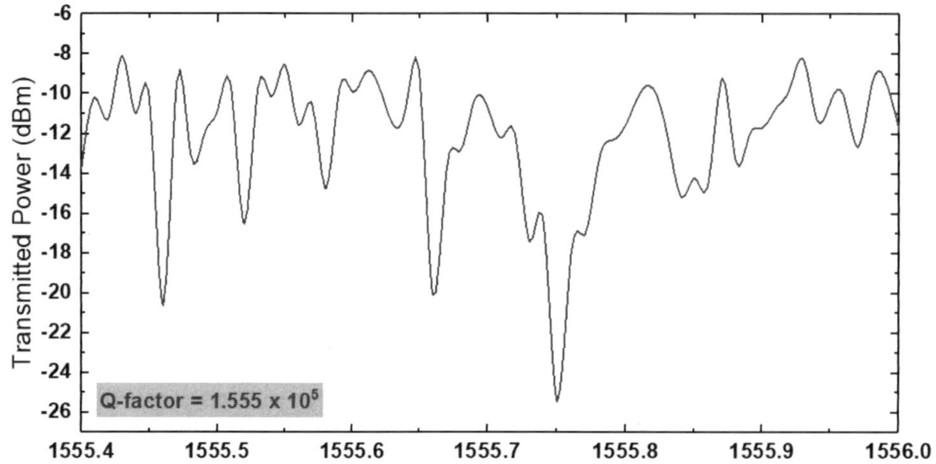

Figure 2. The transmitting spectral of the MBR and the 2 μm tapered microfiber with numbers of WGM resonating.

3. Performance of MBR Temperature Sensor

Figure 3 shows the setup of the experiment used to investigate the performance of MBR as temperature sensors. The vacuum oven (GALLENKAMPN 840.3 FT18) used to monitor temperature level with the MBR and the microfiber are placed inside the oven. The MBR is placed at the tapered area where one end of microfiber is connected to tuneable laser source and the other end to OPM to measure transmitted power. First, the wavelength was set at 1555.750 nm, where the most depth resonated wavelength was recorded. The temperature was then increased from 40^0C to 100^0C and output power were recorded. To investigate the repeatability of the temperature sensor, the experiment was repeated three times on the same day. The transmission spectra at different temperature levels were recorded for 60 seconds to investigate the stability of the sensor. To ensure the best performance, this experiment was compared with different experiment setup.

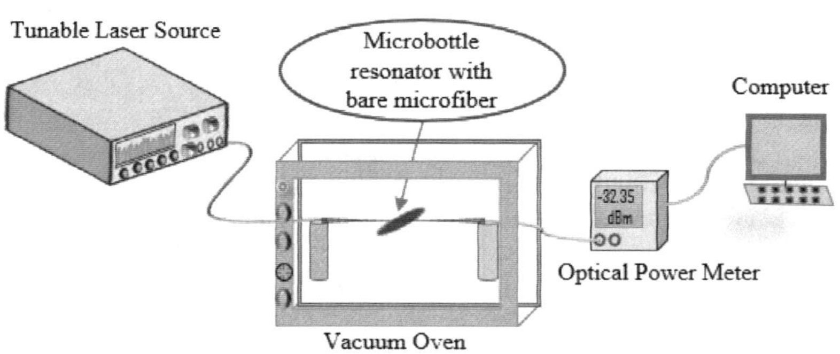

Figure 3. The MBR was attached with tapered fiber ($2\mu m$) for temperature sensing. Optical power meter connected to one end of fiber for final data collection while tuneable laser source supplied wavelength at 1555.750 nm.

The tuneable laser source transmitted a wavelength between the range of 1555.750 nm to 1555.760 nm. The optical power meter and the computer were used to monitor and record the changes of wavelength produced upon changes in temperature. The proposed MBR produce a shift in transmitted spectral wavelength from 1555.750 nm to 1555.758 nm. This is due to the thermo-optic coefficient concept where the MBR conductivity changes drastically with the change of temperature. This work proved that MBR can be used to increase the conductivity of material such as temperature sensing.

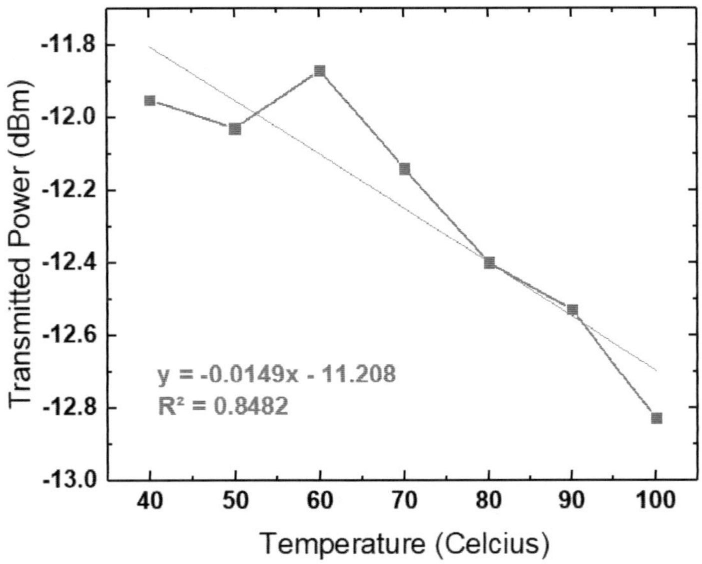

(a)

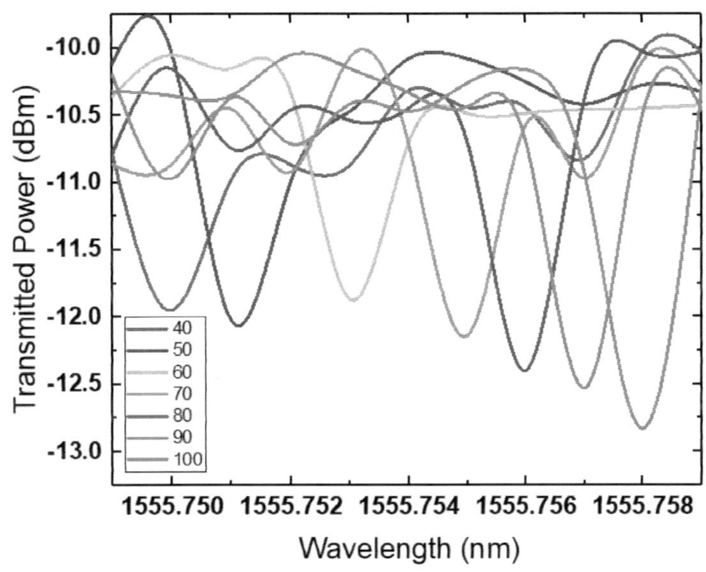

(b)

Figure 4. Analysis of temperature sensing performances (a) and wavelength shift of the MBR microfiber temperature sensor (b).

Figure 4(a) shows the performance of the transmitted spectra of the MBR with the changes in temperature. **Figure 4(b)** shows the value of transmitted power plotted by the peak resonant value against wavelength shift. As the temperature changes, the peak value of the resonant wavelength was captured and recorded in the same graft. Based on the plotting, the value of transmitted power (dBm) decreased per unit rise of temperature level. As mentioned in equation (6) where the thermal expansion influenced the sensor performance where transmitted power value decreases with the increase of temperature.

The sensitivity, linearity and standard deviation are listed parameter used to determine the performance of the MBR towards temperature sensing. However, the MBR significantly showed promising performance on all parameter, which is recorded in **Table 1**. In the sensitivity performance study, the MBR showed 0.0149 dB/^{0}C with the linearity of 94% respectively for transmitted power over temperature. However, by the analysis from the wavelength shift, the sensitivity is observed at 1.3 pm/^{0}C.

Table 1. The analysis results of the MBR microfiber performance in temperature sensing activity

Parameters	MBR
Linearity (%)	94%
Sensitivity (dB/^{0}C)	0.0149
Sensitivity (pm/^{0}C)	1.3
Standard deviation (dBm)	0.3234
P-value	3.886 x 10^{-8}
Linear Range (^{0}C)	40 - 100

The experiment was repeated three times to ensure repeatability and as practised by other earlier research [38]. **Figure 5(a)** shows the repeatability of the proposed work where MBR microfiber shows the same downtrend on every cycle. The MBR is observed to have good sensitivity between the range of 0.014 dB/^{0}C to 0.0155 dB/^{0}C with 0.01 interval between data. For linearity performances, the MBR microfiber shows stable performance which is >80%.

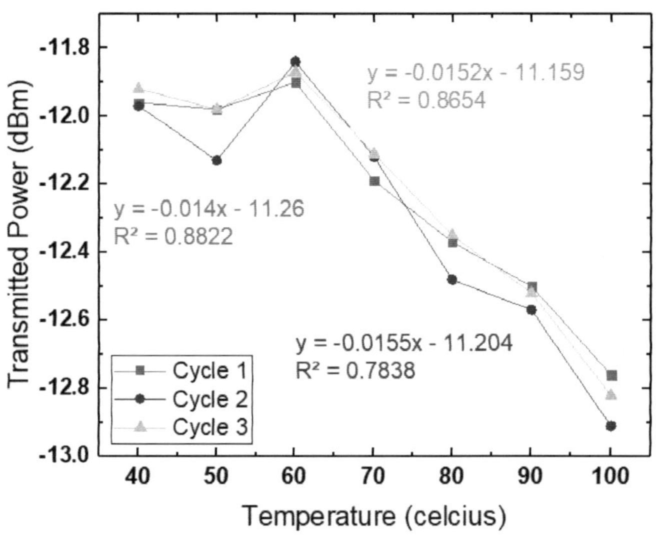

(a)

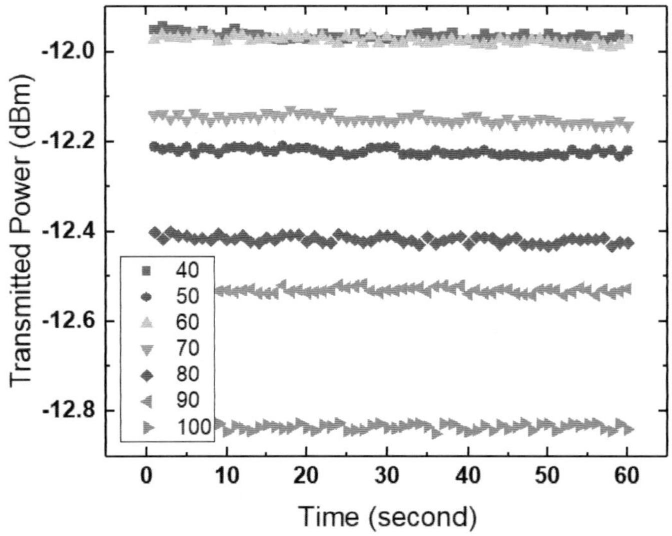

(b)

Figure 5. The (a) repeatability and (b) stability of the MBR microfiber temperature sensor.

The stability test was carried out for the MBR microfiber temperature sensor and recorded in **Figure 5(b)**. The sensing data was taken for 60 seconds where 60 different transmitted power were produced for the changes in each degree of temperature. These results were used as additional information, to observe the condition of the sensor during the experiment. The MBR microfiber showed stable performance while performing temperature sensing, based on the data collected.

4. Conclusions

This proposed work studied the effect of MBR microfiber towards temperature sensor. The method of "soften-and-compress" was applied on SFM-28 silica fiber with several numbers of arc to create bottle structure with a diameter of $D_b = 190\ \mu m$, stem diameter of $D_s = 125$ μm and bottle length of $L_b =$ 182 μm. The characterization of the MBR was defined by the TLS from range of wavelength between 1551 nm to 1560 nm with a step of 0.001 nm via 2 μm optical microfiber. The Q-factor of the MBR was found to have 1.555 x 10^5 which is significantly similar to the previous work. The MBR microfiber performance as a temperature sensor is then compared with no-MBR microfiber. The MBR shows better performance than no-MBR based on linearity, sensitivity and standard deviation analysis. The P-value is defined by calculation and needs to be over than 10^{-5} indicates that the measurement was finely collected. The experiment was repeated three times to ensure stability and accuracy in data collection.

Acknowledgements

The authors would like to acknowledge the University of Malaya, University Technical Malaysia Melaka, INTI International University and Ministry of Education, Malaysia for their financial support.

References

[1] Ahmad, H., A. Zulkhairi, and S. Azzuhri, *Temperature sensor and fiber laser based on optical microfiber knot resonator.* Optik-International Journal for Light and Electron Optics, 2018. **154**: p. 294-302.
[2] Bianucci, P., *Optical microbottle resonators for sensing.* Sensors, 2016. **16**(11): p. 1841.
[3] Gil-Santos, E., et al., *Scalable high-precision tuning of photonic resonators by resonant cavity-enhanced photoelectrochemical etching.* Nature communications, 2017. **8**: p. 14267.
[4] Longhi, S. and L. Feng, *Unidirectional lasing in semiconductor microring lasers at an exceptional point.* Photonics Research, 2017. **5**(6): p. B1-B6.
[5] Jali, M., et al. *Optical characterization of microfiber loop resonator towards humidity sensing.* in *Journal of Physics: Conference Series*. 2019. IOP Publishing.

[6] Johari, M., et al. *Microbottle resonator formaldehyde sensor*. in *Journal of Physics: Conference Series*. 2019. IOP Publishing.

[7] Little, B.E., et al., *Microring resonator channel dropping filters.* Journal of lightwave technology, 1997. **15**(6): p. 998-1005.

[8] Absil, P.P., *Microring resonators for wavelength division multiplexing and integrated photonics applications.* 2000.

[9] Kokubun, Y., *Wavelength selective integrated device by vertically coupled microring resonator filter.* Photonics Based on Wavelength Integration and Manipulation IPAP Books, 2005. **2**: p. 303-316.

[10] Armani, D., et al., *Ultra-high-Q toroid microcavity on a chip.* Nature, 2003. **421**(6926): p. 925.

[11] Armani, D., et al., *Electrical thermo-optic tuning of ultrahigh-Q microtoroid resonators.* Applied physics letters, 2004. **85**(22): p. 5439-5441.

[12] Gérard, J., et al., *Enhanced spontaneous emission by quantum boxes in a monolithic optical microcavity.* Physical review letters, 1998. **81**(5): p. 1110.

[13] Sarma, J. and K. Shore, *Electromagnetic theory for optical disc resonators.* IEE Proceedings J (Optoelectronics), 1985. **132**(6): p. 325-330.

[14] Srinivasan, K., et al., *Experimental demonstration of a high quality factor photonic crystal microcavity.* Applied Physics Letters, 2003. **83**(10): p. 1915-1917.

[15] Lee, P.-T., et al., *Investigation of whispering gallery mode dependence on cavity geometry of quasiperiodic photonic crystal microcavity lasers.* Applied physics letters, 2006. **89**(23): p. 231111.

[16] Johari, M.A.M., et al., *WHISPERING GALLERY MODES ON OPTICAL MICRO-BOTTLE RESONATOR FOR HUMIDITY SENSOR APPLICATION.* Optik, 2019.

[17] Jali, M.H., et al., *Optical characterization of different waist diameter on microfiber loop resonator humidity sensor.* Sensors and Actuators A: Physical, 2019. **285**: p. 200-209.

[18] 18.Lee, P.-T., et al., *High quality factor microcavity lasers realized by circular photonic crystal with isotropic photonic band gap effect.* Applied physics letters, 2007. **90**(15): p. 151125.

[19] Johari, M.A.M., et al., *EFFECT OF SIZE ON SINGLE AND DOUBLE OPTICAL MICROBOTTLE RESONATOR HUMIDITY SENSORS.* Sensors and Actuators A: Physical, 2018.

[20] Johari, M., et al., *Microbottle resonator for formaldehyde liquid sensing.* Optik, 2018. **173**: p. 180-184.

[21] Chiasera, A., et al., *Spherical whispering-gallery-mode microresonators.* Laser & Photonics Reviews, 2010. **4**(3): p. 457-482.

[22] Matsko, A., et al., *Review of applications of whispering-gallery mode resonators in photonics and nonlinear optics.* IPN Progress Report, 2005. **42**(162): p. 1-51.

[23] Berini, P., *Long-range surface plasmon polaritons.* Advances in optics and photonics, 2009. **1**(3): p. 484-588.

[24] Nemova, G. and R. Kashyap, *A compact integrated planar-waveguide refractive-index sensor based on a corrugated metal grating.* Journal of lightwave technology, 2007. **25**(8): p. 2244-2250.

[25] Nemova, G. and R. Kashyap, *Theoretical model of a planar integrated refractive index sensor based on surface plasmon-polariton excitation.* Optics Communications, 2007. **275**(1): p. 76-82.

[26] Jali, M.H., et al., *Applied microfiber evanescent wave on ZnO nanorods coated glass surface towards temperature sensing.* Sensors and Actuators A: Physical, 2018. **277**: p. 103-111.

[27] Yusof, H.H.M., et al., *Detection of Formaldehyde Vapor Using Glass Substrate Coated With Zinc Oxide Nanorods.* IEEE Photonics Journal, 2019. **11**(1): p. 1-9.

[28] Johari, M.A.M., et al. *Corn Oil Concentrations Detection for Food Industry Research Development by Using Application of Fiber Optic Liquid Sensor Concept*. in *MATEC Web of Conferences*. 2017. EDP Sciences.

[29] Johari, M.A.M., et al. *Development of Acetone Liquid Concentration Detection Sensor by Using Fiber Optic for Diabetic Level Detection*. in *MATEC Web of Conferences*. 2017. EDP Sciences.

[30] Murugan, G.S., J.S. Wilkinson, and M.N. Zervas, *Selective excitation of whispering gallery*

modes in a novel bottle microresonator. Optics express, 2009. **17**(14): p. 11916-11925.

[31] Kakarantzas, G., et al., *Miniature all-fiber devices based on CO 2 laser microstructuring of tapered fibers.* Optics Letters, 2001. **26**(15): p. 1137-1139.

[32] Ward, J.M., et al., *Heat-and-pull rig for fiber taper fabrication.* Review of scientific instruments, 2006. **77**(8): p. 083105.

[33] Warken, F., A. Rauschenbeutel, and T. Bartholomaus, *Fiber Pulling Profits from Precise Positioning-Precise motion control improves manufacturing of fiber optical resonators.* Photonics Spectra, 2008. **42**(3): p. 73.

[34] Murugan, G.S., et al., *Hollow-bottle optical microresonators.* Optics express, 2011. **19**(21): p. 20773-20784.

[35] Lim, K., et al., *Fabrication and applications of microfiber*, in *Selected Topics on Optical Fiber Technology*. 2012, InTech.

[36] Nasir, M.N.M., G.S. Murugan, and M.N. Zervas. *Broadly tunable solid microbottle resonator.* in *Photonics Conference (IPC), 2016 IEEE*. 2016. IEEE.

[37] Cai, M., O. Painter, and K.J. Vahala, *Observation of critical coupling in a fiber taper to a silica-microsphere whispering-gallery mode system.* Physical review letters, 2000. **85**(1): p. 74.

[38] Isa, N.M., et al., *Polyaniline Doped Poly (Methyl Methacrylate) Microfiber for Methanol Sensing.* IEEE Sensors Journal, 2018.

Journal of Physics: Conference Series **1371** (2019) 012007 doi:10.1088/1742-6596/1371/1/012007

Q-Switched dual-wavelength erbium-doped fiber laser using graphene as a saturable absorber

R A M Yusoff[1], N Kasim[1], A A A Jafry[1], Y Munajat[1], S W Harun[2], M I Suliman[1]

[1] Department of Physics, Faculty of Science, Universiti Teknologi Malaysia, 81310 Skudai, Johor, Malaysia.
[2] Department of Electrical Engineering, University of Malaya, 50603 Kuala Lumpur, Malaysia.

r.addayusoff92@gmail.com; k.nabilah@utm.my

Abstract. We proposed stable Q-switched dual-wavelength initiated by graphene as saturable absorber inside erbium-doped fiber laser cavity. Continuous-wave and pulse spectrum of erbium-doped fiber laser (EDFL) was successfully demonstrated and analysed where the lasing starts at a pump power of 50.4 mW. Threshold pump power generates continuous-wave at 1560 nm which is then shifted to 1558.7 nm as the SA is inserted. Q-switched EDFL generates a maximum repetition rate of 115.9 kHz which corresponds to a pulse width of 41 μs. The generation of dual-wavelength at two different wavelengths are also demonstrated in these works.

1. Introduction

Last year, distinguished scientists in the field of laser physics celebrated their discovery of chirped optical pulses and optical tweezers which make ultrashort pulses as an important scientific tool in the world. Since then, many researchers start to pay more attention to producing efficient ultrafast laser. Recently, most works in laser physics start to focus more on all-fiber laser system as it provides more advantages than solid-state laser. Its reliability, flexibility, compactness and maintenance-free operation make it more useful than other laser systems. There are two techniques to generate pulses in all-fiber laser system, which are active and passive means. The latter offers more flexibility and feasibility as it did not require additional mirror or lenses inside the system [1]. Therefore, the passive technique is frequently used with the implementation of material inside the laser cavity. Among materials, 2-dimensional material attracts more research attention as it demonstrates excellent confinement of electron and holes inside the atom [2].

In 2003, Geim and Novoselov had discovered a promising carbon allotrope, graphene [3]. This Nobel prizes winning material had drawn vast research attention due to its excellent condensed matter properties as well as its dynamics charge carriers density (massless Dirac Fermion) [4]. Mobility of graphene seems unique as no scattering occurs even after thousands of interatomic distances. In fact, an atomically thin carbon precursor possessed strong interatomic bonds which prevent lattice dislocations and crystal defects even at high temperature [5]. Thus, Bao *et al.* demonstrated a functional picosecond laser using graphene by manipulating its saturable absorption properties in erbium-doped fiber laser (EDFL). This ability is attributed to low saturation intensity and high photocarrier density

Journal of Physics: Conference Series **1371** (2019) 012007 doi:10.1088/1742-6596/1371/1/012007

[6]. This zero-gap semimetal exhibit ultrafast recovery time (~200 fs) which are able to operate at broad operation regimes (1-2 μm). In this work, we demonstrate a Q-switched dual-wavelength operates at 1.55 μm region by utilizing the erbium-doped fiber (EDF) cavity implemented with graphene as saturable absorber.

2. Experimental setup

In this experiment, the fiber laser consists of a 980 nm laser pump. The EDFL is utilized as a lasing medium since it produces laser in a wavelength of 1550 nm. The ring configuration gives an advantage to loop the beam into the system. The laser diode is connected to a wavelength-division multiplexer (WDM). The input of WDM is linked to the patch cable via 980 nm port with a standard type of connector. After that, the output of WDM is connected into the fiber with doping material, erbium. Then the optical coupler is connected to the end side of the EDF and divided into two sections which are 10% and 90% output power. 90% output is used to loop back the signal into WDM while 10% output of optical coupler is connected to measuring instruments for analysis purposes. Measuring tools used are optical power meter and optical spectrum analyser to observe the laser spectrum. Next, 10% output is connected to fiber Bragg grating (FBG) to operates at 1550 nm to observe the dual wavelength result. Figure 1 shows the overview schematic diagram for analysing the laser erbium doped setup.

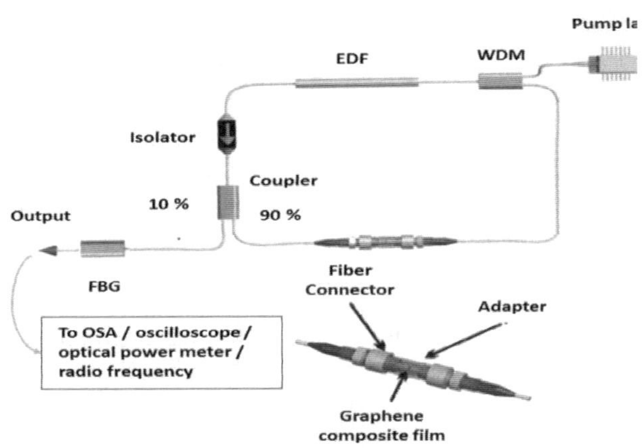

Figure 1. Erbium-doped fiber laser setup.

3. Results and Discussion

Figure 2 compares the output spectrum for the EDF with and without the graphene as saturable absorber when the pump power is fixed at a current of 400 mA. We can observe the operating wavelength of the laser shifts from 1560 nm to 1558.7 nm with the incorporation of the SA. This is attributed to the cavity loss of the resonator. To compensate for the loss, the laser is operated at a shorter wavelength near the peak absorption wavelength of the EDF to acquire more gain. Without the SA, the EDF operates in continuous mode.

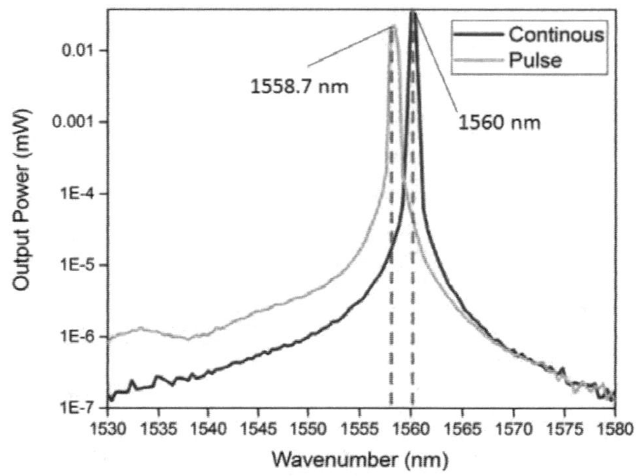

Figure 2. Output spectra from both EDFLs configured with and without the SA at a pump power of 400 mW.

The 10% of the light was coupled to the oscilloscope and the 90% port was coupled to the saturable absorber. The current supply to the laser diode pump starts from 260 mA until 490 mA. Those values correspond to the pump power of 115.83 mW with a maximum pump power of 231.61 mW. The production power growth as the input power increased. The repetition rate and pulse width of the Q-switched pulses generated from this fiber laser taken against the laser diode pump power are shown in figure 3. It can be seen from the figure that the repetition rate increases almost linearly against the pump power from 68.12 kHz at a pump power of 75.9 mW, which is the Q-switching threshold, to a maximum value of 115.9 kHz at a pump power of 136.6 mW. It must be noted that the repetition rate of the EDFL is not limited to this value, and higher repetition rates can be obtained if a pump laser diode with a higher output power is used. By adjusting the pump power from 115.83 mW to 231.61 mW, pulse width displayed reading of 41 μs to 20 μs. However, this work is limited by a maximum available pump power of 137 mW. Figure 4 depicts the output pulse train of 300 cm EDF Q-switched fiber laser, taken from the oscilloscope that was connected to the photodiode. It operates at the same pump power of 115.8 mW, with the repetition rate value of 24.39 kHz. This corresponds to a time interval of 41 μs between the pulses in the pulse train. Δt in the figure denotes the pulse width value obtained at 115.8 mW pump power. The intensity of the peaks is almost constant at 7.8 mV, indicating that the output of the laser is adequately stable.

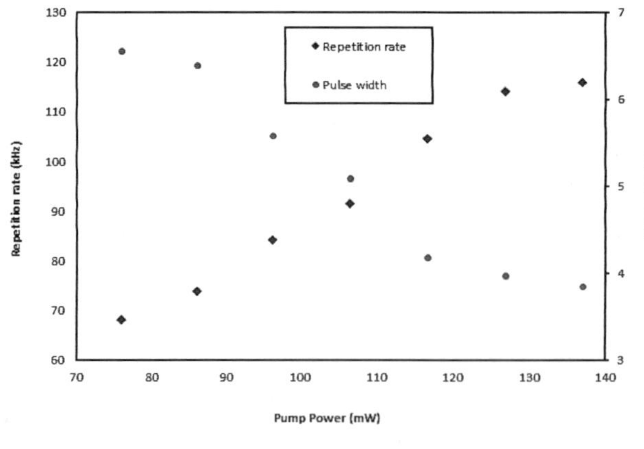

Figure 3. Pulse repetition rate and pulse width against pump power.

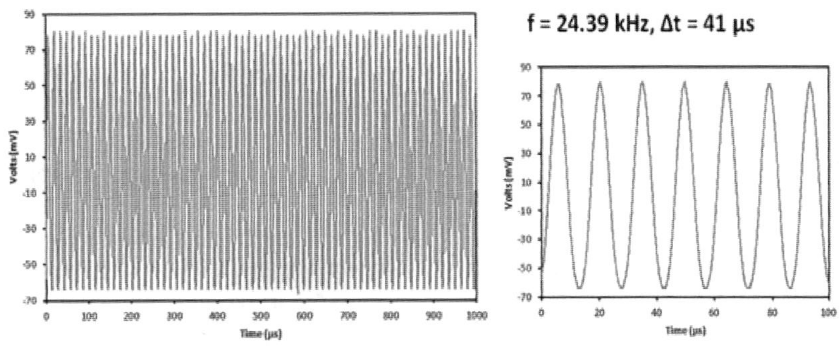

Figure 4. An output pulse train of the graphene-based Q-switched EDFL

The experimental setup for the dual-wavelength lasing spectrum of graphene EDFL is shown in Figure 1. The system was constructed using Fiber Bragg grating as a wavelength selective mirror. FBG used 95.78% reflection of the laser back to the cavity. The peak of the reflectivity is about 1550 nm. By considering only 4% of the transmission FBG as the output laser, wide-band wavelength spectrum is visible confirming the existence of FBG in the system. Figure 5 shows the dual wavelength spectra that oscillate at two wavelengths which are 1547.3 nm and 1554.6 nm. However, the result gained is not strong as there is no obvious result for dual wavelength to appear.

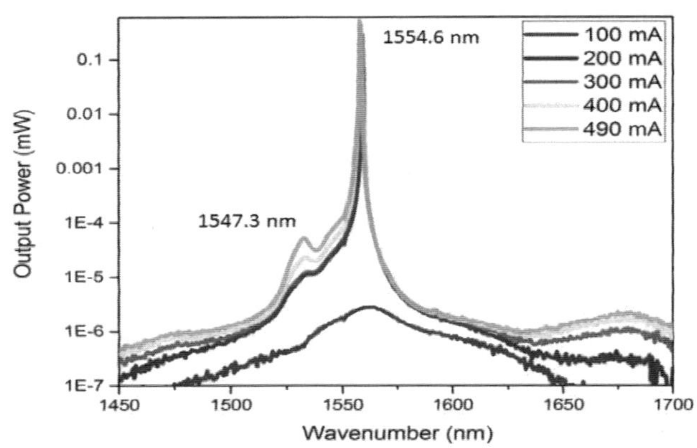

Figure 5. A typical dual-wavelength lasing spectrum of graphene Q-switched EDFL

4. Conclusions

The experiment was successfully conducted for the passively Q-switching, dual wavelength EDFL. Up to this point, the objective of exploring and investigating graphene as saturable absorber has been fulfilled. However, there are many other avenues in which this research could be continued and expanded. One of them is to explore and investigate the superiority of graphene oxide in place of graphene as the saturable absorber. Tunable Q-switched fiber lasers can also find many potential applications in various fields. Thus, in this work, three types of wavelength selective elements are suggested to improve the data from the previous result in generating tunable Q-switched fiber lasers.

Acknowledgements

This research is fully supported by FRGS grant, R.J130000.7826.4F928. The authors fully acknowledged the Ministry of Education (MOE) Malaysia and Universiti Teknologi Malaysia for the approved fund which makes this important research viable and effective.

References

[1] Jafry AAA, Arof H, Kasim N, Munajat Y, Apsari R, Harun SW. Q-switched Erbium-doped Fiber Laser Using MoS_2 Deposited Side-polished D-shape Fiber. *Nonlinear Optics Quantum Optics-Concepts in Modern Optics*. **49(3-4)**:285-293.

[2] Mahyuddin MBH, Latiff AA, Rusdi MFM, Irawati N, Harun SW. Quantum dot cadmium selenide as a saturable absorber for Q-switched and mode-locked double-clad ytterbium-doped fiber lasers. *Optics Communication*. **397**:147-52.

[3] Novoselov KS, Geim AK, Morozov SV, Jiang D, Zhang Y, Dubonos SV, et al. Electric field effect in atomically thin carbon films. *Science*. **306(5696)**:666-9.

[4] Geim AK, Novoselov KS. The rise of graphene. *Nature Materials*. **6(3)**:183-91.

[5] Bonaccorso F, Sun Z, Hasan T, Ferrari AC. Graphene photonics and optoelectronics. *Nature Photonics*. **4(9)**:611-22.

[6] Bao QL, Zhang H, Wang Y, Ni ZH, Yan YL, Shen ZX, Loh KP, Tang DY. Atomic-Layer Graphene as a Saturable Absorber for Ultrafast Pulsed Lasers. *Advanced Functional Materials*. **19(19)**:3077-83.

Q-switched Thulium-doped fibre laser using Bismuth (III) Telluride-based saturable absorber

M F M Rusdi[1], X S Cheng[2], A A Latiff [3], A H A Rosol[1], M T Ahmad[1], M F A Rahman[1,4], SW Harun[1]

[1]Photonics Engineering Laboratory, Department of Electrical Engineering, Faculty of Engineering, University of Malaya, 50603 Kuala Lumpur, Malaysia
[2]School of Engineering, KDU University College, Shah Alam, Selangor Darul Ehsan, Malaysia
[3]Faculty of Electronic and Computer Engineering, Universiti Teknikal Malaysia Melaka, 76100 Hang Tuah Jaya, Melaka, Malaysia
[4]Fakulti Teknologi Kejuruteraan Elektrik & Elektronik, Universiti Teknikal Malaysia Melaka, 76100 Hang Tuah Jaya, Melaka, Malaysia.

xs.cheng@kdu.edu.my

Abstract. We demonstrate a passive Q-switched thulium-doped fibre laser (TDFL) pulse generation using bismuth (III) telluride (Bi_2Te_3) thin film as saturable absorber. The Bi_2Te_3 was fabricated by a simple processing technique; simply by embedding the Bi_2Te_3 into the polyvinyl alcohol (PVA) film. By sandwiching 1 mm x 1 mm Bi_2Te_3 thin film between two fibre ferrules in a TDFL ring cavity, a stable Q-switching pulses train was generated. The repetition rate and pulse width were tuneable from 10.72 kHz to 39.9 kHz and 13.08 μs to 5.98 μs, respectively, as the pump power increased from 309 mW to 454 mW. The slope efficiency was 4.51%, while, the maximum pulse energy was calculated as 0.17 μJ.

1. Introduction

Passively Q-switched pulsed fibre laser operating at 2-micron wavelength region have attracted growing attention in recent years as it can be used in various applications such as military, spectroscopy, medical surgery, and light detection and ranging (LIDAR) [1, 2]. Compared to an active Q-switched laser, the passive technique is more popular in the recent years, because of its dimension which is small in nature, cheap and easy in fabrication process, as well as can provide considerably high laser efficiency [3]. The passively Q-switched pulses can be generated by using a saturable absorber (SA), which modulates a continuous wave (CW) into a repeated pulses train. Up to date, many passive SAs have been demonstrated including semiconductor saturable absorber mirrors (SESAMs), single-wall carbon nanotube (SWCNT), carbon nanotubes (CNT), and two dimensional (2D) materials such as graphene, transition metal dichalcogenides (TMD), topological insulator (TI) and black phosphorus (BP) as well as other types of SA [4-12]. Among these materials, graphene has attracted much research interest due to its small band gap, sufficient modulation depth and efficiency saturable properties [13, 14].

Previously, passively Q-switched fibre lasers were demonstrated by using Bismuth (III) Selenide (Bi_2Se_3) and Bismuth (III) Telluride (Bi_2Te_3) SAs at 1.55 μm region [15-19]. In this paper, we demonstrated a passively Q-switched Thulium-doped fibre laser (TDFL) by incorporating Bi_2Te_3 thin film SA, The Bi_2Te_3 film SA was prepared by dissolving Bi_2Te_3 powder into PVA host solution. The formed thin film was then integrated into a ring cavity fibre laser for generating a self-pushed Q-switching pulses train.

2. Preparation and Characterization of Bi_2Te_3 SA

The Bi_2Te_3-SA was formed by a simple method process; where, 14 mg of Bi_2Te_3 powder with a molecular weight of 800.76 g/mol was dissolved in 5 ml of the polyvinyl alcohol (PVA) (Sigma Aldrich) solution using a magnetic stirrer for three hours. Then, the Bi_2Te_3-PVA solution was placed in the ultrasonic bath for about 10 minutes to ensure all the Bi_2Te_3 powder fully dissolved within the PVA. This process was repeated for a few times until the Bi_2Te_3-PVA solution fully dissolved. The prepared solution was then poured onto the petri dish. It was left to dry at a room temperature for 48 hours to form a Bi_2Te_3-PVA composite film. The Bi_2Te_3-PVA thin film was then cut into 1 mm x 1 mm dimension and sandwiched between two FC/PC fibre ferrules. The nonlinear optical profile of the Bi_2Te_3 film was investigated via a twin-balance detector measurement technique. A stable self-mode-locked fibre laser with a repetition rate of 1 MHz was used as an input laser source. A reference power for normalization was recorded as a function of incident intensity by varying the input power. Figure 1 shows the data obtained for the nonlinear absorption curve of Bi_2Te_3, where these data were fitted based on a simple two-level SA model; $T(I) = 1 - \alpha_o \,.\exp(-I/I_{sat}) - \alpha_{ns}$,where $T(I)$, α_o I, I_{sat} and α_{ns} stand for the transmission, modulation depth, input intensity, saturation intensity and non-saturable absorption, respectively [20]. The modulation depth and non-saturable absorption were measured to be 30% and 55%, respectively.

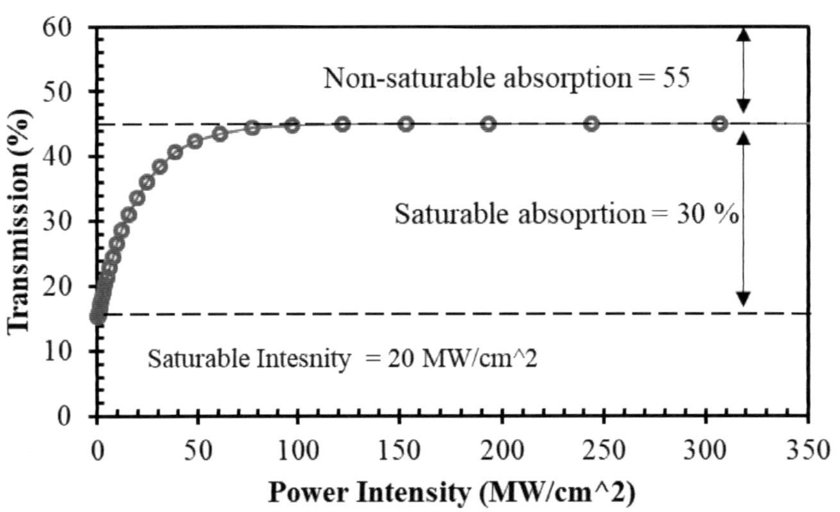

Figure 1. Nonlinear absorption curve of the Bi_2Te_3 thin film SA.

3. Q-switched TDFL Experimental setup

The TDFL experimental setup is shown in Figure 2. The laser cavity consists of 1552 nm erbium ytterbium co-doped fibre laser (EYDFL) pump power, 1550/2000 nm wavelength division multiplexer (WDM), 5 m TDF as a gain medium and the optical coupler of 90/10. TDF used (Nuferm SM-TSF-9/125) has a core and cladding diameter of 9 μm and 125 μm respectively. The total cavity length of TDFL ring cavity is about 13.1 m and composed of single mode fibre (SMF) with the group velocity dispersion (GVD) of 80 ps^2/km at 1990 nm. To minimize the insertion loss in the cavity, all connections of fibre are directly spliced except for a joint, where to put the SA. The Bi_2Te_3 film SA was cut into 1 mm and 1 mm and then sandwiched between two FC/PC fibre ferrules. The 90/10 optical coupler splits the output laser into two, where 90% is circulated back into the ring cavity, while, 10% is used as the measurement output. The measurement devices used in this experiment consists of an optical spectrum analyzer (Yokogawa AQ6370B); to measure the optical spectra of the laser output, while, the pulses train and radio frequency (RF) spectra were measured via an oscilloscope (LeCroy Wavejet 352A) and RF spectrum analyzer (Anritsu MS2683A), respectively.

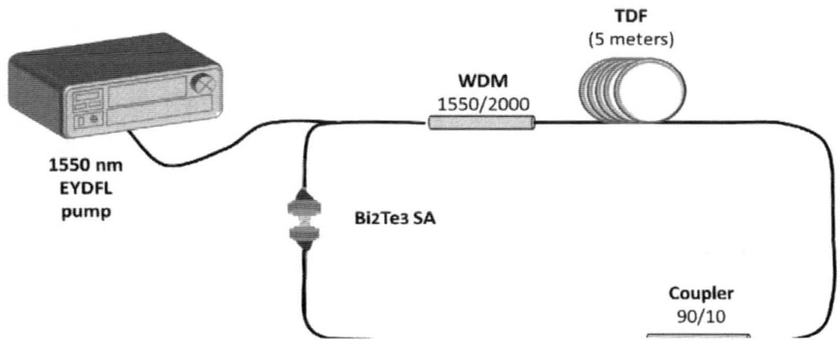

Figure 2. The proposed TDFL ring cavity experimental setup.

4. Result and discussion

Figure 3 shows the performance of Q-switched TDFL using the Bi_2Te_3 thin film as SA. A self-generated Q-switched TDFL appears as the pump power increase from the threshold of 309 mW to the maximum pump power of 454 mW. Figure 3(a) shows the optical spectrum of Q-switched TDFL at the threshold pump power. As illustrated, the blue color of the spectrum is a continuous-wave (CW) spectrum, while, the red color is a Q-switched TDFL spectrum, with each of them has a central wavelength of 1940.33 nm and 1968.99 nm, respectively. After the SA integrated into the TDFL ring cavity, the spectrum shifted to the left and this phenomenon is called as a blue shifted. Figure 3(b) shows the characteristic of the output power, that increases from 0.3 mW to 6.6 mW, when the pump power is increased. As shown, the pulse energy also increases from 0.03 μJ to 0.17 μJ. The slope efficiency obtained from the output power profile is measured as 4.51%. Figure 3(c) shows the pulse train of the Q-switched TDFL at the maximum pump power of 454 mW. The repetition rate is observed as 39.5 kHz. The insert of Figure 3(c) shows the enlargement image of the pulse train, where it can be seen, that the measured peak-to-peak is 26 μs. The fundamental frequency of the Q-switched TDFL with several harmonics is shown in Figure 3(d). At the maximum pump power of 454 mW, the signal-to-noise ratio (SNR) is measured as 38.73 dB. The fundamental frequency is obtained as 39.9 kHz, which confirms the validity of the obtained pulses train in the time domain (Figure 3(c)).

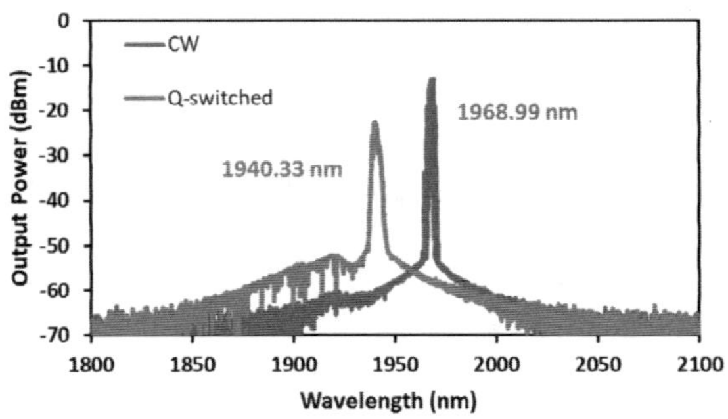

(a)

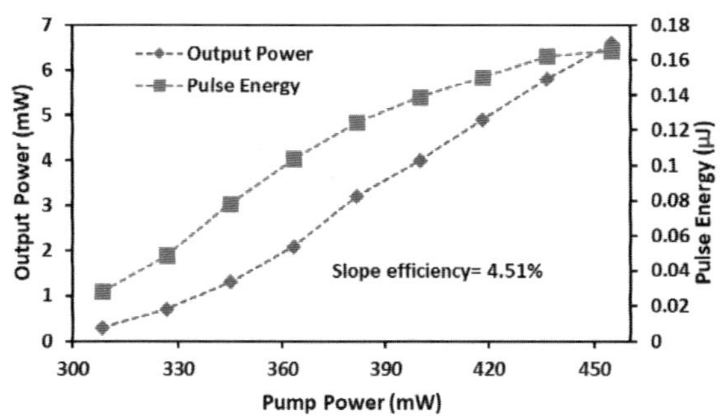

(b)

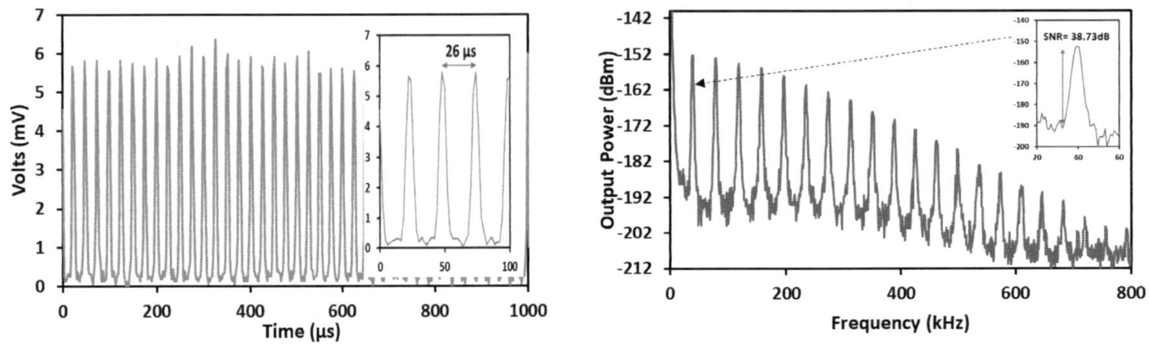

(c) (d)

Figure 3. Performance of the Q-switched TDFL by using Bi_2Te_3 thin film SA. (a) Optical spectrum. (b) Output power and pulse energy (c) Pulse train in the time domain (d) RF spectrum.

5. Conclusion

We have demonstrated a passively Q-switched TDFL using Bi_2Te_3 thin film SA, which was obtained via a simple processing technique; by embedding Bi_2Te_3 elements into the PVA. The Q-switched TDFL self- generated at a threshold pump power of 309 mW and was centred at 1940.33 nm wavelength. The repetition rate was tuneable within a range of 10.72 - 39.9 kHz, as the pump power varied from 309 mW to 454 mW. This allowed the formation of a pulse width within a range of 13.08 - 5.98 μs. The maximum pulse energy was calculated as 0.17 μJ, while, the corresponding slope efficiency was obtained as 4.51%.

Acknowledgment

This work is financially supported by the KDU University College (Grant No: KDURG/2017/1/011).

Reference

[1] Gong, W., T.H. Chyba, and D.A. Temple, *Eye-safe compact scanning LIDAR technology.* Optics and lasers in engineering, 2007. **45**(8): p. 898-906.

[2] Walsh, B.M., *Review of Tm and Ho materials; spectroscopy and lasers.* Laser physics, 2009. **19**(4): p. 855.

[3] Rahman, M.F.A., et al., *An 8 cm long holmium-doped fiber saturable absorber for Q-switched fiber laser generation at 2-μm region.* Optical Fiber Technology, 2018. **43**: p. 67-71.

[4] Zirngibl, M., et al., *1.2 ps pulses from passively mode-locked laser diode pumped Er-doped fibre ring laser.* Electronics Letters, 1991. **27**(19): p. 1734-1735.

[5] Zhang, M., et al., *Tm-doped fiber laser mode-locked by graphene-polymer composite.* Optics express, 2012. **20**(22): p. 25077-25084.

[6] Hisyam, M.B., et al., *Generation of Mode-locked Ytterbium doped fiber ring laser using few-layer black phosphorus as a saturable absorber.* IEEE Journal of Selected Topics in Quantum Electronics, 2016. **23**(1): p. 39-43.

[7] Zakaria, U., et al., *Q-switched hafnium bismuth erbium-doped fiber laser with bismuth (III) telluride based saturable absorber.* Chalcogenide Letters, 2018. **15**(4): p. 181-186.

[8] Rusdi, M., et al., *Q-switched and mode-locked thulium doped fiber lasers with nickel oxide film saturable absorber.* Optics Communications, 2019. **447**: p. 6-12.

[9] Baharom, M., et al., *Lutetium oxide film as a passive saturable absorber for generating Q-switched fiber laser at 1570 nm wavelength.* Optical Fiber Technology, 2019. **50**: p. 82-86.

[10] Rahman, M.F., et al., *Q-switched ytterbium-doped fiber laser via a thulium-doped fiber saturable absorber.* Applied Optics, 2018. **57**(22): p. 6510-6515.

[11] Rahman, M., et al., *Q-switched and mode-locked thulium-doped fiber laser with pure Antimony film Saturable absorber.* Optics Communications, 2018. **421**: p. 99-104.

[12] Rahman, M., et al., *An 8 cm long holmium-doped fiber saturable absorber for Q-switched fiber laser generation at 2-μm region.* Optical Fiber Technology, 2018. **43**: p. 67-71.

[13] Kasim, N., et al., *Short-pulsed Q-switched Thulium doped fiber laser with graphene oxide as a*

saturable absorber. Optik, 2018. **168**: p. 462-466.

[14] Rosol, A., et al. *Graphene Oxide saturable absorber for generating eye-safe Q-switched fiber laser*. in *IOP Conference Series: Materials Science and Engineering*. 2017. IOP Publishing.

[15] Koo, J., et al., *Passively Q-switched 1.56 μm all-fiberized laser based on evanescent field interaction with bulk-structured bismuth telluride topological insulator*. JOSA B, 2014. **31**(9): p. 2157-2162.

[16] Ahmad, H., et al., *A passively Q-switched ytterbium-doped fiber laser based on a few-layer Bi2Se3 saturable absorber*. Laser Physics, 2015. **25**(6): p. 065102.

[17] Haris, H., et al., *Passively Q-switched and mode-locked Erbium-doped fiber laser with topological insulator Bismuth Selenide (Bi2Se3) as saturable absorber at C-band region*. Optical Fiber Technology, 2019. **48**: p. 117-122.

[18] Rahman, M.F.A., et al., *Passively Q-switched fibre laser utilizing erbium-doped fibre saturable absorber for operation in C-band region*. Journal of Modern Optics, 2019. **66**(2): p. 235-239.

[19] Rahman, M., et al., *Q-Switched Thulium-Doped Fiber Laser Using Vanadium Oxide Saturable Absorber*. Nonlinear Optics, Quantum Optics: Concepts in Modern Optics, 2018. **48**(4).

[20] Tian, Z., et al., *Mode-locked thulium fiber laser with MoS2*. Laser Physics Letters, 2015. **12**(6): p. 065104.

A study of relative humidity sensor on micro-ball resonator

U U M Ali[1], M A M Johari[1], Z Jusoh[2], H A Rahman[3], S W Harun[1]

[1] Department of Electrical Engineering, University of Malaya, 50603 Kuala Lumpur, Malaysia.
[2] Faculty of Electrical Engineering, University Teknologi Mara (Terengganu), 23000 Dungun, Terengganu, Malaysia
[3] Faculty of Electrical Engineering, Universiti Teknologi Mara, 40450 Shah Alam, Selangor, Malaysia

umairah.ali@gmail.com

Abstract. We investigate the use of a class of whispering gallery mode optical resonators, namely, optical micro-ball resonator (MBR) as a relative humidity (RH) sensor. The MBR is fabricated using fusion splicer at the tip of standard optical fibers; single-mode fibers (SMF-28). The MBR is then optically excited by using an $8\mu m$ optical microfiber and was found to have a Q-factor of >10^5. The MBR was then employed as a humidity sensor with an RH range of between 40% to 100%. The MBR RH sensor was found to have a sensitivity 0.284 dB/%, linearity >90% and is superior to the bare microfiber in all measured parameters. The MBR RH sensor was also found to have good repeatability and stability compared to bare microfiber.

1. Introduction

Microresonators (MRs) have been widely studied due to their advantage can confining the light with high-quality factor (Q) and in a small modes volume [1-2]. The MRs have many geometries such as microring, microdisk and microtoroid and this MRs is supports whispering gallery mode (WGM) [3-11]. MR operates by creating continuous internal reflection at specific resonant wavelengths.

A WGM MRs is typically associated with circular-path resonant cavities, formed e.g. by micro-ball [12]. The most commonly used WGM microresonators for sensing are micro-ball, due to their straightforward fabrication and very high Q-factors. In micro-ball resonators, WGMs are resonant modes with small mode volumes and high-quality factors Q. In terms of geometrical optics, a WGM can be represented by an optical ray transmitted exclusively near the microresonator surface due to grazing-angle total internal reflection [13]. This transmission can occur if a light is evanescently coupled into the tapered fiber. A very narrow resonance dip with a full width half maximum (FWHM) on a level of pm appears in the transmission spectrum of the taper due to coupling light into the WGM.

Optical fiber-based relative humidity (RH) sensors have been widely studied due to their advantages such as the feasibility of long-distance sensing, real-time monitoring, and immunity to electromagnetic interference. Various techniques including tapered optical fiber, heterocore optical fiber, and fiber grating have been reported previously [16–20]. In this work, we proposed and experimentally demonstrated a novel differences RH fiber sensor by using a micro-ball resonator (MBR) and bare fiber. Whispering gallery modes (WGM) resonators, in particular, have been extensively studied due to several advantages including ease of fabrication, high Q factors and low intrinsic losses [12,13]. The proposed resonator also compact in size, fast in response, and low in cost.

2. Experimental set-up

In this paper, we microresonator namely micro-ball resonator (MBR) and this MBR was fabricated at the tip of standard optical fibers; single-mode fibers (SMF-28). A single-mode fibre (SMF) was placed in a manually controlled fibre fusion splicer (Furukawa Electric Fitel S178A) of the fibre was heated through plasma arcing. One of the two clamped-ends compressed the fibre in the direction of the arcing-region. This softening and-compressing procedure yields increasing pronounced bulge along the fibre with each number of arcs performed [2]. **Figure 1** shows the fabricated of MBR resonator with 10 times arc. As shown in Figure 1, the diameter of MBR Db=213 μm and stem diameter Ds = 125 μm.

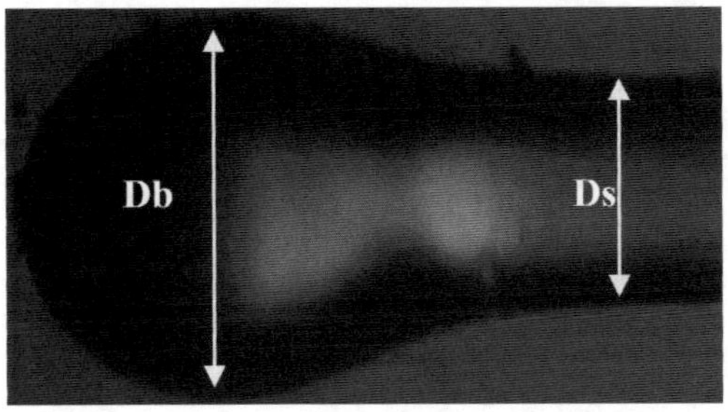

Figure 1. Fabricated micro-ball resonator

In these experiments, a taper with a waist diameter (W) of 8 μm, a waist-length (L1) of 0.1 mm, and a length of the stretching (L2) of 1 mm (see **Figure 2**) was fabricated. The insertion losses of this taper were 13 dB for l λ =1310 nm (a standard communication wavelength). **Figure 3** shows the image of the MBR coupled to a tapered fiber at the centre position with 1550 nm lasing light launched through the tapered fiber.

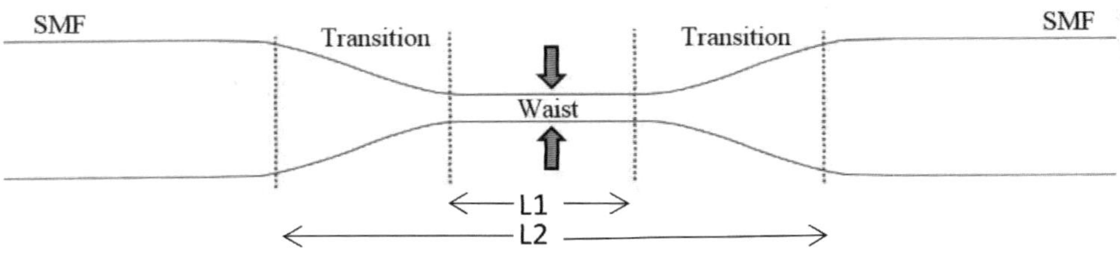

Figure 2. Illustration of tapered fiber

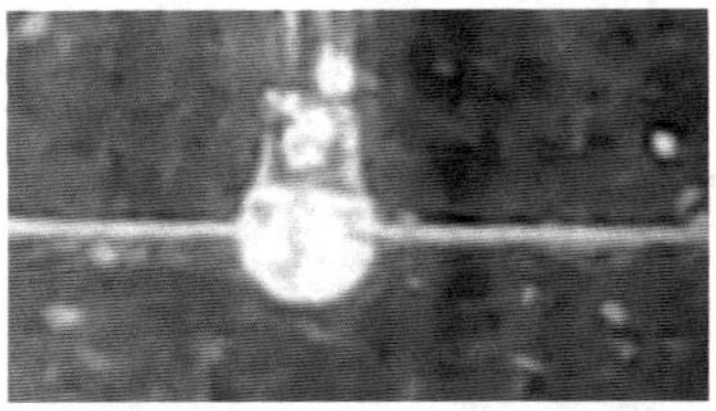

Figure 3. Tapered fiber coupled at the center of MBR

3. Results and discussion

The light source in this investigation is a tuneable laser source (TLS) which operating wavelengths in the range of 1520nm to 1620nm with an average output power of 1dBm. This TLS is launched into the tapered fiber 8μm coupled to a micro-ball resonator. The laser was adjusted wavelength between 1520.2 nm to 1520.3 nm with wavelength interval 0.001nm and the transmitted power is collected by using an optical power meter (THORLABS S145C). **Figure 4** shows the transmission spectrum for a tapered fiber couple with micro-ball resonator (MBR). The insertion loss for MBR around 27.5dBm to 33 dBm, the Q-factor loss could be effective by varying the position or space in between MBR with bare microfiber. The Q-factor identified by $\Delta\lambda/\lambda$, whereas $\Delta\lambda$ is the full width half maximum (FWHM) resonant wavelength and λ is resonant of the wavelength, is discovered to be miniature compared to the previous work [20-22]. The Q-factor for micro-ball resonator is 7.6012×10^5dBm. The fact behind this due to the non-adiabaticity microfiber which amenities notably towards the insertion loss of the full micro-resonators ensemble.

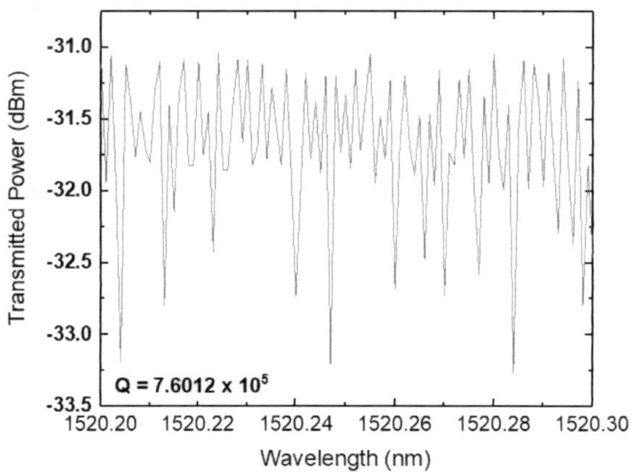

Figure 4. Transmitted Power of micro-ball resonator varies with wavelength

Figure 5 shows the experimental setup of the MBR used to detect relative humidity using the tapered fiber 8μm. The setup consists of a TLS, the proposed resonator, RH measurement meter (humidity), and optical power meter (OPM). Both ends of the tapered micro-fiber are connected to the TLS and OPM, respectively. The TLS is launched into the tapered fiber coupled with MBR placed in a sealed chamber with a dish filled with a saturated salt solution.

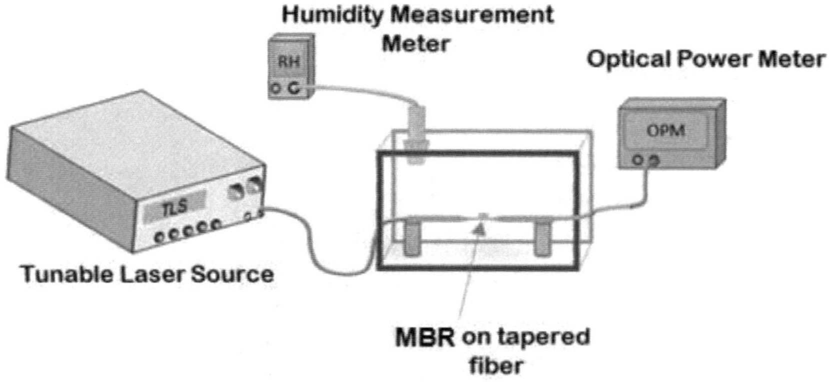

Figure 5. Experimental setup of micro-ball resonator coupled with tapered fiber 8 μm

Figure 6 shows the variation of the transmitted light against the relative humidity for bare micro-fiber and MBR on the tapered fiber. It is observed that the intensity of the transmitted light through the bare micro-fiber and MBR on the tapered fiber increases as relative humidity increases from 40% to 100%. The performance characteristics of the bare micro-fiber and MBR are summarized in **Table 1**. The value sensitivity, linearity, standard deviation and p-value of the MBR are significantly better as compared to the bare micro-fiber. For the bare micro-fiber, the sensitivity is obtained 0.152 dB/%, with slope linearity of 38.99% and limit of detection of 20.197% while the tapered fiber coupled with MBR shows sensitivity 0.2840 dB/% with slope linearity of more than 97.80% and a limit of detection of 19.132%. The outcome of the investigation is that MBR has much better efficiency regarding humidity sensing than bare micro-fiber. One of the cause is surface absorption in transmitted insertion loss of the resonator whereas the losses increasing along with the higher humidity levels. Although losses in the resonator were noticed due to the light circulation inside the resonators, power loss with each pathway of the circle thus significantly increased the sensitivity.

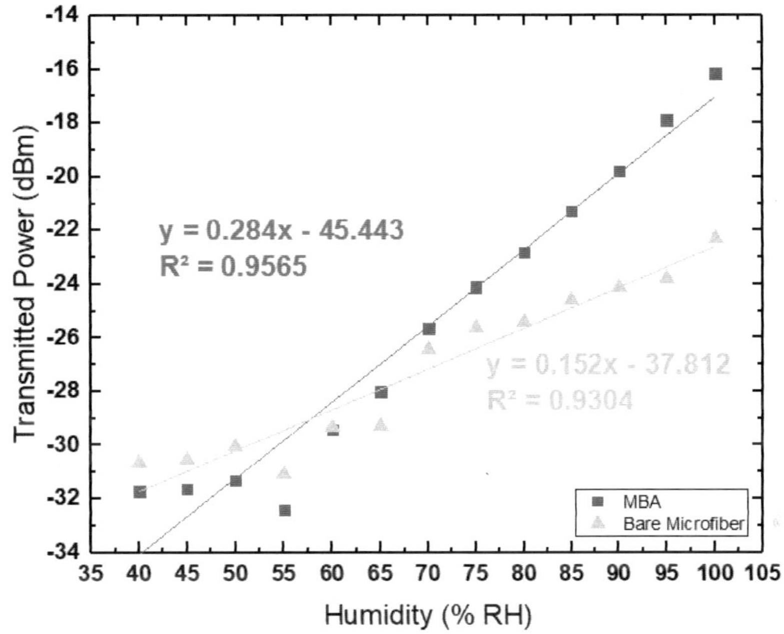

Figure 6. Performances of bare micro-fiber and MBR for the different humidity level

The repeatability of the setup was studied by repeating the experiment three times for MBR and bare microfiber [22]. As showed in **Figure 7**, the results were consistent for MBR and bare microfiber, with similar values of sensitivity – more than 0.25 dB/% for MBR and less than 0.2dB/% for the bare microfiber, respectively. The linearity values for MBR microfiber was 97.8%, as depicted in **Table 1**. Therefore, in general, the MBR performed much better as a humidity sensor as compared to the bare microfiber.

Table 1. Performance of the resonator in humidity sensing activity.

Parameters	Bare micro-fiber	MBR
Linearity (%)	38.99%	97.80%
Sensitivity (dBm/%RH)	0.1520	0.2840
Standard deviation (dBm)	3.07	5.433
Linear Range Humidity (%)	40-100	40 - 100
Limit of detection	20.197	19.132

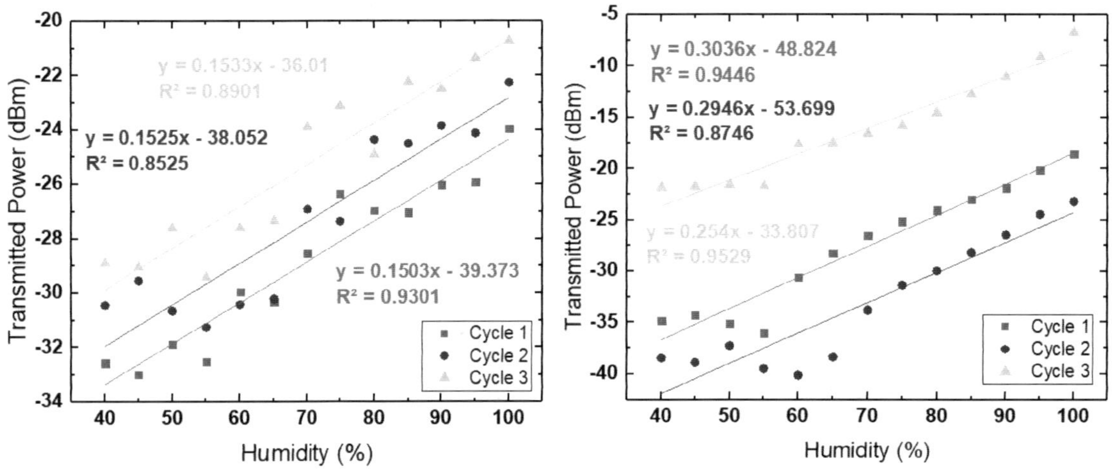

(a) (b)

Figure 7. Repeatability performance of (a) bare micro-fiber and (b) MBR varies with a humidity level

4. Conclusion

We have investigated the use of a class of whispering gallery mode optical resonators, namely, optical micro-ball resonator (MBR) as a relative humidity (RH) sensor. The MBR was employed as a humidity sensor with an RH range of between 40% to 100%. The MBR RH sensor was observed to have a sensitivity of 0.284 dB/%, linearity >90% and showed relatively good repeatability and stability.

References

[1] Chiasera, A., et al., *Spherical whispering-gallery-mode microresonators.* Laser & Photonics Reviews, 2010. **4**(3): p. 457-482.

[2] Murugan, G.S., J.S. Wilkinson, and M.N. Zervas, *Selective excitationof whispering gallery-modes in a novel bottle microresonators.* Optica express, 2009. **17**(14): p. 11916-11925.

[3] Little, B.E., et al., *Microring resonator channel dropping filters.* Journal of lightwave technology, 1997. **15**(6): p. 998-1005.

[4] Absil, P.P., *Microring resonators for wavelength division multiplexing and integrated photonics applications.* 2000.

[5] Kokubun, Y., *Wavelength selective integrated device by vertically coupled microring resonator filter.* Photonics Based on Wavelength Integration and Manipulation IPAP Books, 2005. **2**: p.

303-316.

[6] Armani, D., et al., *Ultra-high-Q toroid microcavity on a chip.* Nature, 2003. **421**(6926): p. 925.

[7] Armani, D., et al., *Electrical thermo-optic tuning of ultrahigh-Q microtoroid resonators.* Applied physics letters, 2004. **85**(22): p. 5439-5441.

[8] Gérard, J., et al., *Enhanced spontaneous emission by quantum boxes in a monolithic optical microcavity.* Physical review letters, 1998. **81**(5): p. 1110.

[9] Sarma, J. and K. Shore, *Electromagnetic theory for optical disc resonators.* IEE Proceedings J (Optoelectronics), 1985. **132**(6): p. 325-330.

[10] Srinivasan, K., et al., *Experimental demonstration of a high quality factor photonic crystal microcavity.* Applied Physics Letters, 2003. **83**(10): p. 1915-1917.

[11] Lee, P.-T., et al., *Investigation of whispering gallery mode dependence on cavity geometry of quasiperiodic photonic crystal microcavity lasers.* Applied physics letters, 2006. **89**(23): p. 231111.

[12] Marcatili, E.A.J., "Bends in optical dielectric waveguides", Bell System Techn. J. 48, 2103-2132 (1969).

[13] Braginsky, V.B., Gorodetsky, M.L., and Ilchenko, V.S., "Quality-factor and nonlinear properties of optical whispering-gallery modes", Phys. Lett. A 137, 393-396 (1989).

[14] Lee, P.-T., et al., *High quality factor microcavity lasers realized by circular photonic crystal with isotropic photonic band gap effect.* Applied physics letters, 2007. **90**(15): p. 151125.

[15] Bianucci, P., *Optical microbottle resonators for sensing.* Sensors, 2016. **16**(11): p. 1841.

[16] X. Wang, J. Zhang, Z. Zhu, J. Zhu, Humidity sensing properties of Pd2+ doped ZnO nanotetrapods, Appl. Surf. Sci. 253 (2007) 3168.

[17] L. Zhang, F. Gu, J. Lou, X. Yin, and L. Tong, "Fast detection of humidity with a subwavelength-diameter fiber taper coated with gelatin film," *Optics Express*, vol. 16, no. 17, pp. 13349–13353, 2008.

[18] X. Dong, T. Li, Y. Liu, Y. Li, C. Zhao, and C. Chan, "Polyvinyl alcohol-coated hybrid fiber grating for relative humidity sensing," *Journal of Biomedical Optics*, vol. 16, no. 7, Article ID 077001, 2011.

[19] Q. Wu, Y. Semenova, J. Mathew, P. Wang, and G. Farrell, "Humidity sensor based on a single-mode hetero-core fiber structure," *Optics Letters*, vol. 36, no. 10, pp. 1752–1754, 2011.

[20] T. Li, X. Dong, C. Chan, C. Zhao, and P. Zu, "Humidity sensor based on a multimode-fiber taper coated with polyvinyl alcohol interacting with a fiber bragg grating," *IEEE Sensors Journal*, vol.12, no. 6, pp.2205–2208, 2012.

[21] Murugan, G.S., et al., *Hollow-bottle optical microresonators.* Optics express, 2011. **19**(21): p. 20773-20784.

[22] Isa, N.M., et al., *Polyaniline Doped Poly (Methyl Methacrylate) Microfiber for Methanol Sensing.* IEEE Sensors Journal, 2018.

Q-switched in figure of 8 by using graphite flakes as saturable absorber

Mofaq Alsaady*[1], NA Awang[1] and Thoalfiqar A Zaker[2]

[1]Faculty of applied science & Technology (FAST), department of physics, Universiti Tun Hussein Onn Malaysia, Pagoh, Johor, Malaysia.
[2] Faculty of Science, department of physics Al-hamdaniya University, Iraq

haidermofaq51@gmail.com

Abstract. In this study, a successful experiment has been reported on pulse compression in the Q-switched fiber laser by utilizing graphite flakes as a saturable absorber (SA). The sputter deposition method was used to deposited graphite flakes as SA, incorporated into the bore of the laser in order to compress the width of the pulse in the operation of Q-switching. The small repetition rate at 29.6 kHz was mentioned as the maximum power of the pump with the single-to-noise ratio (SNR) ~30dB was required to compress the pulse. Accordingly, graphite flakes have the ability to be utilized as an effective SA in producing pulse compression and Q-switching mode, that will be may fixedly lead to further enhancement in the fiber laser pulsed.

Keywords: Q-switching fiber laser, graphite flakes, saturable absorber.

1. Introduction

The short pulse laser has been widely used in many fields as medicine, remote sensing, laser treatment, and telecommunications [1,2]. The two common using methods: Mode locking [3], and Q-switching [4]. However, using these methods, the energy of pulse limited to the Nano-Joule level, and the rate of repetition pulse is fixed in many applications given its operating mechanism such as laser machining and marking, the OTDR (optical time domain reflectometry) and laser ranging [5,6], spectroscopy, biomedical diagnoses, fiber communication [7], environmental sensing, microwave optics, metrology and range finding [8]. Ultrafast pulses are normally Q-switched or mode-locked and are enabled by applying active modulation techniques [9] and other methods in order to get short pulses. The Q-switched fiber lasers can be classified depending on the Q-factor type which divided into two types passive and active Q-switched. The passive Q-switched fiber lasers have regained more attention compared to others due to low cost, simple configurations, and their compactness. The Q-switched fiber lasers actively fiber lasers are requiring many additional switching electronics compared with the passive [10,11,12].

On the other hand, SA is considered one of the important items for passively Q-switched fiber laser, and SA is also very important to the output of Q-switching execution which has attracted immense interest and attention of many researchers. In order to achieve a Q-switched fiber laser and passively mode-locking, a SA is required in the laser cavity. Although, the time response of the SA does not necessary to be shorter than the bore round-trip time, but it needs to be shorter when it compared to the time of the iron life in the upper levels of the media given the time of the duration pulse is measured by the time of the gain depletion following the saturation [13]. In addition, the ideal saturable absorber for the operation of Q-switching wants to have high damage threshold, good thermal stability, and large modulation depth, to maintain the operation of the stable pulse with high pulse energy [5]. The SESAM (semiconductor saturable absorber mirror) is one of the most commonly SA which is used for the Q-switched fiber laser passively despite it has various negatives such as small response bandwidth and having a low damage threshold, which makes it very limits to use in the applications of passively Q-switched fiber lasers [11,14]. SESAM is a well-determined method to produce pulses with the fast ultra in Er-doped (Erbium-doped) fiber lasers [1], however, the operation of the SESAMs fabrication is an expensive and complex process.

Graphene has attracted the attention of researchers recently as a new material used in the Q-switching pulse lasers as a saturable absorber [16,11]. In the theoretical side, Graphene is a highly absorbable spectrum and Q- Switch from visible to distant infrared spectroscopy, which differs from classic SA Cr^{2+}: ZnSe, SESAM, etc., [17], graphene has a rapid recovery factor of up to ~200 fs compared with other materials which gave less than this amount [8], In addition, requiring simple fabrication and large response bandwidth [17,18,21,22]. Moreover, graphene has a wavelength independent SA, due to its band gap property zero unique [15] allowing nanoparticle graphene SA to be formed, with its inclusion deepness enhanced between 11% and 20% [5].

Since the discovery of graphene, there have been many studies done using graphene SA in the passively Q-switched fiber lasers [1,11,23-29] which are implausibly systems in this field. Graphene has been provided thermal conductivity and a large response bandwidth compared with the SESAM, these properties are making graphene useful to the produce [11]. Many studies have been demonstrated to support this fact using graphene SA in the tunable Q-switched fiber lasers [2,10-12,30]. In 2011, a Q-switched fiber laser has been developed using a tunable filter and graphene-film SA which wrapped a tunable wavelength range of 32 nm [11]. After two years 2013, Armed et al, have been reported that a graphene Q-switched fiber laser tunable based on Bragg grating fiber tunable, of 10 nm [2], which leads to an increase the complicated the system of the laser. The graphene-film SA which is used in the notified Q-switched fiber laser tunable compared to the graphene SAs based on sharpened fibers give's more benefits such as simplicity to modularize and large interaction length, having a high damage threshold and package [31]. Furthermore, due to the intervention multimode, sharpened fibers can be provided in addition to that can use as tunable refineries in tunable fiber lasers and wavelength-dependent transmissions [32,33]. Therefore, it is expected that peaked fiber coated with graphene use as both SA and a setting item in the tunable Q-switched fiber laser passively, thus leading to production the costly and complicated.

In this study, we report the use of graphite flakes as both a SA and a tunable filter. The graphite SA is collected of a fiber which has been plated with graphite which is a good show for a graphite flakes-based Q-switched fiber laser without the need for any additional setting item.

2. Preparation of the Graphite Flakes

In preparing the graphite flakes, a mechanical exfoliation approach was adopted to fabricate the graphite flake samples (see Figure 1(a)), which is similar to the approach used in preparing Black Phosphorus-based SA for applications of ultra-short pulse laser [34]. The technique we used has advantageous to give its reliability and simplicity, and where the whole process of fabrication is set free from difficult chemical procedures and the use of expensive instruments. First, relatively thin flakes were peeled off from pieces of commercially obtained graphite flakes in which small pieces were placed onto adhesive tape (Figure 1(b)) attached between the end surfaces of two fiber ferrules (Figure 1(c)). This procedure took about 2 minutes to complete as graphite flake material can be easily damaged and must be handled carefully. Next, the graphite flakes-SA was inserted into the laser cavity by applying it between two fiber ferrules with a fiber connector as the display in Figure 1(d).

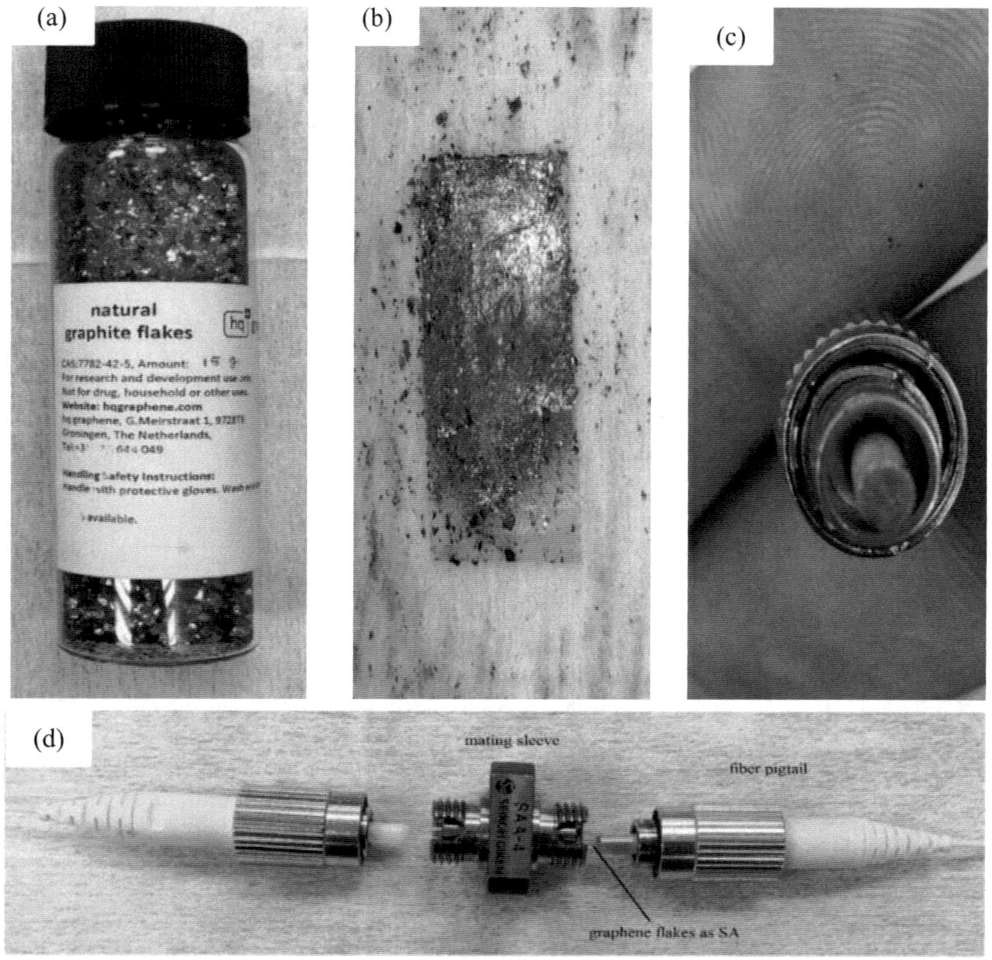

Figure 1.(a) Graphite flakes. (b) Crushed graphite flakes. (c) Graphite flakes deposited on fiber ferrule. (d) Sandwiching the crushed graphite flakes between two ferrules by using a mating sleeve.

3. Experimental Setup

As displayed in Figure 2, a passively Q-switched fiber laser using the graphite flakes as SA was structured using an Oclaro LC 96A74P-20R laser diode with a 980 nm peak wavelength of pump. The laser diode was joint to the laser cavity via a 980/1550 wavelength division multiplexer (WDM). The 2 m length of erbium-doped fiber (EDF, Fibercore) as the gain medium, with a 5.09 dB/m absorption coefficient of a 979 nm wavelength, 5.8 μm mode field diameter (MFD) with a 0.23 numerical aperture (NA), was then connected to the 1550 arm of the WDM. The EDF was next spliced with a 2*2 optical coupler (optical ratio=50:50, operating wavelength and bandwidth=1550∓20) where the output of the coupler was applied to an independent polarisation isolator (ISO) to ensure unidirectional operation when the other one going back to the WDM. A polarisation controller (PC) used after the isolator to control the cavity polarisation. Then to 2 x 1 optical coupler (optical ratio=90:10, operating wavelength and bandwidth=1550±20) was used in the second ring, 90 was back to the 50/50 OC and The fiber (10%) of the OC (optical ratio=10:90, operating wavelength and bandwidth=1550±20), then connected to a 50/50 OC to an optical spectrum analyzer (OSA) and oscilloscope. The laser cavity length is around 11 m. The laser execution was simultaneously detected using (Tektronix MDO 4104) a digital oscilloscope with (PD, Thorlabs DET08CFC/M) 5 GHz photodetector, and (OSA Yokogawa AQ6370D) optical spectrum analyzer. While graphite flakes were used as a saturable absorber in the cavity, minimal graphite grinding occurred on the tape which was applied between the two ferrule fibers.

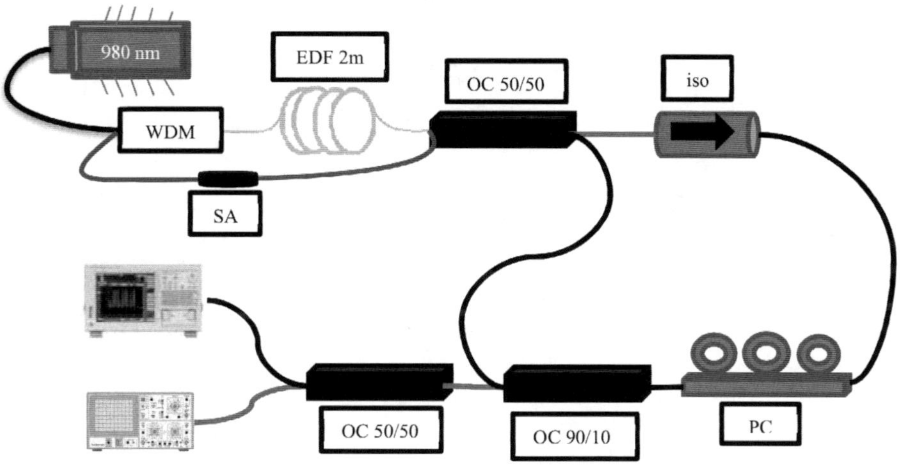

Figure 2. Experimental setup of the passively Q-switched fiber laser using the graphite flakes as SA.

4. Results and Discussion

Figure 3 displays the laser pulse train have taken from the oscilloscope via a 5 GHz bandwidth of the photodetector we used at the (P_p) of 16.6 mW. The pump power (P_p) at 70 mA (16.6 mW), a stable Q-switched pulse was achieved as can be seen in Figure 3(a). With increasing P_p, the laser continues to appear the Q-switched shape, with an increased repetition rate and the P_p. The last pulse appeared at 370 mA (194.5 mW) of the pump power (Figure 3(b)) after that the Q-switched pulse disappeared. However, when the P_p reaches 194.5 mW, the Q-switched pulse train has lost and could not be fixed even with arranging full modification of the PC. For the minimum and maximum finding of the Q-switch, Figure .4. showing a pump power at 16.6 mW the output spectrum listed by the OSA when the cavity was embedded with the graphite flakes.

Journal of Physics: Conference Series **1371** (2019) 012010 doi:10.1088/1742-6596/1371/1/012010

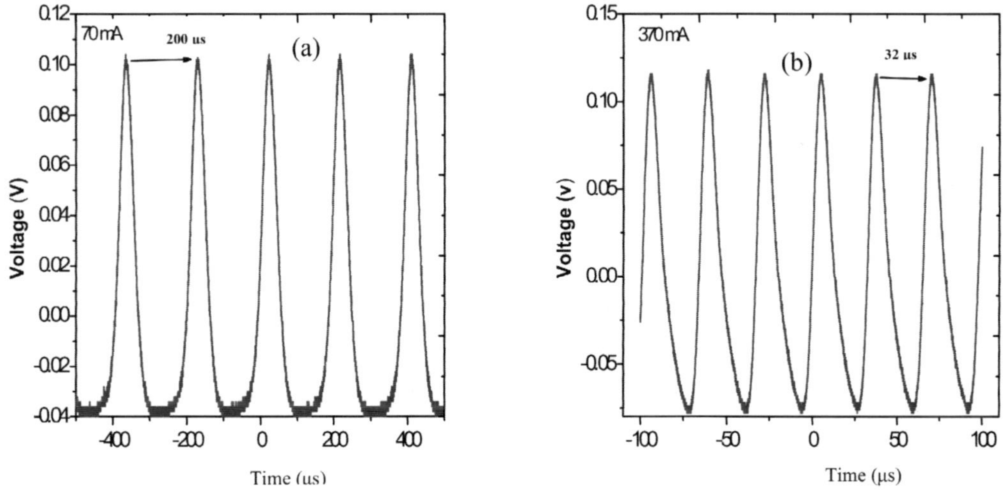

Figure 3. Train pulse of Q-switched at pump powers of (a) 70, (b) 370 (mA).

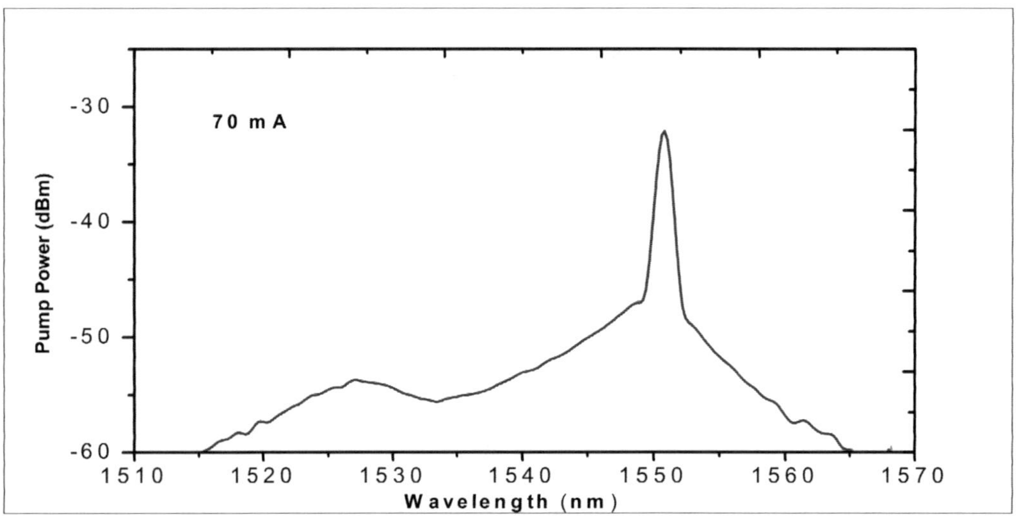

Figure 4. Output spectrum at P_p 16.6 mW.

The laser spectrum was also observed having a broad laser bandwidth at the first reading Q-switched signal, as shown in Figure 4. above. which covered the wavelength range between 1547 nm to 1560 nm. One of the factors that contributed to the broadening of the laser bandwidth was due to the large normal dispersion of the fiber cavity. It is also supposed that this grapheme flakes-SA could be applied to other wavelengths as well. Also, with a small repetition rate, pulse compression can be done which happened when gaining energy from the pump laser. This is because of those non-thermal electrons de-excite via e-e collisions to a thermal bath which deposits its energy into the surrounding lattice or atoms via electron-phonon collisions [35].

In this situation, the free electrons are accelerated in the laser field, where the electron density is finally rising sufficiently to cause pulse compression in the laser cavity to occur [36][37]. Therefore, this demonstrates that graphite flakes have the ability to compress the pulse, which is similar to the metal pulse compressor in the bulk laser system [38]. Figure 5 illustrates a linear increment of the repetition rate from 0 kHz to 29.6 kHz as the increment of the (P_p) from 16.6 mW to 200.4 mW. Whereas, linear decreasing of the pulse width from 0 μs to 6 μs, displaying the normal characteristics of the Q-switching operation [39].

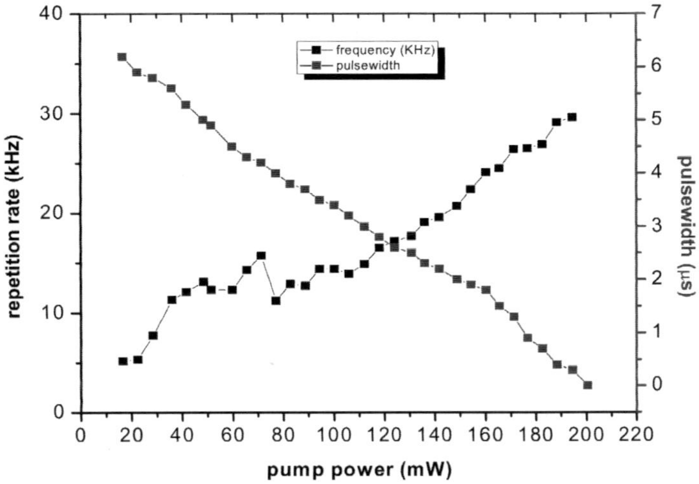

Figure 5. Repetition rate and pulse width against pump

Figure 6 shows the radio frequency (RF) spectrum for generating pulse obtained at 29.7 kHz resolution bandwidth (RBW) and observed having a strong signal peak with a 29.6 kHz repetition rate, and 30 dB signal-to-noise ratio (SNR). Also, with a single repetition rate appearing in the RF spectrum, this suggests that the waveform in the oscilloscope is a cosine wave pattern that corresponds to the broad pulse width.

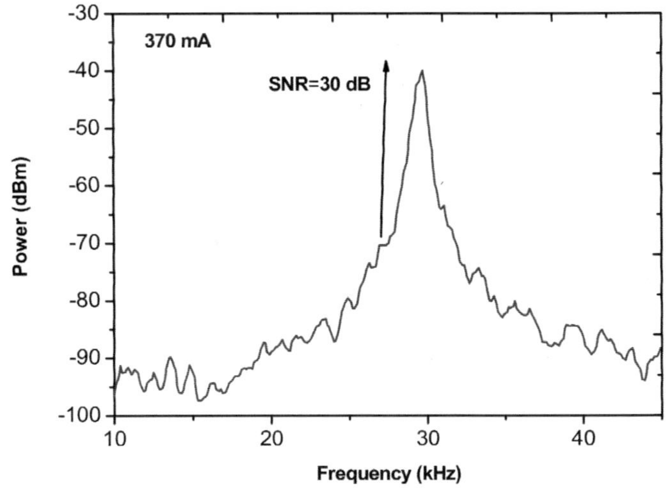

Figure 6. Radio frequency spectrum at 194.5 mW of pump power.

5. Conclusion

The passive Q-switching of an EDFs using graphite flakes as a SA was investigated in this study where this SA generated pulse repetition rates between 0 kHz and 29.6 kHz at the maximum pump power of laser system, with SNR 30 dB. It was shown that a small repetition rate was required to compress the pulse, which has been done by the deployment of graphene SA in the laser cavity. Accordingly, this finding may lead to further research to be undertaken in the photonics field on the capability of the graphene flakes-SA in compression of pulses.

Acknowledgement
We would like to thank Universiti Tun Hussein Onn Malaysia for providing the TIER 1 grant of H139 for funding this project.

References

[1] Wei, L., Zhou, D. P., Fan, H. Y., & Liu, W. K. (2012). Graphene-based Q-switched erbium-doped fiber laser with wide pulse-repetition-rate range. *IEEE Photonics Technology Letters*, *24*(4), 309-311.

[2] Ahmad, H., Zulkifli, M. Z., Muhammad, F. D., Zulkifli, A. Z., & Harun, S. W. (2013). Tunable graphene-based Q-switched erbium-doped fiber laser using fiber Bragg grating. *Journal of Modern Optics*, *60*(3), 202-212.

[3] Haus, H. A. (2000). Mode-locking of lasers. *IEEE Journal of Selected Topics in Quantum Electronics*, *6*(6), 1173-1185.

[4] Wang, Y., & Xu, C. Q. (2007). Actively Q-switched fiber lasers: Switching dynamics and nonlinear processes. *Progress in Quantum Electronics*, *31*(3-5), 131-216.

[5] Wang, Z., Zhan, L., Qin, M., Wu, J., Zhang, L., Zou, Z., & Qian, K. (2015). Passively Q-switched Er-doped fiber lasers using alcohol. *Journal of Lightwave Technology*, *33*(23), 4857-4861.

[6] Dong, Z. Z., Li, H. P., Xia, H. D., Liu, Y., Wang, Z. G., & Chen, Y. F. (2012, November). Passively Q-switched erbium-doped fiber laser using a graphene saturable absorber. In *2012 Asia Communications and Photonics Conference (ACP)* (pp. 1-3). IEEE.

[7] Harun, S. W., Sabran, M. B. S., Azooz, S. M., Zulkifli, A. Z., Ismail, M. A., & Ahmad, H. (2015). Q-switching and mode-locking pulse generation with graphene oxide paper-based saturable absorber. *The Journal of Engineering*, *2015*(6), 208-214.

[8] Wang, Z. T., Chen, Y., Zhao, C. J., Zhang, H., & Wen, S. C. (2012). Switchable dual-wavelength synchronously Q-switched erbium-doped fiber laser based on graphene saturable absorber. *IEEE Photonics Journal*, *4*(3), 869-876.

[9] Ahmad, H., Thambiratnam, K., Muhammad, F. D., Zulkifli, M. Z., Zulkifli, A. Z., Paul, M. C., & Harun, S. W. (2014). Q-Switching and Mode-Locking in Highly Doped Zr _ {2} O _ {3}–Al _ {2} O _ {3} –Er {2} O _ {3} E-Doped Fiber Lasers Using Graphene as a Saturable Absorber. *IEEE Journal of Selected Topics in Quantum Electronics*, *20*(1), 9-16.

[10] Zhou, D. P., Wei, L., & Liu, W. K. (2012). Tunable graphene Q-switched erbium-doped fiber laser with suppressed self-mode locking effect. *Applied optics*, *51*(14), 2554-2558.

[11] Popa, D., Sun, Z., Hasan, T., Torrisi, F., Wang, F., & Ferrari, A. C. (2011). Graphene Q-switched, tunable fiber laser. *Applied Physics Letters*, *98*(7), 073106.

[12] Ahmad, H., Muhammad, F. D., Zulkifli, M. Z., & Harun, S. W. (2013). Wideband tunable Q-switched fiber laser using graphene as a saturable absorber. *Journal of Modern Optics*, *60*(18), 1563-1568.

[13] Keller, U., Weingarten, K. J., Kartner, F. X., Kopf, D., Braun, B., Jung, I. D., ... & Der Au, J. A. (1996). Semiconductor saturable absorber mirrors (SESAM's) for femtosecond to nanosecond pulse generation in solid-state lasers. *IEEE Journal of selected topics in QUANTUM*

ELECTRONICS, *2*(3), 435-453.

[14] Kang, Z., Liu, M., Li, Z., Li, S., Jia, Z., Liu, C. & Qin, G. (2018). Passively Q-switched erbium doped fiber laser using a gold nanostars based saturable absorber. *Photonics Research*, *6*(6), 549-553.

[15] Yang, Y., & Chow, K. K. (2013, December). Optical pulse generation using multi-layered graphene saturable absorber. In *2013 9th International Conference on Information, Communications & Signal Processing* (pp. 1-4). IEEE.

[16] Bao, Q., Zhang, H., Wang, Y., Ni, Z., Yan, Y., Shen, Z. X., & Tang, D. Y. (2009). Atomic-layer graphene as a saturable absorber for ultrafast pulsed lasers. *Advanced Functional Materials*, *19*(19), 3077-3083.

[17] Zhang, X., Liu, J., Shen, D., Yang, X., Tang, D., & Fan, D. (2013). Efficient graphene Q-switching of an in-band pumped polycrystalline Er: YAG ceramic laser at 1617 nm. *IEEE Photonics Technology Letters*, *25*(13), 1294-1296.

[18] Xing, G., Guo, H., Zhang, X., Sum, T. C., & Huan, C. H. A. (2010). The physics of ultrafast saturable absorption in graphene. *Optics express*, *18*(5), 4564-4573.

[19] Martinez, A., & Sun, Z. (2013). Nanotube and graphene saturable absorbers for fibre lasers. *Nature Photonics*, *7*(11), 842.

[20] Sun, Z., Hasan, T., Torrisi, F., Popa, D., Privitera, G., Wang, F., & Ferrari, A. C. (2010). Graphene mode-locked ultrafast laser. *ACS nano*, *4*(2), 803-810

[21] Ma, J., Lu, S., Guo, Z., Xu, X., Zhang, H., Tang, D., & Fan, D. (2015). Few-layer black phosphorus based saturable absorber mirror for pulsed solid-state lasers. *Optics express*, *23*(17), 22643-22648.

[22] Jiang, X., Liu, S., Liang, W., Luo, S., He, Z., Ge, Y.,& Li, J. (2018). Broadband nonlinear photonics in few-layer MXene Ti3C2Tx (T= F, O, or OH). *Laser & Photonics Reviews*, *12*(2), 1700229.

[23] Luo, Z., Zhou, M., Weng, J., Huang, G., Xu, H., Ye, C., & Cai, Z. (2010). Graphene-based passively Q-switched dual-wavelength erbium-doped fiber laser. *Optics letters*, *35*(21), 3709-3711.

[24] [24] Ahmad, H., Zulkifli, M. Z., Muhammad, F. D., Jemangin, M. H., Dimyati, K., Pal, B. P., & Harun, S. W. (2012). Passively Q-switched 11-channel stable brillouin erbium-doped fiber laser with graphene as the saturable absorber. *IEEE Photonics Journal*, *4*(5), 2050-2056.

[25] Wang, J., Luo, Z., Zhou, M., Ye, C., Fu, H., Cai, Z., & Qi, W. (2012). Evanescent-light deposition of graphene onto tapered fibers for passive Q-switch and mode-locker. *IEEE Photonics Journal*, *4*(5), 1295-1305.

[26] Wang, Z., Zhu, S. E., Chen, Y., Wu, M., Zhao, C., Zhang, H., & Wen, S. (2013). Multilayer graphene for Q-switched mode-locking operation in an erbium-doped fiber laser. *Optics Communications*, *300*, 17-21.

[27] Wang, Z. T., Zou, Y. H., Chen, Y., Wu, M., Zhao, C. J., Zhang, H., & Wen, S. C. (2013). Graphene sheet stacks for Q-switching operation of an erbium-doped fiber laser. *Laser Physics Letters*, *10*(7), 075102.

[28] Tang, Y., Yu, X., Li, X., Yan, Z., & Wang, Q. J. (2014). High-power thulium fiber laser Q switched with single-layer graphene. *Optics letters*, *39*(3), 614-617.

[29] Wu, D., Luo, Z., Xiong, F., Zhang, C., Huang, Y., Chen, S., & Xu, H. (2014). Passive Synchronization of 1.06-and 1.53-\(\mu\) m Fiber Lasers Q-switched by a Common Graphene SA. *IEEE Photonics Technology Letters*, *26*(14), 1474-1477.

[30] Wu, D., Xiong, F., Zhang, C., Chen, S., Xu, H., Cai, Z., & Luo, Z. (2014). Large-energy, wavelength-tunable, all-fiber passively Q-switched Er: Yb-codoped double-clad fiber laser with mono-layer chemical vapor deposition graphene. *Applied optics*, *53*(19), 4089-4093.

[31] Wu, K., Chen, B., Zhang, X., Zhang, S., Guo, C., Li, C., & Chen, J. (2018). High-performance mode-locked and Q-switched fiber lasers based on novel 2D materials of topological

insulators, transition metal dichalcogenides and black phosphorus: review and perspective. *Optics Communications*, *406*, 214-229.

[32] Cassidy, D. T., Johnson, D. C., & Hill, K. O. (1985). Wavelength-dependent transmission of monomode optical fiber tapers. *Applied optics*, *24*(7), 945-950.

[33] Kieu, K., & Mansuripur, M. (2006). Tuning of fiber lasers by use of a single-mode biconic fiber taper. *Optics letters*, *31*(16), 2435-2437.

[34] Latiff, A. A., Rusdi, M. F. M., Hisyam, M. B., Ahmad, H., & Harun, S. W. (2016, November). Black phosphorus as a saturable absorber for generating mode-locked fiber laser in normal dispersion regime. In *Second International Seminar on Photonics, Optics, and Its Applications (ISPhOA 2016)* (Vol. 10150, p. 101500U). International Society for Optics and Photonics.

[35] Yuzaile, Y. R., Awang, N. A., Zalkepali, N. U. H. H., Zakaria, Z., Latif, A. A., Azmi, A. N., & Hadi, F. A. (2019). Pulse compression in Q-switched fiber laser by using platinum as saturable absorber. *Optik*, *179*, 977-985.

[36] Boyd, R. D., Britten, J. A., Decker, D. E., Shore, B. W., Stuart, B. C., Perry, M. D., & Li, L. (1995). High-efficiency metallic diffraction gratings for laser applications. *Applied optics*, *34*(10), 1697-1706.

[37] Alessi, D. A., Carr, C. W., Hackel, R. P., Negres, R. A., Stanion, K., Fair, J. E., ... & Britten, J. A. (2015). Picosecond laser damage performance assessment of multilayer dielectric gratings in vacuum. *Optics express*, *23*(12), 15532-15544.

[38] Poole, P., Trendafilov, S., Shvets, G., Smith, D., & Chowdhury, E. (2013). Femtosecond laser damage threshold of pulse compression gratings for petawatt scale laser systems. *Optics Express*, *21*(22), 26341-26351.

[39] Wang, X. D., Luo, Z. C., Liu, H., Zhao, N., Liu, M., Zhu, Y. F., & Xu, W. C. (2015). Gold nanorod as saturable absorber for Q-switched Yb-doped fiber laser. *Optics Communications*, *346*, 21-25.

Journal of Physics: Conference Series **1371** (2019) 012011 doi:10.1088/1742-6596/1371/1/012011

Graphene in chitin based passive Q-switcher

S N F Zuikafly[1], W M F Wan Nawawi[2], L H Ngee[3], H Yahaya[1], W J Yahya[1] and F Ahmad[1]

[1]Malaysia-Japan International Institute of Technology, Universiti Teknologi Malaysia, Jalan Sultan Yahya Petra, 54100, Kuala Lumpur, Malaysia
[2] Department of Biotechnology Engineering, International Islamic University Malaysia, P.O. Box 10, 50728 Kuala Lumpur, Malaysia
[3]Department of Chemistry, Faculty of Science, Universiti Putra Malaysia, Serdang, 43400 Seri Kembangan, Selangor, Malaysia

fauzan.kl@utm.my

Abstract. Many materials have been utilized as saturable absorbers in generating pulsed lasers. By incorporating them with host polymers such as polymethylmethacrylate (PMMA), polyvinyl alcohol (PVA) and polyethylene oxide (PEO), the process of saturable absorber integration into the laser cavity is made easier. This work reported the generation of Q-switched pulsed lasers based on graphene embedded in chitin saturable absorber. Graphene: chitin with a concentration of 1:1 and 1.5:1 are integrated into a fiber laser cavity and the laser performance of both SAs are recorded and compared. The graphene chitin SA of 1:1 concentration produces a higher repetition rate and shorter pulse duration of 111.77 kHz and 1.328 μs, respectively. However, the SA with higher graphene content operated at a larger range of input pump power yielding a higher instantaneous peak power and pulse energy of 7.29 mW, and 14.37 nJ, respectively. This work introduces chitin as a potentially excellent natural host polymer as an alternative to conventional ones such as polyvinyl alcohol (PVA) and polyethylene oxide (PEO).

1. Introduction

As instruments of many uses in different fields of science and technology, fiber lasers have become the go-to choice for its flexibility and ease of use. The Q-switched temporal regime of a fiber laser helps provide the laser in pulsed form. The gain modulation of the Q-switched laser is now most commonly controlled by a saturable absorber (SA). This type of approach is called a passive method with the active ones requiring external control elements causing a much-complicated laser system.

With the discovery of real SAs in the early 1990s, many materials have since been investigated as SAs including semiconductor saturable absorber mirrors (SESAMs), carbon nanotubes (CNTs), topological insulator (TI), and black phosphorus (BP) among many others. The Nobel winning uncovering of graphene in 2009 has led to many new and exciting applications of graphene in industrial uses and researches, particularly in optoelectronics. Graphene's wideband operation ranging from the visible to the near-infrared region is mainly due to its gapless band and linear dispersion of Dirac electrons which also allows for a constant 2.3% light absorption [1-2]. Compared to its precursor, CNTs, graphene has the advantages of ultrafast recovery time, low saturation fluence, and broadband saturable absorption [3].

In photonics, the combination of an SA material and a host polymer to form a polymer composite based SAs is widely used by researchers. The composite form of the SA allows for better handling and mechanical tolerance [1], especially in the laser generation discipline. SAs from many classes of materials have been fabricated by using this technique with no exception to graphene. Polyvinyl alcohol (PVA) and polyethylene oxide (PEO) are the two most commonly used host matrix when fabricating a polymer composite SAs. A graphene-PVA film was reported by Popa *et al.* [4] realizing a tunable Q-switched laser operation from 1522 nm to 1555 nm. Kasim *et al.* [5] generated a Q-switched pulsed laser with the shortest pulse duration of 2.68 μs and a pulse energy of 5.8 nJ. A graphene nanoplatelets (GnPs)-PVA film SA was also demonstrated yielding a maximum repetition rate and pulse width of 91.5 kHz and 2.42 μs, respectively [6]. As an alternative to a man-made polymer, this work is introducing chitin, a biopolymer originally derived from mushrooms [7], in graphene SA fabrication and its performance in a pulsed fiber laser.

2. Graphene Chitin SA Fabrication

The chitin nanofiber was prepared from a cultivated mushroom Agaricus bisporus as reported by Wan Nawawi *et al.* [8]. The process involved the mushrooms being blended for 5 minutes before undergoing hot water extraction and alkali treatment. Graphene Oxide (GO) was synthesized via a modified Hummer's method where graphite flakes were oxidized by Sulfuric acid (H_2SO_4) and Potassium permanganate ($KMnO_4$). The mixture obtained was then washed with a diluted Hydrochloric acid (HCl) solution, using a centrifuge, followed by repeated washing with deionized water until a constant pH of 4–5 was obtained. The graphene oxide powder was mixed thoroughly with the chitin nanofiber for 6 hours aided by a magnetic stirrer before dried at room temperature. After 36 hours, graphene chitin films with different concentrations were formed. 5 mg of graphene oxide powder was added to 5 ml chitin nanofiber to produce a graphene chitin composite of 1:1 ratio and 7.5 mg graphene oxide powder to 5 ml chitin was mixed to produce an SA with 1.5:1 concentration ratio.

3. Experimental Setup

The experimental setup of the Q-switched EDFL with the graphene chitin films as SA is illustrated in figure 1. A 3 m long Erbium-doped fiber (EDF) is used as the gain medium. The EDF is forward-pumped by a 980 nm laser diode via a 980/1550 nm wavelength division multiplexer (WDM). The SA is placed in the cavity by sandwiching it in between two ferrules which are connected with a connector. An optical isolator is also spliced in the cavity to ensure unidirectional propagation of the light. The output of the laser is observed and obtained through an optical spectrum analyser (OSA), an optical power meter (OPM), a radio frequency spectrum analyser (RFSA), and an oscilloscope which are connected through a 3 dB optical coupler.

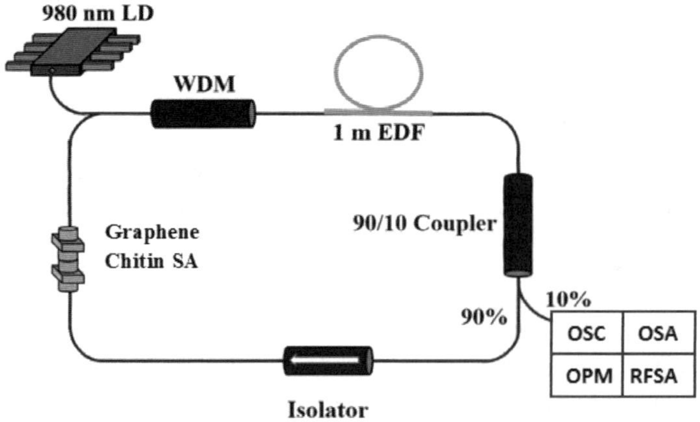

Figure 1. Experimental setup of the Q-switched EDFL based on graphene chitin SA.

Journal of Physics: Conference Series **1371** (2019) 012011 doi:10.1088/1742-6596/1371/1/012011

4. Pulsed Laser Generation

Figure 2 shows the OSA trace of both the graphene chitin with 1:1 and 1.5:1 concentration. The graphene chitin of concentration 1:1 operated at 1560.03 nm with a 3 dB spectral bandwidth of 0.79 nm while the graphene chitin of concentration 1.5:1 oscillated at 1530.51 nm with a 3 dB spectral bandwidth of around 1.41 nm. Graphene chitin of concentration ratio of 1.5:1 shifted to an even shorter wavelength at around 1533.53 nm as compared to the one of 1:1 ratio. This is due to the greater amount of graphene in the SA that induced a greater loss in the cavity. The spectrums were obtained at a maximum input pump power of 28.70 mW and 42.29 mW, for 1:1 and 1.5:1 concentration ratio, respectively.

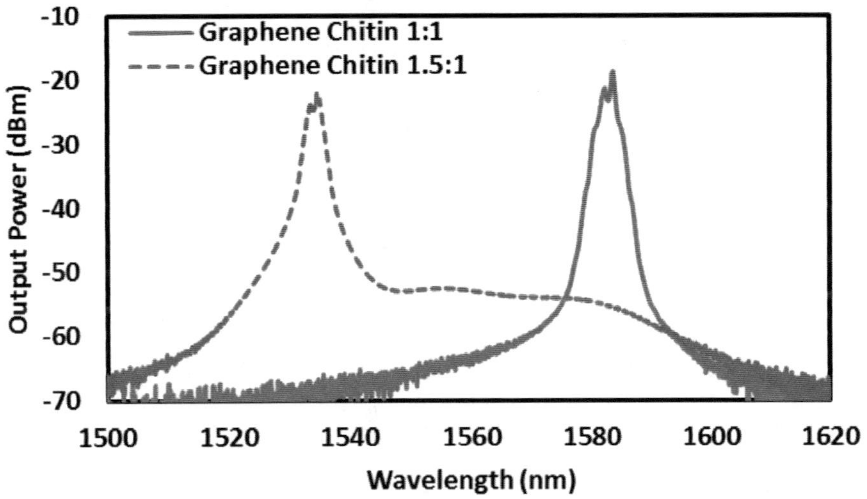

Figure 2. The optical spectrum of Q-switched EDFL based on graphene chitin SAs.

The graphene chitin SA with a concentration ratio of 1:1 was first employed in the cavity. The proposed ring laser cavity generated a Q-switched lasing as the pump power was tuned over 10.58 mW. The lasing was maintained up to the pump power of 28.70 mW before it started to lose stability. By reversing the pump tuning, the lasing can be re-achieved with the pump power range kept within the 10.58 mW to 28.70 mW span. No such lasing can be observed without the SA integration even when the pump power was tuned over a wide range indicating that the SA was responsible for the acquired Q-switched lasing. The same goes for the SA of ratio 1.5:1 with a different input pump power range from 15.11 mW to 42.49 mW. The slightly larger range of tunable input pump power may be attributed to the higher amount of graphene present in the SA, allowing for a higher degree of light saturation and absorption during the laser generation process.

The pulse train repetition rates of the Q-switching operations are shown in figure 3. In the case of the SA with a 1:1 concentration ratio, the highest repetition rate of 111.77 kHz was obtained with a corresponding shortest pulse width of 1.33 μs. 95.24 kHz of repetition rate and 1.95 μs of pulse width were subsequently obtained for the SA ratio of 1.5:1. Both of the pulse trains showed a stable Q-switching operation with little to no amplitude modulation. The insets show the single pulse envelope of the pulse train repetition rate with the corresponding shortest pulse widths.

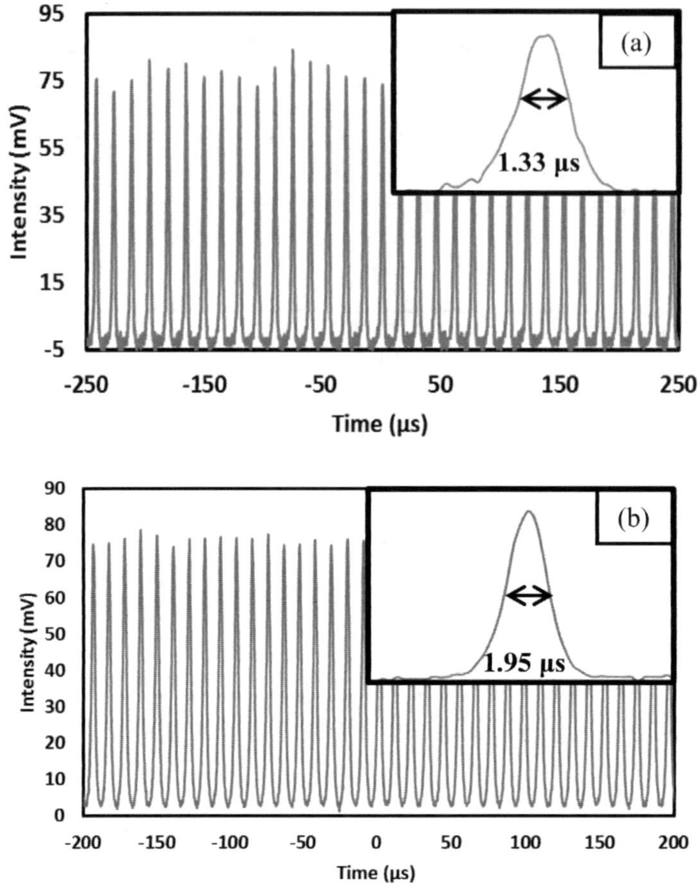

Figure 3. Pulse train repetition rate of the Q-switched EDFL based on graphene chitin SAs on concentration ratio (a)1:1 and (b) 1.5:1 and corresponding pulse envelope

When tuning the input pump power from 10.58 mW to 28.70 mW, the repetition rate increased almost linearly as also seen from past reported works on Q-switching operation [9-11], from 20.76 kHz to 111.77 kHz. The same trend was also observed in the pulse laser using the graphene chitin of concentration 1.5:1 where the repetition rate increased from 33.33 kHz to 95.24 kHz. This can be seen in figure 4. The increase in the repetition rate also contributed to the increase in the pulse energy as the pulse energy was calculated by dividing the average power with the repetition rate. The maximum pulse energy of 7.45 nJ and 15.60 nJ was obtained for SA concentration of 1:1 and 1.5:1, respectively. In the rather small range of the tunable pump power, the highest repetition rate and pulse energy obtained is higher than most of the reported works based on graphene polymer composite SAs [5-6][12].

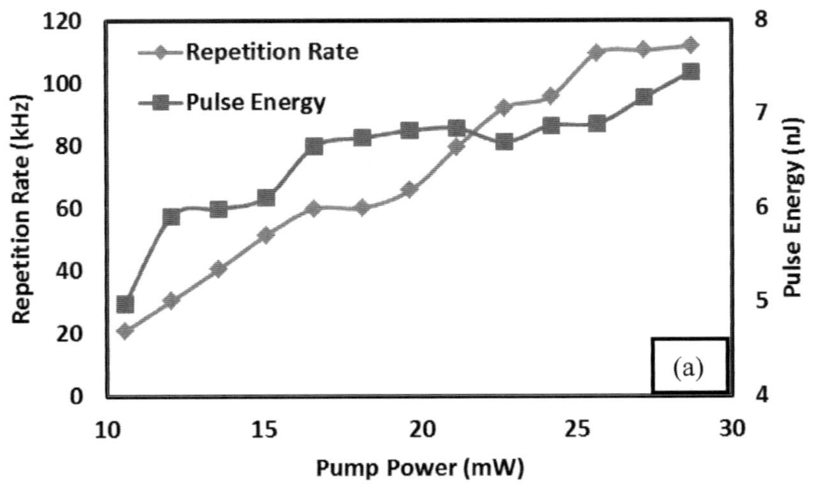

(a)

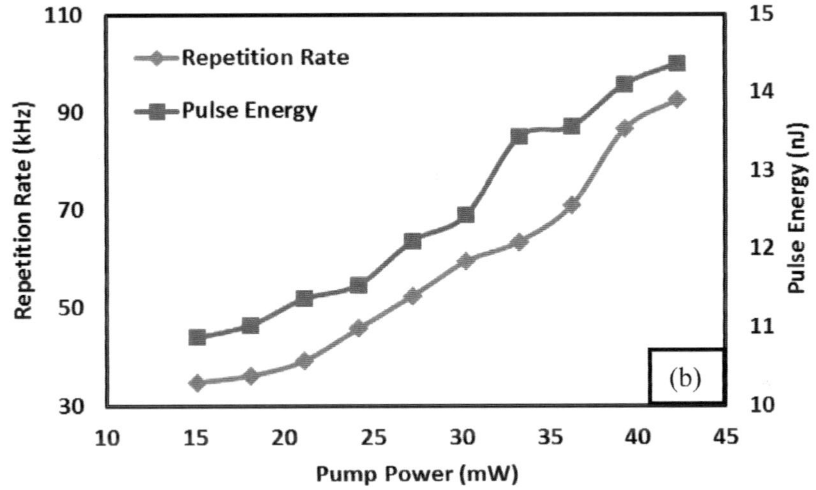

(b)

Figure 4. Repetition rate and pulse energy against input pump power of the Q-switched EDFL based on graphene chitin SAs of concentration ratio (a) 1:1 and (b) 1.5:1.

As a contrast to the repetition rate, the pulse width does not share the same relationship with the increasing input pump power. Demonstrating a typical trend of Q-switched lasers [9-11], the pulse width was reduced from 4.21 μs to 1.33 μs when using the graphene chitin SA with a concentration ratio of 1:1. Thereafter the SA with a 1.5:1 ratio demonstrated the same trend with decreasing pulse duration from 3.43 μs to 1.97 μs, as shown in figure 5. The shortest pulse width obtained was lower than other reported works using graphene-based SAs with other host polymers [5-6][12]. Both pulse widths obtained are lower if not comparable to many SAs based on other materials including black phosphorus, copper, and nickel oxide [13-16]. The peak power increases with the increasing pump power to a maximum of 5.61 mW and 7.29 mW, for each concentration ratios, respectively.

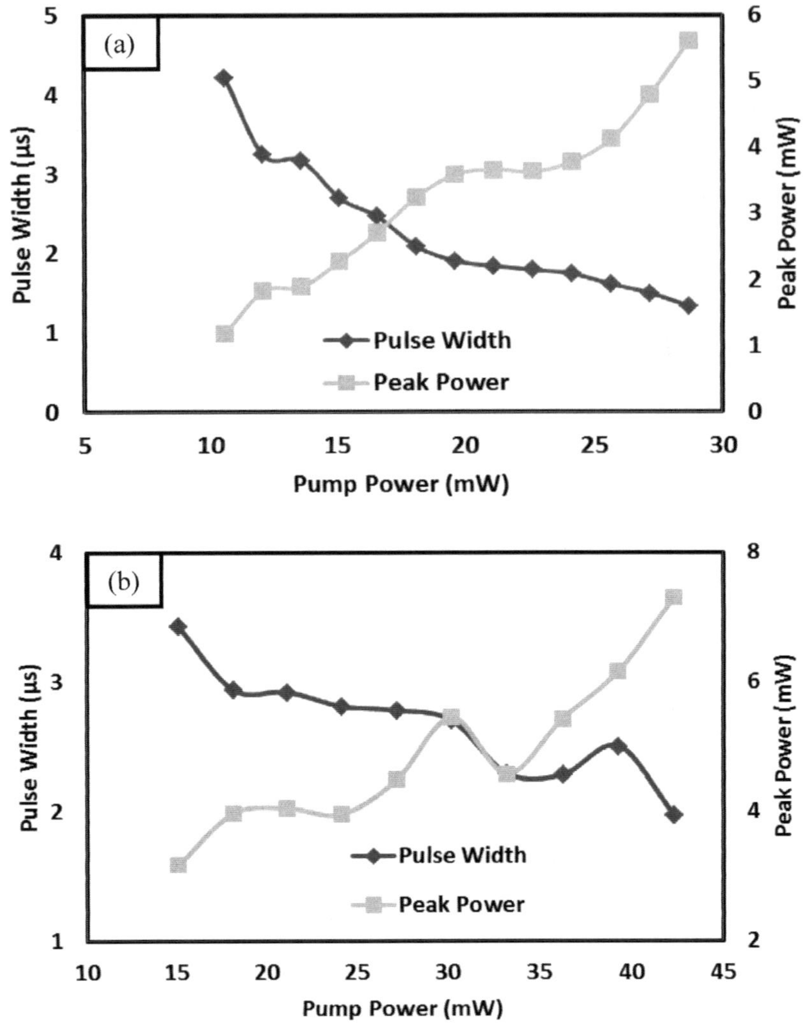

Figure 5. Pulse width and peak power against input pump power of the Q-switched EDFL based on graphene chitin SAs of concentration ratio (a) 1:1 and (b) 1.5:1.

Both of the SA concentrations yield a very high signal-to-noise ratio (SNR) of above 50 dB indicating both lasers generation was operated stably. The RFSA spectrums with a 1000 kHz span are depicted in figure 6. At maximum input pump power, the first beat node at 111.77 kHz was 51.71 dB for the graphene chitin SA of 1:1 concentration ratio. The SA with 1.5:1 concentration ratio yielded a marginally higher SNR of around 56.21 dB.

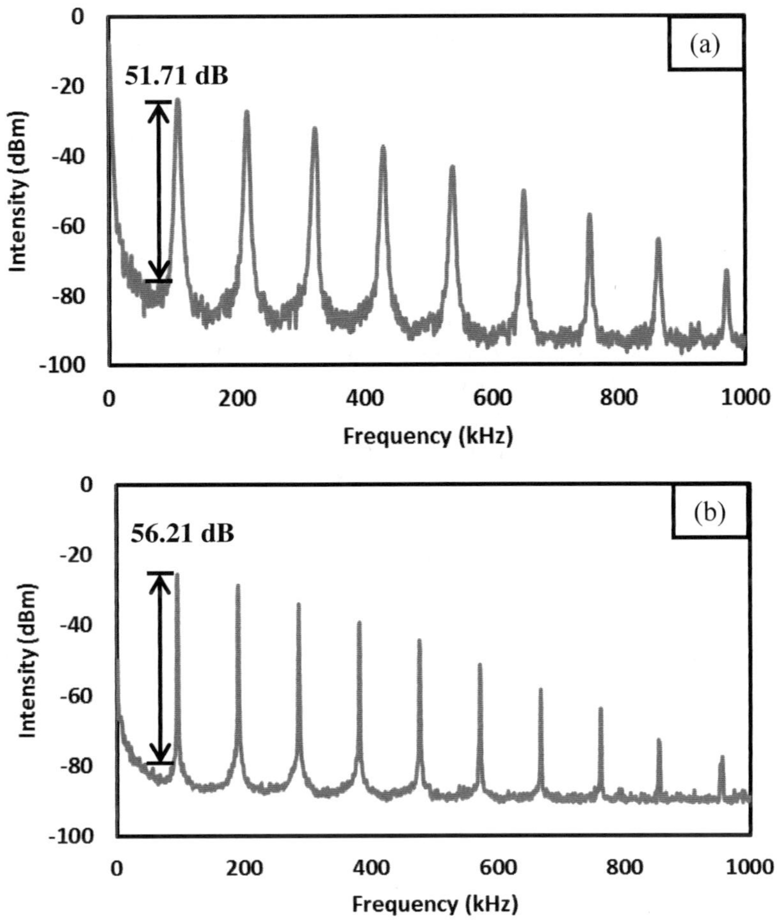

Figure 6. RFSA measurement of the Q-switched EDFL based on graphene chitin SAs of concentration ratio (a) 1:1 and (b) 1.5:1.

5. Conclusion

A Q-switched EDF laser was successfully demonstrated by using graphene composited with a chitin biopolymer in two different concentration ratios. The combination of the SA material with a biopolymer is the first ever to be reported in pulsed laser generation. A high repetition rate and shortest pulse width of 111.77 kHz and 1.33 μs, respectively, were obtained. The high laser stability was also proven by high SNR value for both concentrations of the SAs.

References

[1] Feng Y, Dong N, Wang G, Li Y, Zhang S, Wang K, Zhang L, Blau W J and Wang J 2015 *Optic Express* **23** 559-569
[2] Hasan T, Scardacl V, Tan P H, Bonaccorso F, Rozhin A G, Sun Z and Ferrari A C 2011 *Nanotube and Graphene Polymer Composites for Photonics and Optoelectronics* (Boston: Springer) p 279-354
[3] Bianchi V *et al.* 2017 *Nature Communications* **8** 15763
[4] Popa D, Sun Z, Hasan T, Torrisi F, Wang F and Ferrari A C 2011 *Appl Phys Lett* **98** 0731066
[5] Kassim N, Anyi C L, Haris H, Ahmad F, Ali N M, Ahmad H, Munajat Y and Harun S W 2014 *Journal of Nonlinear Optical Physics & Materials* **23** 1450009
[6] Hussin N, Ibrahim M H, Ahmad F, Yahaya H, Harun S W 2017 *Telkomnika* **15** 814-819

[7] Park B K and Kim M M 2010 *Int. J. Mol. Sci* **11** 5152-5164
[8] Wan Nawawi W M F, Lee K Y, Kontturi E, Murphy R J and Bismarck A 2019 *ACS Sustainable Chemistry & Engineering* **7** 6492-6496
[9] Fan M, Li T, Li G, Zhao S, Yang K, Zhang S, Zhang B, Xu J and Krankel C 2017 *Optic Express* **25** 12796-12803
[10] Kang Z, Liu M, Li Z, Li S, Jia Z, Liu C, Qin W and Qin G 2018 *Photonics Research* **6** 549-553
[11] Liu H H, Chow K K, Yamashita S and Set S Y 2013 *Opt. Laser Technol.* **45** 713
[12] Ahmad H, Samion M Z, Sharbirin A S, Norizan S F, Aidit S N and Ismail M F 2018 *Laser Physics* **28** 055105
[13] Fauziah C M, Rosol A H A, Latiff A A and Harun S W 2017 *IOP Conference Series: Materials Science and Engineering* **210**
[14] Sadeq S A, Al-Hayali S K, Harun S W and Al-Janabi A 2018 *Results in Physics* **10** 264-269
[15] Muhammad A R, Ahmad M T, Zakaria R, Rahim H R A, Yusoff S F A Z, Hamdan K S, Yusof H H M, Arof H and Harun S W 2017 *Chinese Physics Letters* **34** 034205
[16] Rizman Z I, Arof H, Latiff A A, Retna A and Harun S W 2018 *Nonlinear Optics, Quantum Optics* **49** 311-319

PMMA microball resonator for formaldehyde liquid sensing

M M Sharif[1], H R A Rahim[2], M A M Johari[3], M I M Abdul Khudus[4], S W Harun[1]

[1]Department of Electrical Engineering, Faculty of Engineering, University of Malaya, 50603 Kuala Lumpur, Malaysia
[2]Faculty of Electronic and Computer Engineering, Universiti Teknikal Malaysia Melaka, 76100, Melaka, Malaysia
[3]Faculty of Engineering Technology, Universiti Teknikal Malaysia Melaka, 76100 Melaka, Malaysia
[4]Department of Physics, Faculty of Science, University of Malaya, 50603 Kuala Lumpur, Malaysia

maslynnda.ms@gmail.com

Abstract. This experiment reports the fabrication of the Polymethyl Methaclyrate (PMMA) Microball Resonator (MBlR) as a liquid sensor tested on various level of concentration of the Formaldedhye liquid. The PMMA MBlR is fabricated via the "dipped and twirl" method to create the sphere-shaped ball with diameter Db = 320um. The MBlR is then optically excited using a 9μm PMMA microfiber and was found to have a Q Factor of $>10^4$. The MBlR was then employed as a liquid sensor with the level percentage range between 0% to 4% of Formaldehyde liquid and the performance is compared with a non MBlR microfiber. The MBlR sensor was found to have a sensitivity of 6.94 dBm/%, linearity >90% and P-value of more than 10^5. The PMMA MBlR liquid sensor was also found to have good repeatability and stability over a period of 60 seconds.

1. Introduction

In the complexity of the formaldehyde substance CH_2O, hydrogen, oxygen and carbon are three main compounds that exist in the Formaldehyde. Formaldehyde is a colourless, strong smell gas which is used in the making of many building materials and household products. It is used in pressed-wood products, such as particleboard, plywood, and fiberboard; glues and adhesives; permanent-press fabrics; paper product coatings; and certain insulation materials. The formaldehyde also exists in all substance and can be found in many by-products including plants, fish and human living. Formaldehyde can be broken down generally within hours in the air and easily dissolves in water. Formalin is the name of the substance when it is dissolved in water which is commonly used as an industrial disinfectant and as a preservative in funeral homes and medical labs. The formalin is also used as a preservative in some food products, such as antiseptics, medicine and cosmetics. Formaldehyde can be added as a preservative to food, but it can also be produced as the result of cooking and smoking.

Various sensing configurations of optical fiber sensors have been demonstrated such as fiber Bragg grating sensors, side polished optical fiber sensors, optical microfiber sensors, each of which has

Journal of Physics: Conference Series **1371** (2019) 012012 doi:10.1088/1742-6596/1371/1/012012

specific advantages and disadvantages. One class of optical components, the optical micro-resonators, have also shown to have promising properties for sensing applications. Optical micro-resonators (OMR) are optical components which permit the circulation of light through a micron-length closed path fabricated in structures such as microspheres, microdisks and microrings loop and have employed for a verity of different biological or chemical sensors. One subclass of OMRs, namely the whispering gallery mode (WGM) resonators, which permit specific resonance (modes) of the waves along the 'edges' of the optical resonator, produces evanescent fields of the resonator. The resonating modes are also dependent on the refractive index of the surrounding media. As the light travels through the resonator multiple times, any change in both the absorption and refractive index of the surrounding medium is both amplified and is instantly measurable at the output. Multiple geometries of WGM resonators have been investigated but have primarily focussed on glass-based material.

In conjunction with serious needs, many studies have been reported on the detection of the formaldehyde liquid. Several available methods of detection of formaldehyde in air and water have been reported. For air samples is based on spectrophotometry, is one of the widely used methods, while for aqueous samples high-performance liquid chromatography is a preferred choice. The used of both of the techniques are highly sensitive but are costly. Therefore it is significant to develop a precise and sensitive sensing method for formaldehyde detection due to its implementation and applications is widely used in our daily lives.

2. Experimental Setup

First, a small amount of the Polymethyl Methaclyrate (PMMA) powder was heated on a glass plate at around 150°C – 200°C. The PMMA powder was obtained from Across Organic with an Mw of 35 000. A silica microfiber was then fabricated from a single-mode fiber (SMF-28) using the flame brushing technique to produce a non-adiabatic optical microfiber with a diameter of 9 μm and a waist of 5 mm. The fiber was then broken into two and dipped into the molten PMMA. The two ends of the silica fiber were then lifted from the molten PMMA solution, which then forms a 'bridge' between the broken ends of the optical microfiber. To create a PMMA microfiber, the two ends were then pulled apart, producing a PMMA microfiber with an approximate diameter and length of 9 μm and 2 cm, respectively.

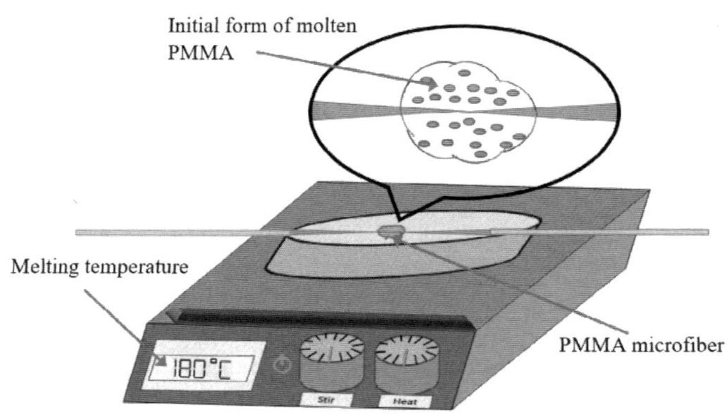

Figure 1. Fabrication of the PMMA Microfiber with a diameter of 9μm and 5mm waist

A PMMA microball resonator was then fabricated using a "dip-and-twirl" technique. This technique involves a tapered fiber tip being dipped into a small amount of molten PMMA and slowly twirled and pulled upwards to form a sphere-like structure. The fabricated structure was then heated by using a torch to form a microball-shaped resonator, shown in Figure 1. The PMMA Microball Resonator (MBIR) has a diameter (D_b) of 320 μm measured using an optical microscope as shown in Figure 2. This use of the PMMA for both the microfiber and MBIR reduces coupling losses due to refractive index differences, potentially increasing the performance of the MBIR sensor.

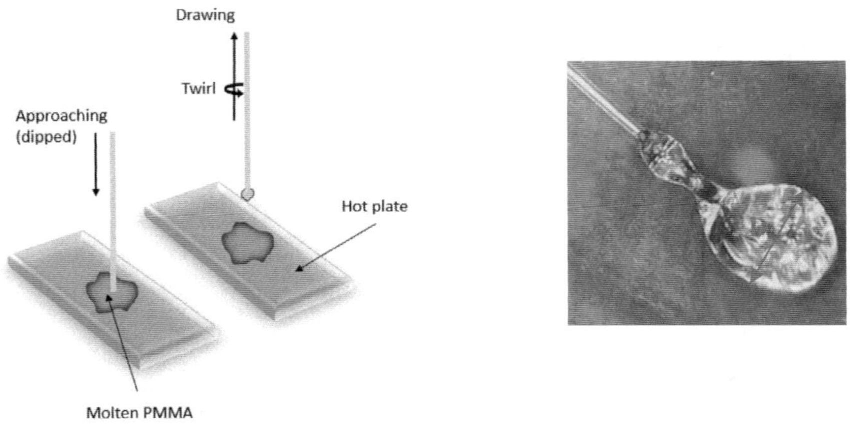

Figure 2. Fabrication of the PMMA Microball Resonator (MBIR)

The PMMA MBIR was characterized by launching light from a tunable laser source (ANDO AQ4321D) with a working wavelength of between 1520 nm and 1620 nm to one end of the broken optical microfiber and collecting the resulting output power by using an optical power meter (THORLABS S145C) connected to the other end of the optical microfiber. The laser was then tuned from 1520.10 nm to 1520.30 nm with a wavelength interval of 1 pm, with the output power recorded. The resulting transmission spectrum is shown in Figure 3. The insertion loss was measured to be approximately 50 dB and includes losses from coupling into the PMMA microfiber as well as the PMMA MBIR. However, as the microfiber and resonator are both made of the same material, it is expected that coupling losses due to Fresnel reflection are relatively small and that the main loss originates from scattering losses due to air bubbles inside the PMMA MBIR as well as from the surface of the resonator. The Q-factor of the MBIR is then calculated from the definition of $Q = \Delta\lambda/\lambda$, where λ is the resonant wavelength and $\Delta\lambda$ is the FWHM of the transmission dip and was found to be 1.689×10^5, similar to silica-based resonators such as silica microbottle and microball resonators.

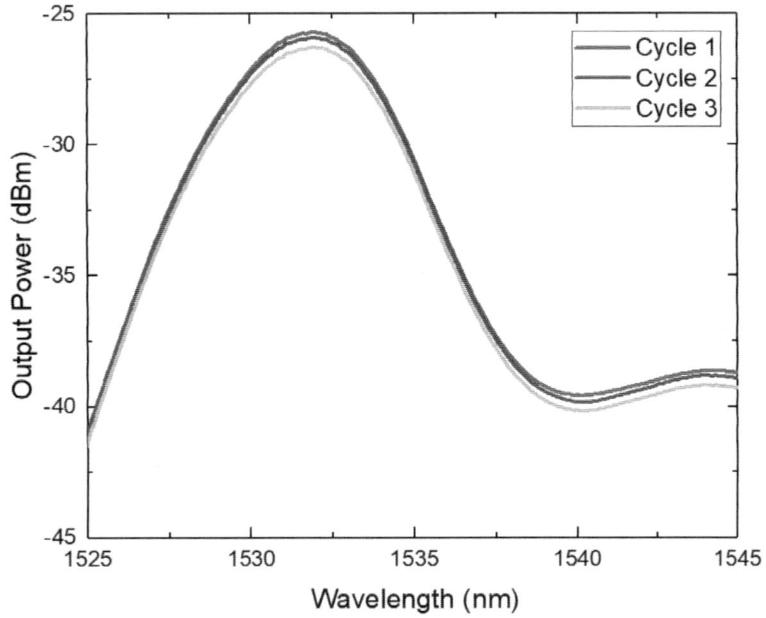

Figure 3. PMMA microfiber 9 μm diameter transmission modes.

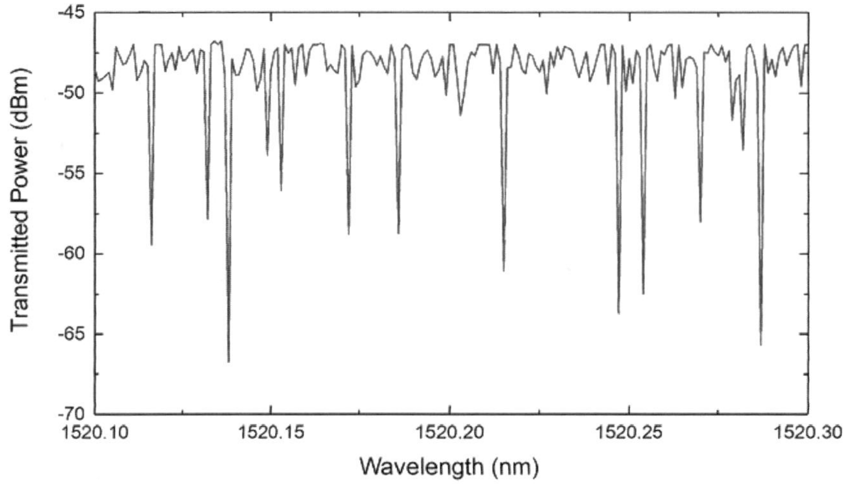

Figure 4. MBIR WGM transmission modes of microball resonator coupled on PMMA microfiber 9 μm diameter.

The performance of the PMMA MBIR as formaldehyde sensor was then investigated by employing the setup as shown in Figure 4. The PMMA MBIR is placed in between the formaldehyde liquid and the microfiber. One end of the microfiber is connected to Tunable Laser Source (TLS) and the other end connected to the Optical Power Meter (OPM) to measure the transmitted power. The formaldehyde liquid concentration was then varied from 0% to 4%. The wavelength was set to λ = 1520 nm and the transmission of the PMMA MBIR at a different percentage of concentration were recorded. The experiment was carried out 3 times to reduce the random errors and investigate the repeatability, the linearity and the stability of the sensor. The stability of the sensor was also studied for 60 seconds by recording the transmission output at a different concentration level.

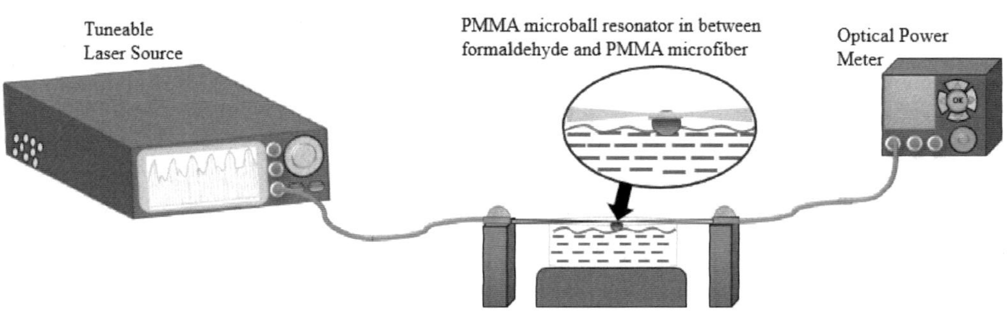

Figure 5. Experimental setup of the PMMA Microball resonator (MBIR) for formaldehyde concentration measurements.

3. Results and Discussions

Figure 6 shows the output signal of the Tunable Laser Source (TLS) transmitted through the PMMA with and without the MBIR at different formaldehyde concentration level. The transmission decreased with the increasing percentage of concentration of the formaldehyde for both with and without the MBIR. The linearity, sensitivity, standard deviation and P-Value of the PMMA MBIR is given in Table 1. The sensitivity of the PMMA MBIR is higher than the one without the resonator at 6.94dBm/% as compared to 5.76dBm/%. From the result, it can be seen that the standard deviation of the bare PMMA microfiber is less than the one with the PMMA MBIR resonator attached to it with the value of 9.2484dBm and 11.5184dBm respectively. This shows that the bare PMMA microfiber data-taking is more stable than the one with attached MBIR resonator. The P-value both of the with and without PMMA MBIR was found to be 2.60 x 10^{-6} and 3.83 x 10^{-5} respectively indicates that the data collection is successfully conducted. This overall reduction may be attributed by the resonator loss, which increases with the increasing formaldehyde percentage of concentration due to the changes of the refractive index and the surface adsorption. However, as the light resonates around the MBIR, the losses accumulate for each round trip, thus increasing overall loss and increasing sensitivity. It can, therefore, be surmised that for liquid formaldehyde sensing, the PMMA MBIR demonstrates better performance compared to the bare PMMA microfiber.

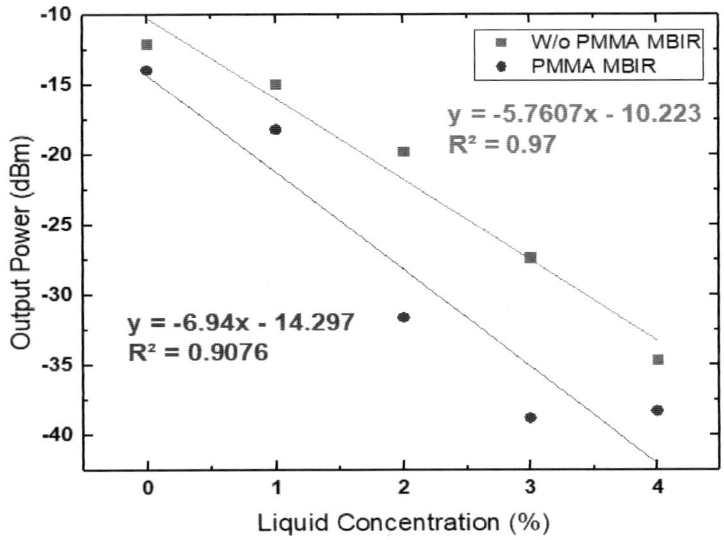

Figure 6. The transmitted power value of the PMMA microfiber with and without the MBIR at varying formaldehyde concentration.

Table 1: Analysis of MBIR and without MBIR microfiber performance in formaldehyde concentration activity.

Parameters	No-MBIR	With PMMA MBIR
Linearity (%)	98.49%	95.27%
Sensitivity (dBm/%)	5.7607	6.9400
Standard deviation (dBm)	9.2484	11.5184
P-value	3.83 x 10^{-5}	2.60 x 10^{-6}
Linear Range (% concentration)	0 - 4	0 - 4

4. Conclusions

This paper discussed the performance of PMMA MBIR and without MBIR microfiber as a liquid sensor tested on the formaldehyde liquid at different concentration level. First, a PMMA microfiber with a diameter of 9 µm and a length of 2 cm is fabricated by dipping two optical microfibers and pulling them apart. Then, a PMMA MBIR is fabricated using the "dip and twirl" method to create the sphere-shaped ball with diameter D_b = 320um. The fabricated PMMA MBIR is then optically excited by a TLS via a 9 um PMMA optical microfiber and characterized at the wavelength of λ = 1520 nm - 1530nm. The PMMA microfiber sensor both with and without the MBIR were then employed for formaldehyde sensing with a concentration of between 1-4%. The sensitivity, the linearity, standard deviation and P-Values were calculated, and the results indicate that the PMMA microfiber with the MBIR is superior in performance. The experiment was then repeated 3 times to investigate the accuracy and stability of the two sensor configurations for different formaldehyde concentrations. In general, the stability of both sensor configurations are comparable, but the sensor with the MBIR has superior sensitivity and repeatability, indicating that the PMMA MBIR is a promising sensor as a liquid formaldehyde sensor.

Acknowledgements
The authors would like to acknowledge the University of Malaya, Kuala Lumpur for their financial support. [Ni: RP039C-18AFR].

References
[1] M.A.M. Johari, A.A.N., M.I.M. Abdul Khudus, M.H. Jali, H.H.M. Yusof, S.W. Harun, M. Yasin, *Microbottle resonator for formaldehyde liquid sensing.* Optik, 2018.
[2] Batumalay, M., et al., *A study of relative humidity fiber-optic sensors.* IEEE Sensors Journal, 2015. **15**(3): p. 1945-1950.
[3] Irawati, N., et al., *Relative humidity sensing using a PMMA doped agarose gel microfiber.* Journal of Lightwave Technology, 2017. **35**(18): p. 3940-3944.
[4] Korpan, Y.I., et al., *Development of highly selective and stable potentiometric sensors for formaldehyde determination.* Biosensors and Bioelectronics, 2000. **15**(1-2): p. 77-83.
[5] Kawamura, K., et al., *Development of a novel hand-held formaldehyde gas sensor for the rapid detection of sick building syndrome.* Sensors and Actuators B: Chemical, 2005. **105**(2): p. 495-501.
[6] Xu, X., et al., *Determination of formaldehyde in beverages using microwave-assisted derivatization and ionic liquid-based dispersive liquid–liquid microextraction followed by high-performance liquid chromatography.* Talanta, 2011. **85**(5): p. 2632-2638.
[7] Ascorbe, J., et al., *Recent developments in fiber optics humidity sensors.* Sensors, 2017. **17**(4): p. 893.
[8] Yoon, H., *Current trends in sensors based on conducting polymer nanomaterials.* Nanomaterials, 2013. **3**(3): p. 524-549.
[9] Zhang, W., et al., *Water content detection in aviation fuel by using PMMA based optical fiber grating.* Sensors and Actuators B: Chemical, 2019. **282**: p. 774-779.
[10] Cui, T., et al., *From monomeric nanofibers to PbS nanoparticles/polymer composite nanofibers through the combined use of γ-irradiation and gas/solid reaction.* Journal of the American Chemical Society, 2006. **128**(19): p. 6298-6299.
[11] Noy, A., et al., *Fabrication of luminescent nanostructures and polymer nanowires using dip-pen nanolithography.* Nano letters, 2002. **2**(2): p. 109-112.
[12] Li, D. and Y. Xia, *Electrospinning of nanofibers: reinventing the wheel?* Advanced materials, 2004. **16**(14): p. 1151-1170.
[13] Nain, A.S., et al., *Drawing suspended polymer micro-/nanofibers using glass micropipettes.* Applied Physics Letters, 2006. **89**(18): p. 183105.
[14] Lim, K., et al., *Fabrication and applications of microfiber*, in *Selected Topics on Optical Fiber Technology*. 2012, InTech.
[15] Ong, Y., et al., *Fabrication of polymer microfiber through direct drawing and splicing of silica microfiber via vapor spray and flame treatment.* Applied Optics, 2015. **54**(13): p. 3863-3867.
[16] Bianucci, P., *Optical Microbottle Resonators for Sensing.* Sensors, 2016.
[17] Tong, L., *Brief introduction to optical microfibers and nanofibers.* Frontiers of Optoelectronics in China, 2010. **3**(1): p. 54-60.
[18] Zhang, Y., et al., *A Multipoint Liquid Level Sensor Based on Two Twisted Polymer Optical Fibers in a Race-Track Helical Structure.* Journal of Sensors, 2018. **2018**.

Optical comb filter generation in micro cavity employing Zinc Oxide based film

H M Burhanuddin[1], S N F Zuikafly[1], A A Ibrahim[2], S Ambran[1], O Mikami[2], C Fujikawa[2] and F Ahmad[1]

[1] Malaysia-Japan International Institute of Technology, Universiti Teknologi Malaysia, Jalan Sultan Yahya Petra, 54100, Kuala Lumpur, Malaysia
[2] Department of Optical and Imaging Science and Technology, Tokai University, Shonan Campus: 4-1-1 Kitakaname, Hiratsuka-shi, Kanagawa 259-1292, Japan.

fauzan.kl@utm.my

Abstract. Optical comb filter generation was proposed and experimentally demonstrated by using Fabry-Perot interferometer in micro cavity. The change of length of Fabry-Perot interferometer modifies the interference pattern with the adjusted length is from 0 μm to 350 μm. The parameter of the generated comb such as free spectral range, extinction ratio, quality factor and finesse was recorded with and without Zinc Oxide based film (nano powder and nano particles). The stable comb filter was generated with a length of 100 μm for both Zin Oxide-based film and 150 μm for the uncoated mirror. At the maximum length of 350 μm, Zinc Oxide nano particles film generates most stable optical comb filter with free spectral range, extinction ratio, quality factor and finesse of, 3 nm, 8 dB, 1298 and 2.5, respectively.

1. Introduction

The study on optical fiber can be traced back to early 1982 [1]. Optical fiber usually comes in two types; single-mode fiber (SMF) and multimode fiber (MMF). SMF was widely used for its ability to transmit a signal over a long distance. The combination of the two types of fiber was successfully demonstrated in generating a simple comb filter as reported by Wang et al. [2] and Zhao et al. [3], using a single mode-multimode-single mode fiber (SMS).

The fiber comb filter as the name implies acts as an optical signal sorter that helps to separate signals from different channels in proximity, thus aiding in reducing chances of crosstalk [4]. Comb filter generations have been reported by Zhou et al. [4] using a cavity consisted of only fibers. The research also made use of the multimode interference and Mach-Zender interferometer effects on the cavity. The generation of comb filter was also usually based on Bragg grating [5] which is becoming less attractive with the surfacing of interferometric cavities. These versatile cavities are higher in sensitivity and can be configured with different designs [6]. A micro cavity generating comb filters is no stranger in the optical sensor world and have been reported by a few in the past few years [6-7]. An optical comb filter was also reported using tapered fiber [8]. The principle is to make use of the interference effect in the ring resonator. This method is especially apparent in sensing applications and is usually approached by tapering the fiber through chemical etching process [9]. The use of fiber allows for ease of disposal and handling. However, the complex and delicate form of the fiber cavity has made the technique less popular than its newer counterparts. Another configuration based on polydimethylsiloxane (PDMS)

Journal of Physics: Conference Series **1371** (2019) 012013 doi:10.1088/1742-6596/1371/1/012013

Fabry-Perot interferometer [6] was reported recently, with the micro cavity is between the SMF surrounded by PDMS.

This work, in particular, used the Fabry-Perot interferometer in the micro cavity to allow for length adjustment between the fiber tip and the mirror. The air gap is the common way to approach this technique, nevertheless, this work will demonstrate the effect of placing a thin film based on ZnO-PVA film (nano powder and nanoparticles) in between the fiber tip and the mirror and the resulting performance is discussed and compared to the ones with air gap micro cavity.

2. Methodology

The experimental setup of the micro cavity is as shown in figure 1. The cavity was pumped by a 1550 nm amplified spontaneous emission (ASE) light source through a 50:50 optical coupler. The microscope was used to observe and accurately adjust the distance between the fiber tip and the mirror. The reflected light by the mirror was measured using an optical spectrum analyzer (OSA) in the form of comb filters.

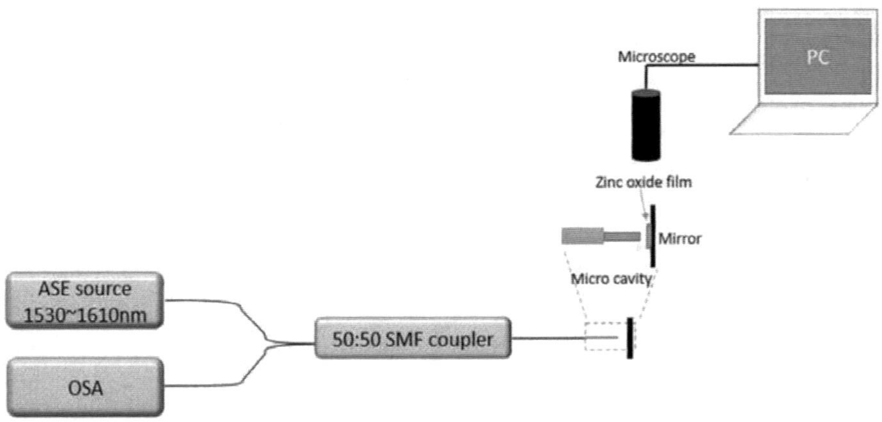

Figure 1. Experimental setup of the micro cavity.

3. Results and Discussions

From the air gap, ZnO nanopowder and ZnO micro cavities, three sets of comb filter were obtained as shown in figure 2. The comb filter with ZnO nanoparticles was observed to have a better and more constant spacing at around 200 μm with highest Q-factor of around 1945 while the other two micro cavities showed more amplitude modulations with unstable bandwidth separation.

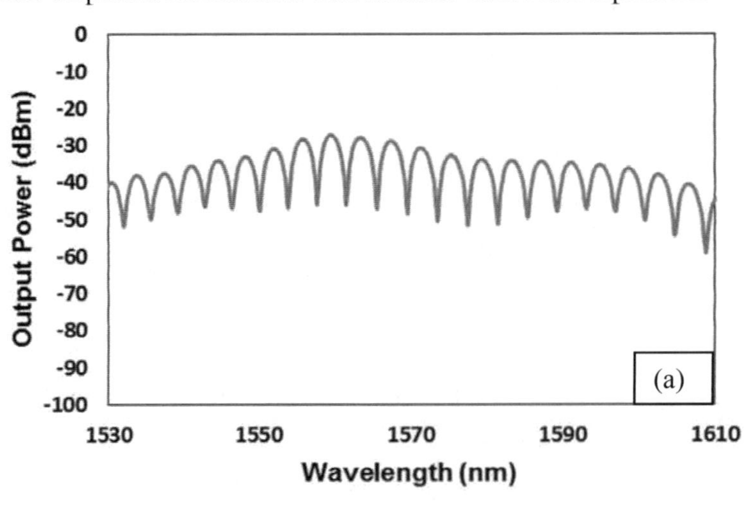

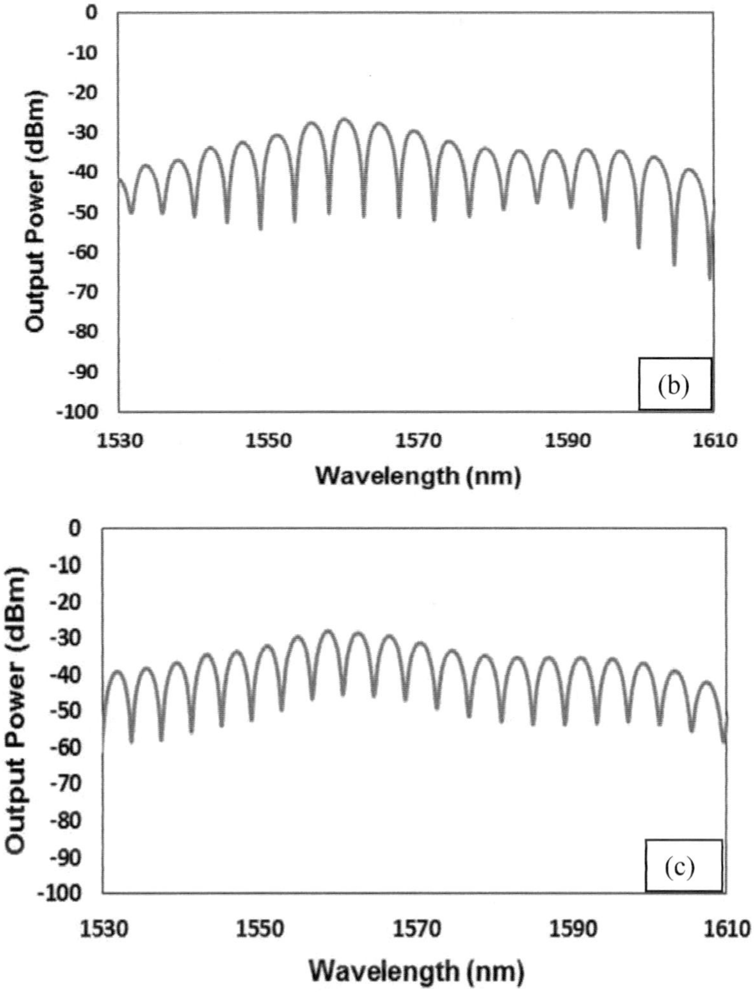

Figure 2. Optical comb filter of (a) air gap, (b) ZnO nanopowder, and (c) ZnO nanoparticles

The comb filter's performances with air gap, ZnO nanopowder, and ZnO nanoparticles were compared in terms of its free spectral range (FSR), extinction ratio (ER), Q-factor, and finesse. The distance between the fiber tip and the mirror was slowly increased to obtain the best distance with the most stable comb filter operation. The distance was increased from 0 to 350 μm with the comb filter started forming at around 100 μm. The relationship between the microcavity performances against the tip-mirror distance in terms of its FSR is illustrated in figure 3. The FSR was seen to decrease with the increasing distance applied. The decrease in FSR was more linear in air gap microcavity. This may be due to lower refraction occurring in the cavity without the ZnO based films.

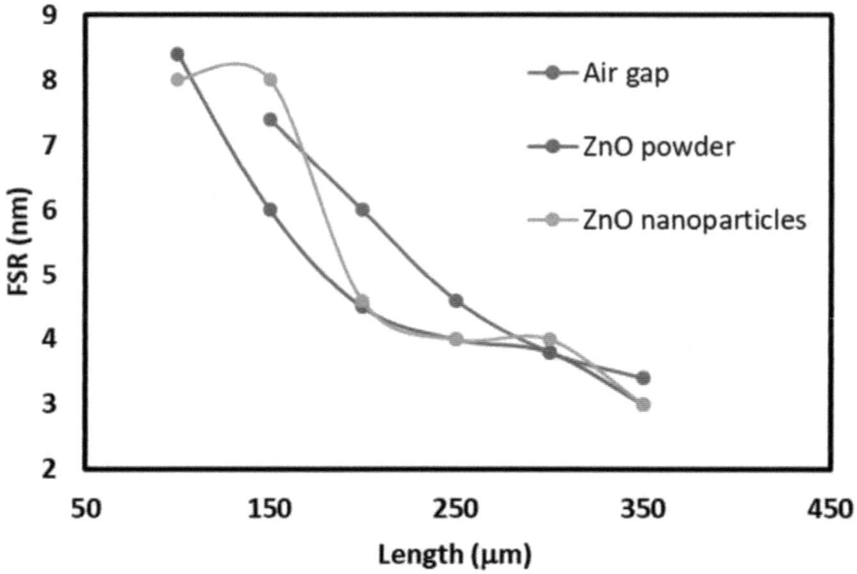

Figure 3. FSR performances comparison against tip-to-mirror distance.

The graph of the ER's relation with regards to the tip-mirror distance is depicted in figure 4. At around 200 µm, all of the tests showed nearly similar value indicating that their minimum and maximum spectrum were in equal high at about 9 dB. ZnO nanoparticles showed higher ER of around 8 dB to 10 dB.

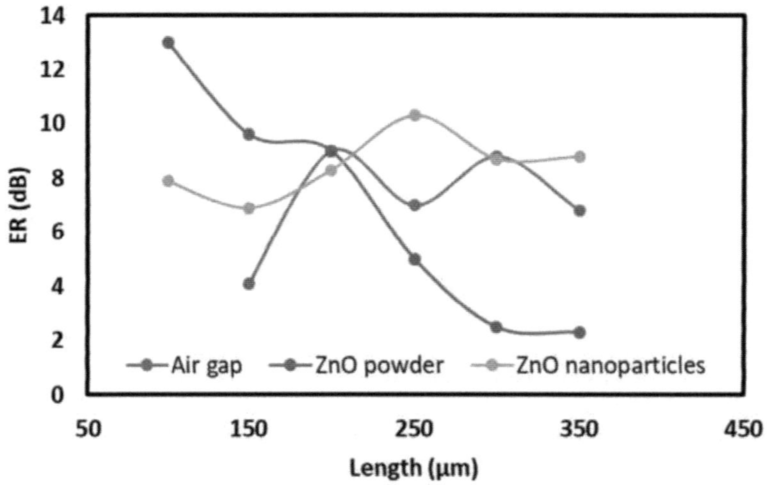

Figure 4. ER performances comparison against tip-to-mirror distance.

The Q-factors of the micro cavities shows an increasing linear trend in figure 5. The Q-factor micro cavity with airgap increases throughout the distance while both of the ZnO films contributed to the slight decrease at 350 µm.

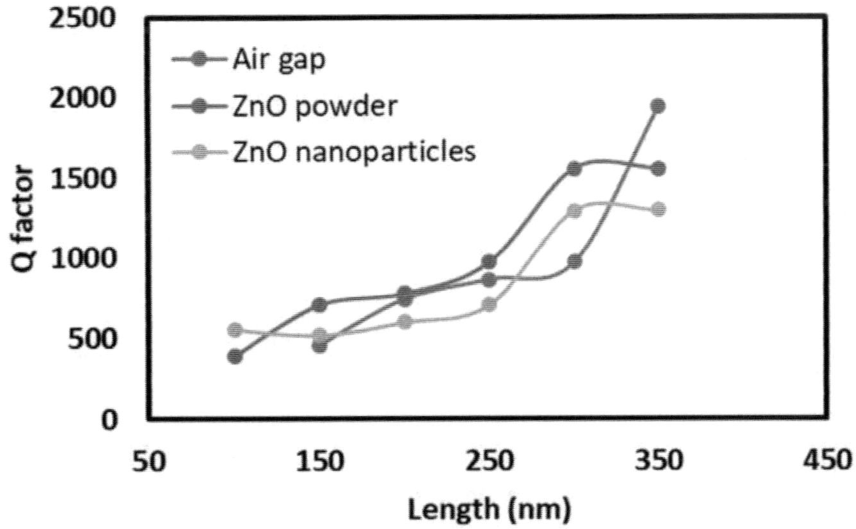

Figure 5. Q-factor performances comparison against tip-to-mirror distance.

The finesse of the micro cavities for the three cases is also plotted against the distance as shown in figure 6. The narrow heaving resonance bandwidth of the cavities causes the wave pattern of the finesse graph [10]. Researchers usually seek for a moderately valued finesse in a comb filter to have a good balance between the stop and pass filter as higher finesse means a wider pass filter [11]. Hence, this particular parameter can be manipulated to get the desired filter effects.

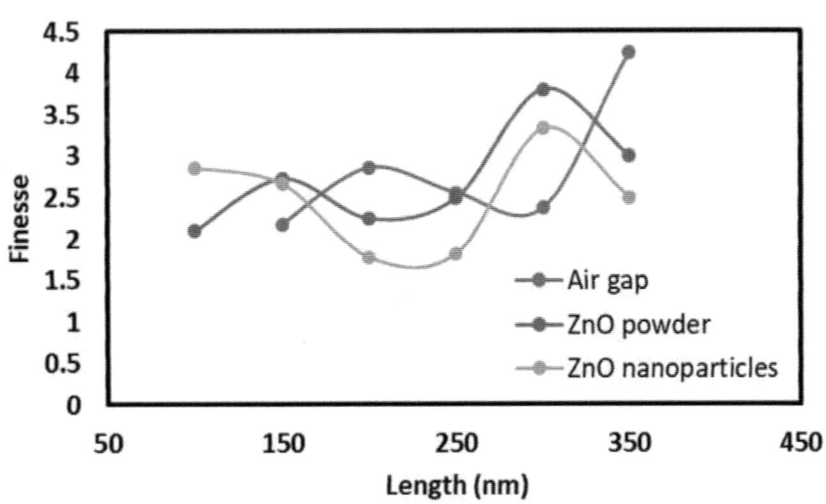

Figure 6. Finesse performances comparison against tip-to-mirror distance.

At the fiber tip and mirror distance of 350 μm, the FSR was recorded at 3.4 nm for the air gap and 3 nm for both ZnO nanopowder and nanoparticles. The ER for the microcavity with ZnO nanoparticles recorded the highest at 8.8 nm. Higher ER is preferable in a sensor system to ease the process of detecting errors through attenuation effects [12]. The ratio of resonance-to-bandwidth of the operating wavelength which defines the Q-factor of the comb filter was calculated at 1945.5, 1555.4, and 1298 for each of the microcavity components, respectively. The finesse of the microcavity with the airgap, ZnO nanopowder, and ZnO nanoparticles stayed within below 5 at 4.25, 3, and 2.5, respectively. The performance comparison is tabulated in table 1.

Table 1. Micro cavities performances comparison

Micro cavity	FSR (nm)	ER (dB)	FHWM (nm)	λ (nm)	Q-factor	Finesse
Airgap	3.4	6.8	0.8	1556.4	1945.5	4.25
ZnO nanopowder	3	2.32	1	1555.4	1555.4	3
ZnO nanoparticles	3	8.8	1.2	1557.6	1298	2.5

4. Conclusion

The ZnO nanoparticles showed the best average performance in terms of comb filter generation and can seriously be considered as a pioneer in replacing micro-airgap cavity comb filter for sensor purposes. The micro cavity with ZnO nanoparticles generated a stable optical comb filter with free spectral range, extinction ratio, quality factor and finesse of, 3 nm, 8 dB, 1298 and 2.5, respectively.

References

[1] Giallorenzi T G et al. 1982 *IEEE Transactions on Microwave Theoru and Techniques* **30** 472
[2] Wang Q, Farrel G and Yan W 2008 **Journal of Lightwave Technology** *26* 512
[3] Zhao Y, Jin Y and Liang H 2011 *IEEE Symposium on Photonis and Optoelectronics*
[4] Zhou G et al. 2018 *Scientific Reports* **8** 11803
[5] Othonos A and Kalli K 1999 *Fiber Bragg Grating: Fundamentals and Applications in Telecommunications and Sensing* (Boston: Artech House)
[6] Kacik D, Martincek I and Tarjanyi N 2018 *Proceedings of SPIE* **10680** 106801X-1
[7] Chen J, Tong Z, Yang H, Zhang W, Xue L and Pan H 2017 *Laser Physics* **27**
[8] Spisser A, Ledantec R, Seassal C, Leclercq J L, Benyattou T, Rondi D, Blondeau R, Guillot G and Viktorovitch P 1998 *IEEE Photonics Technology Letters* **10** 1259
[9] Harith Z, Irawati N, Batumalay M, Rafaie H A, Yun Il G, Harun S W, Nor R M and Ahmad H 2015 *Indian Journal of Science and Technology* **8** 85348
[10] Lokman A et al. 2016 *IEEE Sensors Journal* **16** 312
[11] Yasin M 2012 *Selected Topics on Optical Fiber Technology* (Croatia: InTech) p 473
[12] Sun H,Wang D, Chen C, Long K and Sun X 2018 *Applied Optics* **57** 7344

Journal of Physics: Conference Series **1371** (2019) 012014 doi:10.1088/1742-6596/1371/1/012014

Q-switched erbium-doped fiber laser employing gold thin film saturable absorber

Y R Yuzaile[1], Z Zakaria[1,*], N U H H Zalkepali[1], N A Awang[1], A A Latif[2], A H Ali[1] and N N H E Nik Mahmud[1].

[1]Optical Fiber Laser Technology (OpFLAT) Focus Group, Department of Physics and Chemistry, Faculty of Applied Sciences and Technology, Universiti Tun Hussein Onn Malaysia, 84600 Pagoh, Johor, Malaysia.
[2]Department of Physics, Faculty of Science, Universiti Putra Malaysia, Selangor, Malaysia.

zahariah@uthm.edu.my

Abstract. This research demonstrates an erbium-doped fiber laser (EDFL) which was passively Q-switched by gold thin film saturable absorber (SA). A very simple technique was used to fabricate the SA, in which the gold was directly deposited on the fiber end face through DC magnetron sputtering at 120 mA of sputtering current. The SA was then inserted into the single ring cavity and achieved Q-switching at a threshold pump power of 38.8 mW. By gradually increasing the pump power from 38.8 mW to 119.3 mW, the repetition rate increases from 11.34 kHz to 21.72 kHz. Meanwhile, the pulse duration decreases from 13.91 μs to 7.44 μs. The maximum achievable pulse energy was 13.28 nJ. This research further expands the application of gold thin film SA to the field of Q-switched pulsed fiber laser.

1. Introduction

In recent years, the development of fiber lasers has become a research focus as a possible replacement to the bulky solid-state lasers [1]. Fiber lasers can be developed into pulsed fiber lasers, typically known as Q-switched and mode-locked fiber lasers. These two operations are distinguished by their repetition rates, where kilohertz for Q-switching while megahertz for mode-locking. Q-switching can be achieved by modulating the Q-factor of the laser resonator and its frequency depends on the input power [2]. Meanwhile, mode-locking is achieved by controlling the dispersion and nonlinearity inside the cavity and the repetition rate is influenced by the total cavity length [3].

Q-switched fiber lasers are widely used in the applications that require slower pulse rates but high output powers [4-8], such as 3D optical data storage, tattoo removal and hair removal [9-11]. It emits energetic pulse through active [12] or passive [13] Q-switching techniques. In active Q-switching and mode-locking, some external bulk equipment such as acousto-optics and electro-optics devices are needed to control the absorption wavelength, saturation energy and recovery time. Thus, the passive technique is more preferable since it is simpler, compact and easy to implement.

Passive Q-switching is realized by employing a saturable absorber (SA) inside the laser cavity. The SA plays a vital role in determining the Q-switching performances. Till date, there are several types of saturable absorber have been investigated including carbon nanotubes (CNT) [14-15], graphene [16-18], transition metal dichalcogenides (TMD) such as molybdenum disulphide (MoS_2) [19] and molybdenum diselenide ($MoSe_2$) [20] and topological insulators including bismuth selenide [21]. Yet, only little efforts have been made to discover the potential of metal as saturable absorbers. Recently, platinum has been successfully discovered as saturable absorber to generate Q-switching at 1.5 μm of operating wavelength in our previous work [22]. Ahmad et. al [23] also reported a similar kind of work but using carbon platinum SA on side-polished fiber. Other than platinum, other transition metals such as silver and gold have been explored as Q-switcher due to their unique optical properties in nonlinear optical communications [24]. Besides that, these metals have large third-order nonlinearity and broad absorption governed by localized surface plasmon resonance (SPR) [25].

In this paper, a gold thin film is utilized as SA to induce Q-switching in C-band region. The SA is fabricated by depositing gold on fiber end face via sputtering. It does not need any other materials to act as buffer such as PVA as this technique comprises direct deposition.

2. Experimental details

2.1. Preparation and characterization of gold thin film SA

The fabrication of the gold thin film SA was fully depended on the sputter deposition method. It involves two main processes which are vacuum and deposition. Before depositing the gold target on the fiber end face (substrate), the chamber must be ready in a full vacuum condition. This to allow electrons to strike the Au target in order to eject gold particles. At 5×10^{-3} mBar of vacuum pressure and 120 mA of sputtering current, the deposition process took place. The gold deposited surface is shown in Figure 1(a). The surface was analyzed by SEM and EDS as shown in Figure 1(b) and (c), to ensure the fiber core region was fully covered by the gold thin film. There are 3 peaks of Au element in the EDS graph, confirming the existence of Au at the core area.

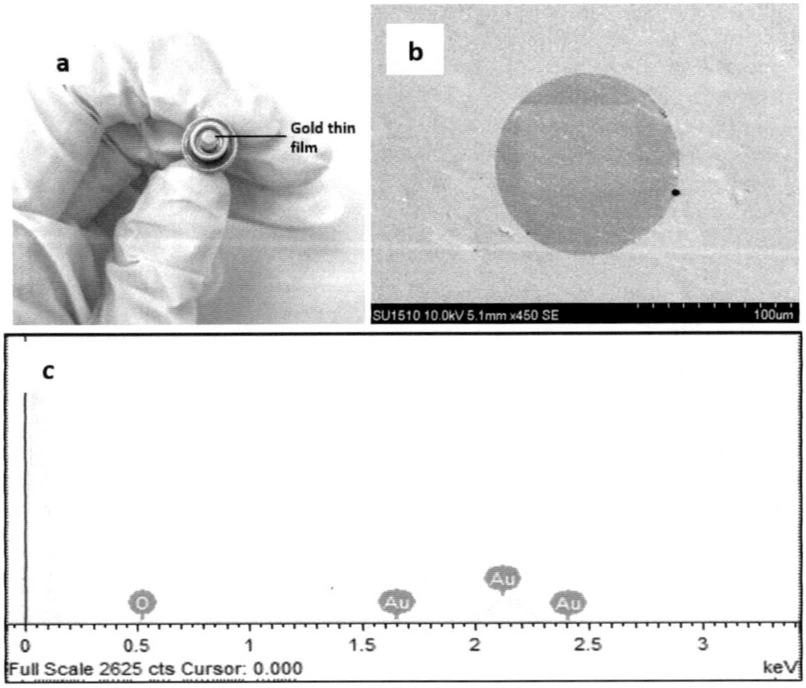

Figure 1. (a) Gold thin film on fiber end face (b) SEM image (c) EDS analysis.

2.2. Q-switched fiber laser setup

The experimental setup of the Q-switched fiber laser is illustrated in Figure 2. The setup was constructed as a single ring cavity, consisted of a 5 m long erbium-doped fiber (EDF) as the gain medium, pumped by a 980 nm laser diode (II-VI LC96Z600-74) via 980/1550 nm wavelength division multiplexer (WDM). The absorption coefficient of the EDF is very low, which is only about 5.09 dB/m. To ensure a unidirectional propagation of light, an isolator was inserted into the cavity. The gold thin film SA was employed to realize passive Q-switching. A 90/10 optical coupler was used to extract 10% of the laser output for data collection while leaving the other 90% to oscillate inside the cavity. The 10% output was tapped by a 50/50 optical coupler, in which the 50% was channeled to the optical spectrum analyzer (OSA, Anritsu MS 9740A) while the other half to the oscilloscope (OSC, Tektronix MDO 3104). A 5 GHz photodetector (PD, Thorlabs DET08CFC/M) was preintegrated with the oscilloscope to detect the pulses. The oscilloscope used in this experiment is able to measure both the time and frequency domain.

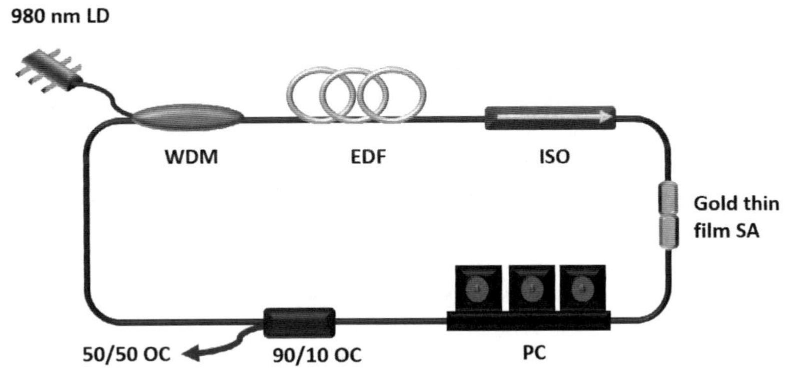

Figure 2. Schematic setup of the Q-switched fiber laser employing gold thin film SA.

3. Results and discussions

The experiment started by varying the pump power to determine the threshold of the Q-switching operation. If the energy inside the cavity reaches a certain saturable value, the Q-switching will occur. This is determined by the SA's recovery time. In this experiment, the Q-switching started to appear at a pump power of 38.8 mW. By further increasing the pump power, the pulse became unstable and disappeared after reaching 119.3 mW. This is because the SA was over-saturated at high incident intensity. Figure 3(a) shows the Q-switched optical spectrum at 119.3 mW of pump power. The peak power and the central wavelength are -7.52 dBm and 1562.4 nm, respectively. The pulse train, which was also taken at the same pump power, is shown in Figure 3(b). The pulse period is measured to be around 46.04 μs, consistent to the repetition rate of 21.72 kHz. The corresponding pulse duration is about 7.44 μs.

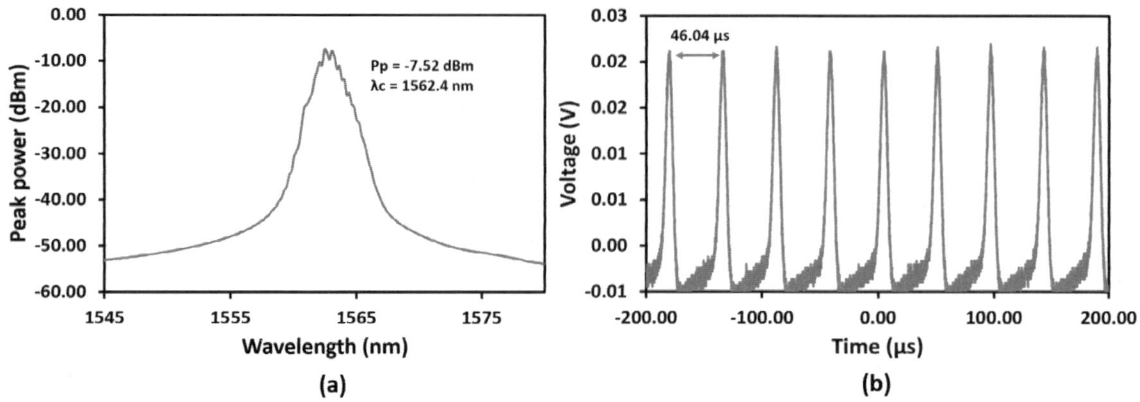

Figure 3. Q-switched fiber laser characteristics at pump power of 119.3 mW. (a) Optical spectrum from the OSA (b) Pulse train from the OSC

The performance of the Q-switched pulses are analyzed at various pump power, presented in Figure 4(a) and (b). The dependence of repetition rate and pulse duration on pump power is illustrated in Figure 4(a). It can be seen that the pulse repetition rate (purple dots) escalated from 11.34 kHz to 12.72 kHz as the pump power was increased from 38.8 mW to 119.3 mW. On the other hand, the pulse duration (orange squares) shrunk from 13.91 μs to 7.44 μs within the same range of pump power variations, representing typical Q-switching characteristics. Figure 4(b) shows the trends of average output power (purple triangles) and pulse energy (orange diamonds) which both were increasing with the pump increment. The pulse energy increased from 3.88 nJ to 13.28 nJ, were obtained by dividing the average output power with the pulse repetition rate at each pump power. The pulse energy signifies the total energy available in a single pulse.

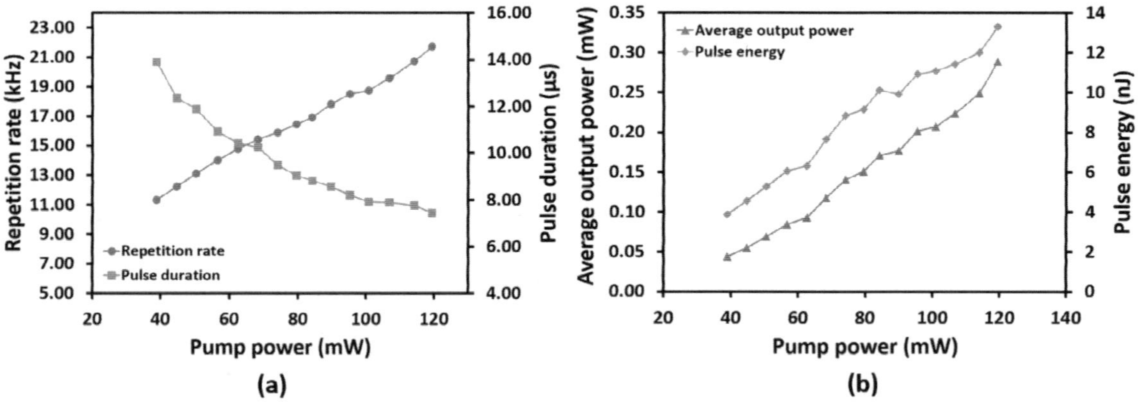

Figure 4. (a) The dependence of pulse repetition rate and pulse duration on pump power (b) Average output power and pulse energy as a function of pump power

Figure 5 indicates the radio frequency (RF) spectra of the generated pulse, taken at 200 Hz of resolution bandwidth. There are 9 harmonics lining up over 200 kHz span. It can be seen that the fundamental frequency peaks at 21.3 kHz, which corresponds well with the measured repetition rate at the same pump power. A high signal-to-noise (SNR) value of about 62.41 dB was obtained, signifying a highly stable pulse.

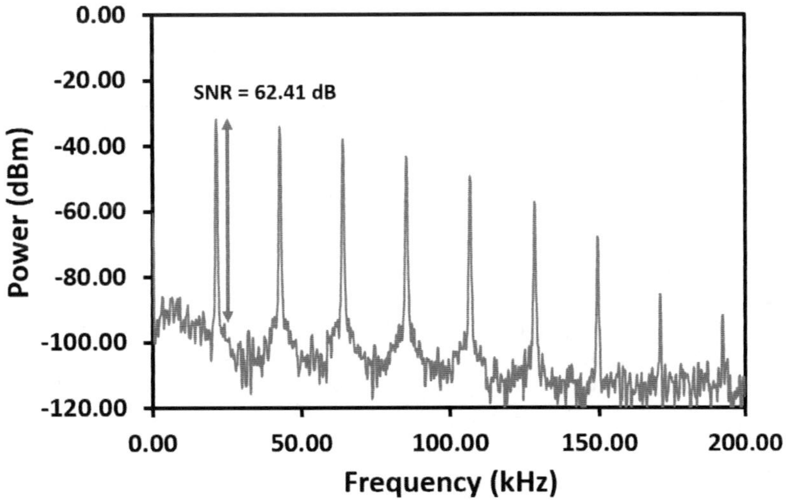

Figure 5. RF spectra at a pump power of 119.3 mW.

4. Conclusion

A passively Q-switched EDFL using gold thin film SA is demonstrated. The SA was prepared through direct deposition of gold on the end surface of a standard FC/PC fiber connector by using a DC magnetron sputter coater. The proposed Q-switched EDFL operates at 1562.4 nm wavelength. The highest repetition rate and pulse energy are 21.72 kHz and 13.28 nJ, respectively. Meanwhile, the smallest pulse duration obtained is 7.44 μs. The results suggest that the gold thin film can be a promising SA for pulsed fiber laser applications.

Acknowledgement

Authors would like to thank Research Management Centre (RMC) of Universiti Tun Hussein Onn Malaysia for supporting this work under TIER 1 Grant: H139 and H133 and GPPS Grant: H065. Authors would also thank Universiti Tun Hussein Onn Malaysia (Pagoh Campus) for providing the facilities and instruments to complete this work.

References

[1] Jackson SD, Sabella A and Lancaster DG 2007 *IEEE J. Sel. Top. Quantum Electron.* **13** 567-72
[2] Morris JA and Pollock CR 1990 *Opt. Lett.* **15** 440-2
[3] Keller U, Knox WH and Roskos H 1990 *Opt. Lett.* **15** 1377-9
[4] Kobtsev S, Kukarin S and Fedotov Y 2008 *Opt. Exp.* **16** 21936-41
[5] Tan SJ, Harun SW, Arof H and Ahmad H 2013 *Chin. Opt. Lett.* **11** 073201
[6] Amiri I, Soltanian M, Alavi S and Ahmad H 2015 *Opt. Quant. Electron.* **47** 2455-64
[7] Soltanian M, Amiri I, Chong W, Alavi S and Ahmad H 2015 *IEEE Photonics J.* **7** 1-11
[8] Grelu P and Akhmediev N 2012 *Nat. Photon.* **6** 84-92
[9] Bäumler W, Eibler ET, Hohenleutner U, Sens B, Sauer J, Landthaler M 2000 *Lasers Surg. Med.* **26** 13-21
[10] Lask G, Elman M, Slatkine M, Waldman A and Rozenberg G 1997 *Dermatol. Surg.* **23** 737-9
[11] Lott J, Ryan C, Valle B, Johnson III JR, Schiraldi DA, Khan J, Singer KD and Weder C 2011 *Adv. Mater.* **23** 2425-9
[12] Andrés MV, Cruz JL, Díez A, Pérez-Millán P and Delgado-Pinar M 2008 *Laser Phys. Lett.* **5** 93-9
[13] Garnov SV, Solokhin SA, Obraztsova ED, Lobach AS, Obraztov PA, Chernov AI, Bukin VV, Sirotkin AA, Zagumennyi YD, Zavartsev YD, Kutovoi SA and Shcherbakov IA 2007 *Laser*

Phys. Lett. **4** 648-51
[14] Haris H, Anyi CL, Muhammad AR, Ahmad F, Nor RM, Zulkepely NR, Ali NM, Harun SW and Arof H 2015 *Microwave Opt. Technol. Lett.* **57** 1897-901
[15] Zarei A, Tiu Z, Ahmad F, Ahmad H and Harun SW 2015 *IET Optoelectron.* **9** (2015) 96-100
[16] Haris H, Anyi CL, Muhammad AR, Ahmad F, Nor RM, Zulkepely NR, Ali NM, Harun SW and Arof H 2015 *J. Optoelectron. Adv. Mater.* **17** 539-44
[17] Luo Z, Zhou M, Weng J, Huang G, Xu H, Ye C and Cai Z 2010 *Opt. Lett.* **35** 3709-11
[18] Yu H, Chen X, Zhang H, Xu X, Hu X, Wang Z, Wang J, Zhuang S and Jiang M 2010 *ACS Nano* **4** 7582-6
[19] Ahmad H, Hassan H, Safaei R, Thambiratnam K, Ismail MF and IS 2017 *Opt. Comm.* **400** 55-60
[20] Ahmad H, Ismail MA, Sathiyan S, Reduan SA, Ruslan NE, Lee CSJ, Zulkifli MZ, Thambiratnam K, Ismail MF and Harun SW 2017 *Opt. Comm.* **382** 93-8
[21] Haris H, Harun SW, Muhammad AR, Anyi CL, Tan SJ, Ahmad F, Nor RM, Zulkepely NR and Arof H 2017 *Opt. Laser Technol.* **88** 121-7
[22] Yuzaile YR, Awang NA, Zalkepali NUHH, Zakaria Z, Azmi AN and Abdul Hadi FS 2019 *Optik* **179** 977-85
[23] Ahmad H, Hassan H, Safaei R, Thambiratnam K and Amiri IS 2017 *Chin. Opt. Lett.* **15** 090601
[24] Liao H, Xiao R, Fu J, Yu P, Wong GKL and Sheng P 1997 *Appl. Phys. Lett.* **70** 1-3
[25] Liao H, Xiao R, Wang H, Wong K and Wong G 1998 *Appl. Phys. Lett.* **72** 1817-19

Influence of concentration on the geometry of ZnO nanostructures prepared by chemical bath deposition

N Rosli[1], M M Halim[1], M R Hashim[1] and W Maryam[1]

[1]Nano-Optoelectronics Research and Technology Laboratory, School of Physics, Universiti Sains Malaysia, 11800 Penang, Malaysia

nurizatirosli@student.usm.my

Abstract. Zinc Oxide (ZnO) nanostructures such as nanorods (NRs), nanowalls (NWs) and nanoflakes (NFs) were successfully grown on a glass substrate via chemical bath deposition (CBD) method by varying solutions concentration (0.01 M, 0.05 M, 0.10 M, 0.15 M, 0.20 M and 0.25 M). Effect of CBD concentration on the growth and morphology of ZnO nanostructures are dignified. The structural and optical properties of ZnO NRs, NWs, and NFs were investigated by Field Emission Scanning Electron Microscope (FESEM), energy-dispersive X-ray (EDX), X-Ray Diffraction (XRD), and UV-Visible (UV-Vis) analyses. Results reveal that as the CBD concentration increases, FESEM images showed differ top-down morphology of ZnO nanostructures from NRs, NWs to NFs. Likewise, the alignment of ZnO cross-section evolved from randomly oriented to vertically oriented. As revealed by EDX, the stoichiometric ratio of Zn:O were mostly towards 1:1. From XRD analyses, the preferred structure was wurtzite hexagonal with *c*-axis orientation along (002) plane. The UV-Vis analysis exposed the optical band gap energy closes to the standard energy gap value of ZnO at 3.37 eV. Thus, this work introduces a variable geometry of ZnO nanostructures (NRs, NWs, and NFs) grown on glass with high structural and optical quality as standard ZnO. The fabricated materials are potential candidates for investigating random lasing in ZnO from different structural geometry.

Keywords: ZnO, nanostructure, nanorods, nanowalls, nanoflakes, CBD

1. Introduction

The ZnO nanostructure has attracted substantial interest in the research community due to several remarkable optical and structural properties such as a wide direct band gap at 3.37 eV, large exciton binding energy at 60 meV, and could be prepared into numerous geometries by different synthesis techniques [1]. These include, (but not limited to) physical synthesis (metal organic chemical vapor deposition-MOCVD, hydrothermal, radio frequency or direct current sputtering), liquid phase synthesis (wet chemical methods, CBD, sol-gel process and colloidal method), and gas phase synthesis (spray pyrolysis and inert gas condensation methods) [2-4]. Amongst all, CBD is known to be one of the most versatile techniques to control growth condition. CBD methods also offer a large semiconductor deposition area by easy procedures, low cost, low processing temperature, no vacuum, catalyst-free and relatively safe (no hazardous chemicals) and do not require any application of external field or surfactant to produce preferred nanostructures with good structural and optical properties [5-8]. Considering these factors, therefore, CBD methods were chosen for the ZnO nanostructures preparation.

2. Methodology

The sample preparation was divided into several steps to grow ZnO nanostructures on top of glass substrates coated with ZnO seed layer. **Figure 1** shows the overall flow chart of each step and process. First, the glass substrates were cleaned using DI, ethanol, and acetone sequentially in the ultrasonic bath at room temperature for 10 minutes. Then, a 200 nm thick ZnO seed layer was grown on the glass substrates using RF magnetron sputtering system with 150 W power, followed by the annealing process at 400 °C for 1 hour. The next step is to grow the ZnO nanostructures by CBD technique. For the CBD solution, six different concentration of $Zn(NO_3)_2{\cdot}6H_2O$ at 0.01 M, 0.05 M, 0.10 M, 0.15 M, 0.20 M, and 0.25 M were prepared by dissolving the concentration in DI water. Separately, HMTA ($C_6H_{12}N_4$) was dissolved similarly with the same concentration. Later, both solutions were combined as 1:1 ratio in one beaker before the prepared substrates were submerged and positioned vertically inside the beaker. The beaker was then heated in an oven at 96 °C for 3 hours. After the growth process, the samples were rinsed with DI water and dried by nitrogen blow. The samples then underwent the second annealing process, again at 400 °C for 1 hour. Physical, structural and optical characterization were done using FESEM, EDX, XRD and UV-Vis respectively.

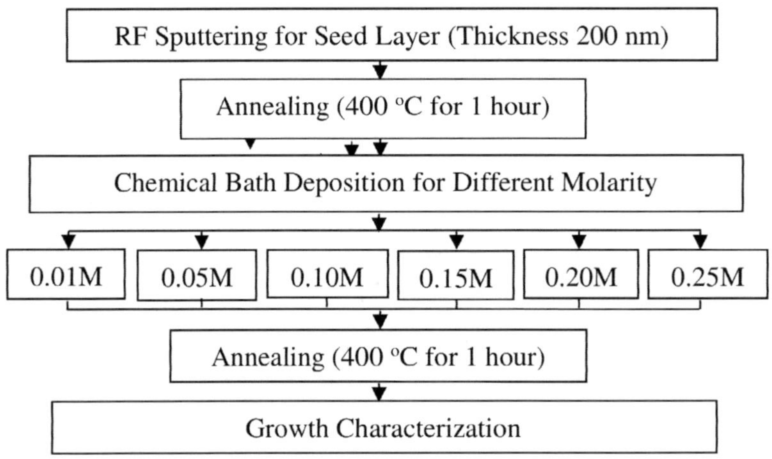

Figure 1. Flow chart for the preparation of ZnO nanostructures.

3. Results and discussion

The growth process of ZnO nanostructures through CBD technique begins with the agglomeration of the HMTA and $ZnNO_3.H_2O$, resulting in the formation of ZnO nanostructures at the elevated temperature. The seed layer reduces the interface energy and provides as the nucleation site for the nanostructures. **Figure 2** shows FESEM images in a planar view and a cross-sectional view of the ZnO nanostructure for six different solution concentrations. The planar view reveals that the surface morphology of the ZnO nanostructures evolves from NRs, to NWs and finally NFs. Information on average diameter, average nanorods height as well as atomic and weight percentage from EDX analysis are listed in **Table 1**. In term of atomic percentage for the ratio between Zn and O, the value was towards 1:1 from 0.05 M onwards. For 0.01 M concentration, the ratio was closely resemblance to the results from 0.15 M.

From the lowest concentration, 0.01 M up to 0.10 M, the nanostructures vary from randomly oriented top and smaller size (22.63 nm average diameter) to sharp hexagonal top, and bigger size (90.86 nm average diameter) with highly vertical oriented. These properties were observed from the cross-sectional view. Beyond 0.10 M, the nanostructures evolved from NRs to NWs (for 0.15 M and 0.20 M) and finally became NFs (for 0.25 M). From these observations, the major variation was in the geometrical shape and the size of the nanostructures. However, the achieved height of the nanostructures was maintained at around 1.00 μm for all samples except for the sample with the lowest concentration, 0.01 M where the nanostructures were in random orientation. It showed here that the solution concentration does not affect the lateral height of the nanostructures, but rather their geometrical shape.

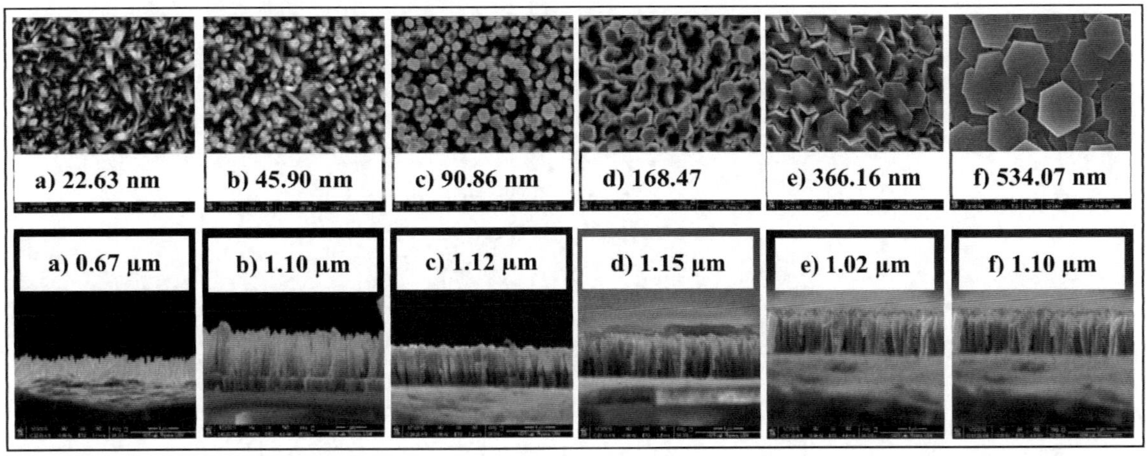

Figure 2. Planar and cross-section morphology for ZnO NRs, NWs to NFs as the CBD concentration increases from 0.01 M to 0.25 M.

Table 1. Average physical dimensions and their atomic and weight percentages of Zn:O as the CBD concentration increases from 0.01 M to 0.25 M.

Concentration, (M)	Mean diameter, (nm)	Height, (μm)	% Atomic	% Weight
0.01 M	22.63	0.667	O – 44.12 Z – 55.88	O – 16.2 Z – 83.8
0.05 M	45.90	1.098	O – 41.21 Z – 58.79	O – 15.86 Z – 84.14
0.10 M	90.86	1.123	O – 43.57 Z – 56.25	O – 15.89 Z – 84.11
0.15 M	168.47	1.153	O – 44.08 Z – 55.92	O – 15.99 Z – 84.01
0.20 M	366.16	1.017	O – 45.09 Z – 54.91	O – 16.73 Z – 83.27
0.25 M	534.07	1.10	O – 47.09 Z – 52.91	O – 17.89 Z – 82.11

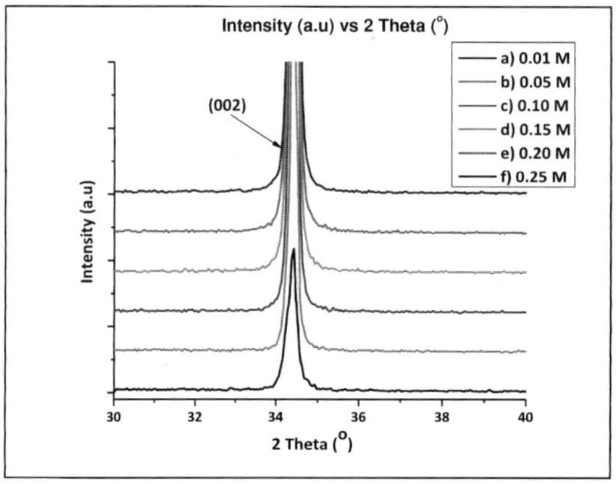

Figure 3. XRD peaks for ZnO NRs, NWs and NFs as the CBD concentration increases, from 0.01 M to 0.25 M.

The structural properties of the ZnO nanostructures were analyzed from XRD results. As can be seen in **Figure 3**, the only clear peak was seen around 34.4°. This corresponds to the preferential growth direction of the nanostructures along (002). To determine the average crystallite size which related to the average diameter (D) of the crystallite columns, the Debye-Scherer formula was employed [9]:

$$D = \frac{0.94\lambda}{\beta \cos\theta} \tag{1}$$

where β is the full width at the half maximum in radians, D (m) is the crystallite size and λ is the X-ray wavelength (1.54×10^{-10} m). In the case where ZnO crystallizes are in the hexagonal wurtzite structure, the oxygen atoms are recognized in a hexagonal close-packed with zinc atoms vacating half the tetrahedral sites and zinc and oxygen atoms are tetrahedrally coordinated to each other. According to Bragg's law [9]:

$$n\lambda = 2d \sin\theta \tag{2}$$

where n is the order of diffraction (first order, $n = 1$ and second order, $n = 2$), λ is the X-ray wavelength, and d is the spacing between planes of given Miller indices h, k, and l. In the ZnO hexagonal structure, the plane spacing d is related to the lattice constants a and c, as $a = b \neq c$. Following the equation of c-axis lattice parameter of the axial height of a unit cell, as edge length, $a = 0$ as only (002) peak appears [10]:

$$\frac{1}{d_{hkl}^2} = \frac{4}{3}\left(\frac{h^2 + hk + k^2}{a^2}\right) + \frac{l^2}{c^2} \tag{3}$$

where d_{hkl} is the interplanar spacing between planes of given Miller indices h, k, and l, while a and c are the lattice constants. The lattice strain of ZnO microstructures synthesized on the PS substrate along c-axis was calculated using the following equation:

$$\%\varepsilon = \frac{c - c_o}{c_o}\ x\ 100 \tag{4}$$

where c_o was the standard values of ZnO single crystals, ($c_o = 0.52098$ nm) which indicated the good quality crystalline in nature of ZnO as be represented in XRD database, c is the lattice constant of the attained ZnO by CBD method. The dislocation density (δ), which represents the amount of defects in the sample is defined as the length of dislocation lines per unit volume of the crystal and is calculated using equation: (5)

$$\delta = \frac{1}{D^2}$$

where D is the crystallite size as from above. All the results from XRD analyses were listed in **Table 2**. Overall, from this XRD analyses, it is confirmed that the ZnO nanostructures grown with c-axis preferred orientation under typical growth conditions due to the lowest surface energy of the (001) or (002) basal plane in ZnO compared to growth in the (101) and (010) direction. This allocated that the most stable morphology of ZnO hexagonal structure elongated along the c-axis, whereas its crystal consists of a nonpolar plane (100), polar basal oxygen terminated plane (001), and a polar face (001) of tetrahedral Zn terminated with O [2,11]. Besides, nonappearance of any other XRD peak specifies that these ZnO nanorods are well aligned along the c-axis [7]. **Table 2** also shows that as the concentration of the reactants increased, the narrowing of FWHM value confirmed that the nanostructures were more aligned and oriented along the c-axis which is related to (002) direction as presented by the FESEM images earlier. Another observation from this analysis, as the concentration increases, it showed that the crystallite size increases. While the dislocation density diminished, which concentrates on the reduction of the lattice imperfections, the lattice misfit also lessening as a result of small changes in tensile stress that acts on the surface.

Table 2. XRD results for ZnO NRs, NWs and NFs as the CBD concentration increases, from 0.01 M to 0.25 M.

Concentration, (M)	1st order, 2θ, (°)	FWHM	D (nm)	D-spacing, d_{hkl}	Lattice, c (nm)	Lattice strain, (%)	Dislocation density, (δ)
0.01 M	34.4156	0.30996 ° 0.00541 rad	28.01	0.2603	0.5206	-0.08	0.0013
0.05 M	34.4244	0.18977 ° 0.00331 rad	48.89	0.2602	0.5204	-0.11	0.0004
0.10 M	34.4258	0.16869 ° 0.00294 rad	51.54	0.2602	0.5204	-0.11	0.0004
0.15 M	34.4398	0.15872 ° 0.00277 rad	54.71	0.2601	0.5202	-0.15	0.0003
0.20 M	34.4400	0.15532 ° 0.00271 rad	55.92	0.2601	0.5202	-0.15	0.0003
0.25 M	34.4413	0.15131 ° 0.00264 rad	57.41	0.2601	0.5202	-0.15	0.0003

UV-Vis investigations were employed to determine the excitation of the photon from the valance band, (VB) to the conduction band, (CB) and to calculate the optical band-gap energy (E_g). From **Figure 4a**, it shows that the UV-visible spectra have high transmittance starting from 380 nm (UV) and getting higher in the visible region. The optical band gap was determined by utilizing Tauc's equation, while the absorption coefficients, α was analysed using Kubelka-Munk (K-M) equation with α is directly proportional to K-M function, and $F(R)$ in terms of diffuse reflectance (R), $F(R)$ is given by [12-14]:

$$F(R) = \frac{(1-R)^2}{2R} \quad (6)$$

where $F(R)$ is the K-M function. The band gap determination based on Tauc's equation is given by [15]:

$$(F(R).hv)^2 = A(hv - E_g) \quad (7)$$

where A is a proportionality constant, hv is the energy of the incident photon and E_g is the band gap energy. The optical energy band gap was measured as a direct band semiconductor and the energy band gap is obtained by linearly extrapolating to the $y = 0$. There are several aspects that can affect the value of the E_g, such as size, the temperature, dopants, defects, tensions, and the functionalization with organic surfactants [16]. In this work, the decreasing in the band gap energy may cause by tensions and stress in the crystalline lattice by highly packed of ZnO grown on the glass substrate as the concentration increases as shown in **Figure 4b**. The band gap achieved is in the range of (3.13 eV < E_g < 3.24 eV) which slightly smaller to the value of intrinsic ZnO, 3.37 eV. The band gap of ZnO nanostructures shows red-shifted trend compared to bulk ZnO as the concentration increases. This shift may be affected by the origination of the structural defects during the chemical synthesis, which nearer the allowed states to the conduction band in the energy band gap. It was suggested that the red-shift may be caused by the increasing size of the synthesized ZnO nanostructures [17], which totally agreed with the observation from FESEM and XRD analysis earlier.

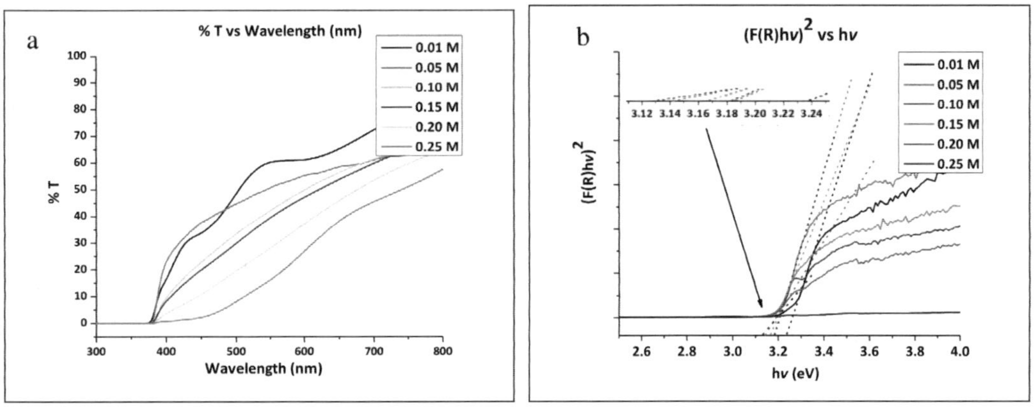

Figure 4. UV-Vis results: a) % Transmittance, and b) $(F(R)\ hv)^2$ vs hv (eV) of ZnO NRs, NWs and NFs as the CBD concentration increases from 0.01 M to 0.25 M.

4. Conclusion

ZnO nanostructures were successfully synthesized with different geometry (NRs, NWs, and NFs) by varying CBD reactants solution ($Zn(NO_3)_2$ and HMTA) at an equimolar ratio. The FESEM, EDX, XRD and UV-Vis were used to investigate the effect of increasing concentration on the structural and optical properties of the nanostructures. FESEM images reveal the structural evolution from NRs to NWs and finally NFs at the highest concentration. The nanostructures also evolved from randomly oriented to highly vertical oriented. XRD indicates that the structural properties of the nanostructures grew along *c*-axis, (002) orientation and getting more aligned and oriented as the concentration increased. The optical band-gap energy revealed by UV-Vis diffused reflectance show smaller and decreasing of energy in the range of (3.13 eV < E_g < 3.24 eV) to the value of intrinsic ZnO, (3.37 eV). Overall, these results show that the increase in reactants concentration leads to a variety of geometrical nanostructure with good growth quality using a simple CBD method. This work aspires a quest for the investigation on random lasing effect from different ZnO nanostructure geometry.

Acknowledgements
We thank Universiti Sains Malaysia (USM) for supporting this work under RUI grant number 1001/PFIZIK/8011092 and totally indebted to the staff from NOR Lab, School of Physics, USM.

References
[1] O. F. Farhat, M. M. Halim, N. M. Ahmed, and M. A. Qaeed, "ZnO nanofiber (NFs) growth from ZnO nanowires (NWs) by controlling growth temperature on flexible Teflon substrate by CBD technique for UV photodetector," *Superlattices Microstruct.*, vol. 100, pp. 1120–1127, 2016.

[2] M. S. Kim, K. G. Yim, S. Kim, G. Nam, D. Lee, Jin Soo Kim, Jong Su Kim, J. Y. Lee., "Growth and characterization of indium-doped zinc oxide thin films prepared by sol-gel method," *Appl. Phys. A Mater. Sci. Process.*, vol. 121, no. 1, pp. 3–6, 2012.

[3] M. Khasif, S. M. U. Ali, H. I. Abduldafour, M. Willander, and Z. Hassan, "Morphological, optical, and Raman characteristics of ZnO nanoflakes prepared via a sol–gel method," *Phys. Status Solidi*, vol. 147, no. 1, pp. 143–147, 2012.

[4] M. Chen, Z. Wang, D. Han, F. Gu, and G. Guo, "Porous ZnO polygonal nanoflakes: Synthesis, use in high-sensitivity NO_2 gas sensor , and proposed mechanism of gas sensing," *J. Phys. Chem. C*, vol. 115, no. 2, pp. 12763–12773, 2011.

[5] O. F. Farhat, M. M. Halim, M. Ahmed Naser, Ammar A. Oglat, A. A. Abuelsamen, M. Bououdina, M. A. Qaeed, "A study of the effects of aligned vertically growth time on ZnO nanorods deposited for the first time on Teflon substrate," *Appl. Surf. Sci.*, vol. 426, pp. 906–912, 2017.

[6] R. Shabannia, H. Abu Hassan, H. Mahmodi, N. Naderi, and H. R. Abd, "ZnO nanorod ultraviolet photodetector on porous silicon substrate," *Semicond. Sci. Technol.*, vol. 28, no. 11, p. 6, 2013.

[7] W. Maryam, N. Fazrina, M. R. Hashim, and H. C. Hsu, "Optically driven random lasing in ZnO nanorods prepared by chemical bath deposition," *Photonics Nanostructures - Fundam. Appl.*, vol. 26, pp. 52–55, 2017.

[8] N. I. Maad Tazri, O. L. Muskens, M. K. Shakfa, and W. Maryam, "Polycrystalline ZnO nanorods for lasing applications," *J. Appl. Phys.*, vol. 125, no. 123103, pp. 1–5, 2019.

[9] H. Sarma and K. C. Sarma, "X-ray peak broadening analysis of ZnO nanoparticles derived by precipitation method," *J. Sci. Res. Publ.*, vol. 4, no. 3, pp. 1–7, 2014.

[10] A. Iqbal, M. Zakria, and A. Mahmood, "Structural and spectroscopic analysis of wurtzite $(ZnO)_{1-x}(Sb_2O_3)_x$ composite semiconductor," *Prog. Nat. Sci. Mater. Int.*, vol. 25, no. 2, pp. 131–136, 2015.

[11] J. Zhao, Z. Jin, T. Li, and X. Liu, "Nucleation and growth of ZnO nanorods on the ZnO-coated seed surface by solution chemical method," *J. Eur. Ceram. Soc.*, vol. 26, pp. 2769–2775, 2006.

[12] R. Bhardwaj, A. Bharti, J. P. Singh, K. H. Chae, N. Goyal, and S. Gautam, "Structural and electronic investigation of ZnO nanostructures synthesized under different environments," *Heliyon*, vol. e00594, pp. 1–21, 2018.

[13] S. D. Hutagalung, M. M. Fadhali, R. A. Areshi, and F. D. Tan, "Optical and electrical characteristics of silicon nanowires prepared by electroless etching," *Nanoscale Res. Lett.*, vol. 12, no. 1, pp. 1–11, 2017.

[14] P. Bossuyt, E. Louis, J.-Y. Mary, S. Vermeire, and Y. Bouhnik, "UV-Vis and PL spectroscopy of ZnO nanorods and nanotubes," *J. Crohn's Colitis*, no. 3, pp. 63–88, 2018.

[15] A. Ahmed, A. Ahmed, Z. A. Talib, M. Z. Hussein, M. H. Flaifel, and N. M. Al-hada, "Influence of ZnFe molar ratio on optical and magnetic properties of ZnO and $ZnFe_2O_4$ nanocrystal as calcined products of layered double hydroxides," *J. Spectrosc.*, vol. 2014, no. 732163, pp. 1-8, 2014.

[16] A. F. Moreno, A. M. Herrera, and G. J. García, "Modification of the crystal lattice and optical band gap of ZnO nanostructures by the polyelectrolytes presence," *J. Mater. Sci. Mater. Electron.*, vol. 29, no. 18, pp. 15604–15612, 2018.

[17] S. S. Kanmani and K. Ramachandran, "Role of aqueous ammonia on the growth of ZnO nanostructures and its influence on solid-state dye-sensitized solar cells," *J. Mater. Sci.*, vol. 48, no. 5, pp. 1-19, 2013.

Fiber Laser Temperature Sensor based on Sagnac Interferometer

M A Omar[1], N A Cholan [1,2*], A M Abdulrahman[1], N A N Jaharuddin[1], R Talib[1] and N H Ngajikin[1]

[1]Faculty of Electrical and Electronics Engineering, Universiti Tun Hussein Onn Malaysia 86400 Parit Raja, Batu Pahat, Johor, Malaysia.
[2] Research Center for Applied Electromagnetics, Universiti Tun Hussein Onn Malaysia 86400 Parit Raja, Batu Pahat, Johor, Malaysia.

noran@uthm.edu.my

Abstract. A fiber laser temperature sensor which is based on Sagnac interferometer is proposed in this work. The fiber laser consists of an erbium-doped fiber amplifier for signal amplification and a Sagnac interferometer for signal filtering. The Sagnac interferometer is made of 30 cm polarization maintaining fiber (PMF) and it acts as the sensing head. Experimental results suggest that the temperature does affect the laser wavelength. As the temperature varies from 30°C to 41°C, the laser wavelength is shifted to the lower wavelength from 1596.5 nm to 1581.8 nm with the recorded sensitivity of 1.1965 nm/°C. The change of the laser wavelength with temperature validates the use of this fiber laser as the temperature sensor.

1. Introduction

In recent years, all-fiber optical sensors especially those with high sensitivity have gained great interest. Such a growing demand for fiber laser sensors stems from their advantages of simple structure, high sensitivity, high signal-to-noise ratio (SNR), and excellent stability [1]. Due to these advantages, fiber laser sensors have been successful in measuring temperature, strain, refractive index, vibration, twist, gas absorption and so on [2]. They have found applications in biomedicine, environmental monitoring, agricultural engineering, health monitoring and aeronautics [3].

Optical fiber sensors could be in the form of wavelength-modulated sensors, intensity modulated sensors and interferometric sensors. The mechanism for wavelength modulated sensors such as fiber Bragg grating (FBG) is based on the wavelength shift [4], [5]. Despite offering a wide dynamic range, the fabrication of the sensor head is rather complex. Intensity modulated sensors meanwhile work based on the change of the light intensity [6]-[8]. They are simpler in fabrication but prone to the instability due to the intensity fluctuation. As for interferometric sensors such as Mach-Zhender, Sagnac and Fabry-Perot sensors, the sensing function is obtained from the interference effect of propagating lights [9]-[14]. The interferometric sensors are basically more stable in comparison to intensity modulated sensors.

In this work, we demonstrate a temperature sensor using a reflective Sagnac loop in a fiber ring laser. The laser wavelength is found to have a linear relationship with temperature. As the temperature varies from 30°C to 41°C, the laser wavelength is shifted to the lower wavelength from 1596.5 nm to 1581.8 nm with the sensitivity of 1.739 nm/°C.

2. Methodology

The experimental setup of the fiber ring laser for temperature detection is shown in Fig. 1. The ring laser consists of a custom-made erbium-doped fiber amplifier (EDFA), a reflective Sagnac loop filter and an output coupler. The EDFA comprises of a 10m erbium-doped fiber (EDF) which is pumped by a 980 nm laser diode through a wavelength division multiplexing (WDM) coupler. The 10m EDF provides optical amplification over the C-band region [15], [16]. Meanwhile, the reflective Sagnac loop is developed by a combination of a 50:50 coupler and a 30 cm polarization maintaining fiber (PMF). The Sagnac loop functions as a comb filter in which the spacing is influenced by the birefringence and length of the PMF. A 90:10 coupler is utilized to tap out 10% of the oscillating light for the laser output. Throughout the experiment, the output spectrum is captured using an optical spectrum analyzer (OSA) with a resolution of 0.05 nm.

Besides filtering, the Sagnac loop in this design also behaves as a temperature sensor head. In order to experience different temperatures, the sensor head is placed in a temperature chamber. As the temperature of the sensing head increases, the birefringence of the PMF in the Sagnac loop also changes as well. This consequently causes a shift of Sagnac loop transmission spectrum and ultimately a shift of laser wavelength.

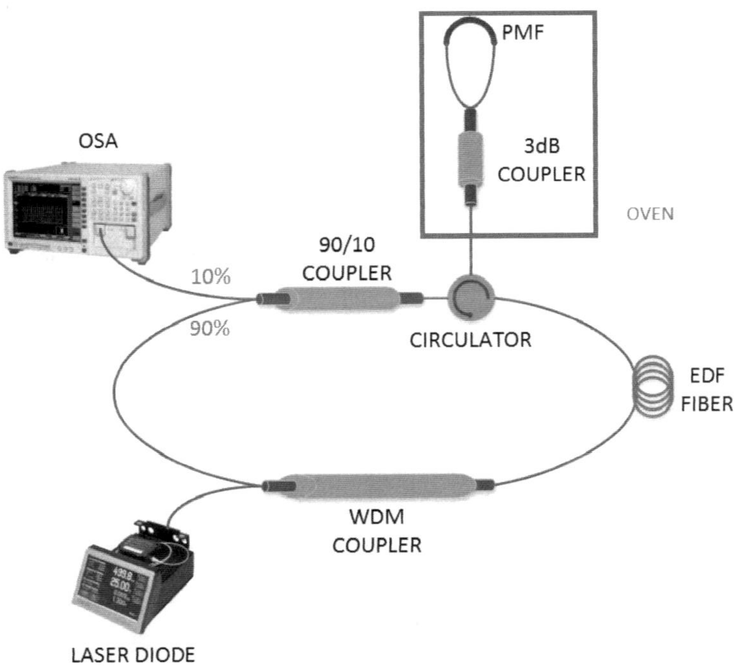

Fig. 1. Experimental setup of fiber laser temperature sensor.

The Sagnac loop structure together with the circulator is shown in Fig. 2. The Sagnac loop consists of a 3-dB coupler and a PMF as the birefringent element. The mechanism of the Sagnac loop as the temperature sensor is as follows; the input light firstly enters the 3-dB coupler via the circulator. In the 3-dB coupler, the signal is split into a clockwise and anti-clockwise direction. These two light beams then pass through the PMF and they get a phase difference due to the birefringence of the PMF. When the two light beams recombined at the output port of the 3-dB coupler, the interference spectrum can be formed. The phase difference ϕ is given by

$$\phi = 2\pi BL/\lambda \tag{1}$$

where, B is the birefringence of the PMF, L is the length of the PMF, and λ is the wavelength of the input light. The output transmission spectrum at the circulator is a periodic function of the wavelength, which is stated as [11]

$$T = (1 - \cos\phi)/2 \tag{2}$$

If $\phi = 2\pi m$, the transmission spectrum will reach its maximum intensity. Here, m is an integer. A change in the phase can be induced by either changing the PMF length ΔL or the birefringence ΔB which is stated as

$$\phi' = 2\pi(B + \Delta B)(L + \Delta L)/\lambda \tag{3}$$

As the PMF is not strained, therefore $\Delta L << L$ and Eqn (3) can be reduced to

$$\phi' = 2\pi(B + \Delta B)L/\lambda \tag{4}$$

Resulting from the change of phase, the phase difference $\Delta\phi$ is defined as

$$\Delta\phi = \phi' - \phi = 2\pi\Delta BL/\lambda \tag{5}$$

which contributes to the shifting of the transmission spectrum. That mean, shifting of the transmission spectrum of Sagnac loop can be induced by changing the temperature of the PMF.

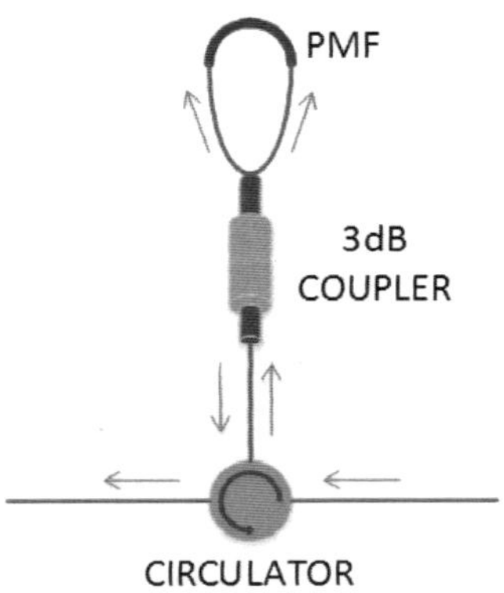

Fig. 2. Sagnac loop configuration

3. Results and discussion

The threshold power of the laser is firstly investigated. The threshold power is defined as the amount of EDF pump power at which the laser output power increases significantly. Figure 3 shows the change of laser output power as compared to the input pump power. In this investigation, input pump power increases from 0 mW to 80 mW with a step of 5 mW. An optical power meter is used to measure the laser output power. It can be seen from Fig. 3 that the threshold pump power is above 20 mW and beyond it, the output power increases linearly with the pump power. Based on the curve fitting, the recorded laser efficiency is x%.

The evolution of the laser output spectrum with input pump power is then examined. Figure 4 shows the output spectrum as the pump power increases from 0 mW to 80 mW with a step of 5 mW. In this observation, the temperature is set at 33°C. Based on Fig. 4, it is observed that the laser spectrum starts to appear at 25 mW. That mean, the input pump power from 0 mW to 20 mW is not enough to generate the laser output spectrum

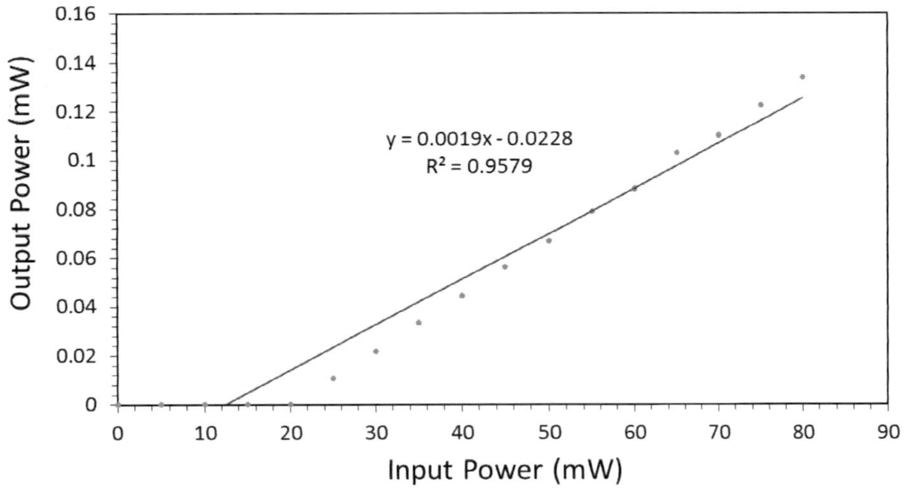

Fig. 3. Laser output power with a variation of pump power

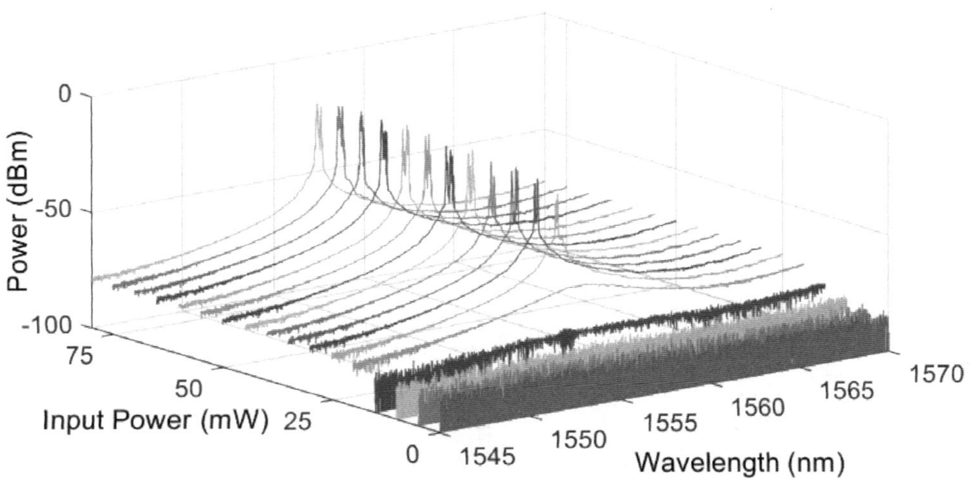

Fig. 4. Laser output spectrum with a variation of pump power

The behavior of the laser output spectrum as the temperature changes is then explored. Figure 5 illustrates the change of the output spectrum as the temperature sensor head is varied from 30°C to 41°C with a step of 1°C. In this observation, the pump power is fixed 200 mW. It is noticeable from Fig. 5 that the laser output spectrum shifts to the shorter wavelength with increasing temperature. When the temperature is set to 30°C, the laser wavelength is at 1570.69 nm and it shifts to the shorter wavelength at 1556.88 nm as the temperature is fixed at 41°C. The blue-shift of the laser wavelength with increasing temperature is due to the change of the PMF birefringence in the Sagnac loop. The birefringence change modifies the interference effect between the clockwise and anti-clockwise light, leading to the blue-shift of the laser output spectrum.

Figure 6 meanwhile plots the laser output wavelength as a function of temperature. It is evident that the laser output wavelength shifts linearly as temperature increases from 30°C to 41°C. Linear fitting is also done to the data in Fig. 6 and the slope of the linear line is found to be 1.1965 nm/°C. This slope basically represents the temperature sensitivity of this fiber laser sensor.

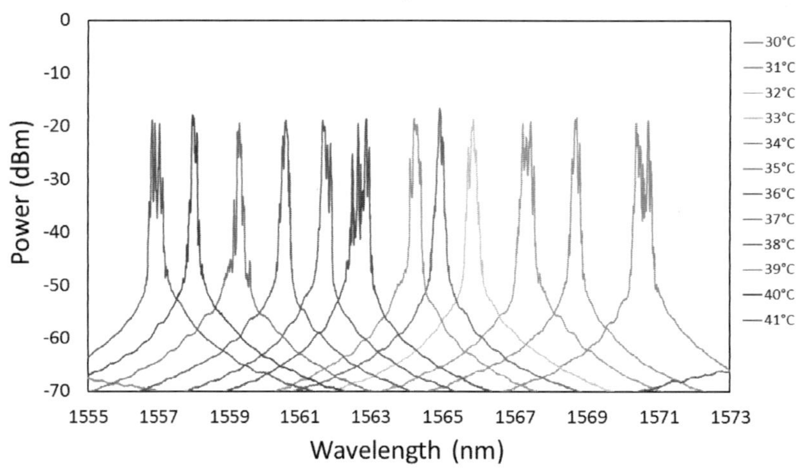

Fig. 5. Laser output spectrum with a variation of temperature

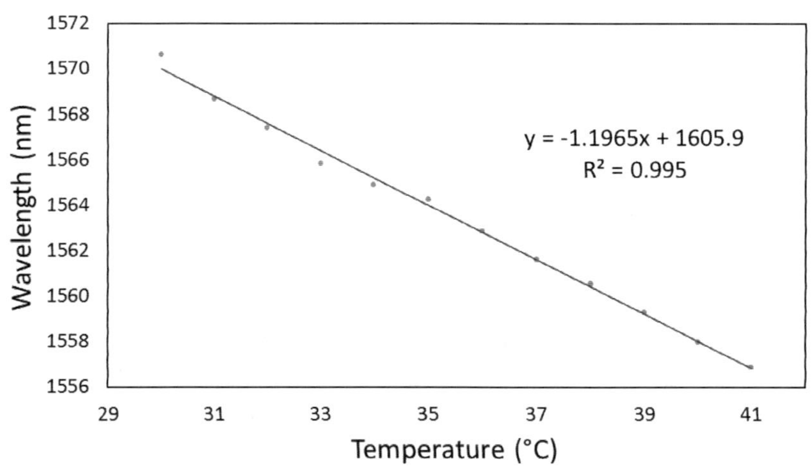

Fig. 6. Laser output wavelength with a variation of temperature

4. Conclusion

In summary, a fiber laser temperature sensor based on Sagnac interferometer is presented. The fiber laser consists of a Sagnac interferometer working as a signal filter and EDFA for signal amplification. The sensing head is made of 30 cm PMF. Based on experimental results, it is evident that temperature indeed affects the laser wavelength. As the temperature varies from 30°C to 41°C, the laser wavelength experiences a blue-shift from 1596.5nm to 1581.8nm. The recorded sensitivity for this fiber laser sensor is 1.1965 nm/°C.

References

[1] Z. Liu, X. Zhang, Z. Gong, Y. Zhang, and W. Peng, "Fiber Ring Laser-Based Displacement Sensor," *IEEE Photonics Technol. Lett.*, vol. 28, no. 16, pp. 1723–1726, 2016.

[2] W. Han, Z. Tong, and Y. Cao, "Simultaneous measurement of temperature and liquid level base on core-offset singlemode-multimode-singlemode interferometer," *Opt. Commun.*, vol. 321, pp. 134–137, 2014.

[3] Y. W. Huang, J. Tao, and X. G. Huang, "Research progress on F-P interference-based fiber-optic sensors," *Sensors (Switzerland)*, vol. 16, no. 9, 2016.

[4] Y.Zhu, P.Shum, C.Lu, M.B.Lacquet, P.L.Swart, A.A.Chtcerbakov and S.J.Spammer, "Temperature insensitive measurements of static displacements using afiber Bragg grat-ing", Optics Express, Vol. 11, pp 1918-1924, (2003).

[5] S.C.Jiang, J. Wang, Q.M.Sui, and Y. Q Cao, "A novel wide measuring range FBG displacement sensor with variable measurement precision based on helical bevel gear", Optoe-lectronics Letter, Vol. 11, pp. 81-83 (2015).

[6] A. Shimamoto and K. Tanaka, "Optical fiber bundle displace-ment sensor using an ac-modulated light source with subna-nometer resolution and low thermal drift," Applied Optics,Vol. 34, Issue 25, pp. 5854-5860 (1995).

[7] E.Fujiwara, M.F.M.Dos Santos and C.K.Suzuki, "Flexible optical fiber bending transducer for application in gloved-based sensors", IEEE Sensors Journal, Vol.14 (10), 3631-3636 (2014).

[8] J.Liu, Y. Hou, P.Jia, S.Su, G.Fang, W.Liu and J.Xiong, "A wide-range displacement sensor based on plastic fiber macro bending coupling", Sensors, MDPI, 17, 196 (2017).

[9] J.P.Chen, J.Zhou and Z.H.Jia, "High sensitivity displacement sensor based on a bent fiber Mach-Zhender interferometer", IEEE Photonics Technology Letters, Vol. 25, no. 23, pp. 2354-2357 (2013).

[10] J.N.Dash, R.Jha, J.Villatoro and S.Dass, "Nano displace-ment sensor based on photonic crystal fiber modal interfer-ometer", Optics Letter, Vol. 40 no. 4, pp. 467-470 (2015).

[11] Q.Rong, X.Qiao, Y.Du, D.Feng, R.Wang, Y.Ma, H.Sun, M.Hu and Z.Feng, "In fiber quasi-Michelson interferometer with a core-cladding-mode fiber end-face mirror", Applied Optics, Vol. 52, no. 7, pp. 1441-1447 (2013).

[12] N.I. Ismail, N.H. Ngajikin, N.F.M. Zaman, M. Awang, A.I. Azmi, N.N.N.A. Malik, N.M. Kassim, "Resolution Improve-ment in Fabry-Perot Displacement Sensor Based on Fringe Counting Method", Telkomnika, Vol. 12, No. 4, pgs. 811-818, (2014).

[13] Y.Bai, F.Yan, S.Liu, and X.Wen, "All fiber Fabry-Perot interferometer for high sensitive micro displacement sensing", Opt. Quantum Electron , Vol. 48, p. 206 (2016).

[14] B.H.Lee, Y.H.Kim, K.S.Park, J.B.Eom, M.J.Kim, B.S.Rho and H.Y.Choi, "Interferometric fiber optics sensors", Sensors, MDPI, vol. 12, pp. 2467-2486 (2012).

[15] N. A. B. Ahmad, S. H. Dahlan, N. A. Cholan, H. Ahmad, I. S. Amiri and Z. C. Tiu, "Dual-Wavelength Thulium Fluoride Fiber Laser Based on SMF-TMSIF-SMF Interferometer as Potential Source for Microwave Generation in 100-GHz Region," IEEE Journal of Quantum Electronics, vol. 54, no. 2, pp. 1600207, April 2018.

[16] N. A. Cholan, M. H. Al-Mansoori, A. S. M. Noor, A. Ismail and M. A. Mahdi, "Self-seeded four-wave mixing cascades with low power consumption ," Journal of Optics (UK), vol. 16, no. 10, pp. 105203, Aug 2014

Journal of Physics: Conference Series **1371** (2019) 012017 doi:10.1088/1742-6596/1371/1/012017

Tapered single-mode optical fiber-based on localized surface plasmon resonance as refractive index sensor

Nurfatin Musa[1], Hazri Bakhtiar[1, 2], Ganesan Krishnan[1, 2], Harith Ahmad[4], Sulaiman Wadi Harun[3], Muhammad Safwan Abd Aziz*[1, 2]

[1] Department of Physics, Faculty of Science, Universiti Teknologi Malaysia, 81310 Skudai, Johor, Malaysia
[2] Laser Center, Ibnu Sina Institute for Scientific and Industrial Research, Universiti Teknologi Malaysia, 81310 Skudai, Johor, Malaysia
[3] Photonics Engineering Laboratory, Department of Electrical Engineering, Faculty of Engineering, Universiti Malaya, 50603 Kuala Lumpur, Malaysia
[4] Photonics Research Centre, University of Malaya, 50603 Kuala Lumpur, Malaysia

safwanaziz@utm.my

Abstract. A simple tapered single-mode optical fiber-based on localized surface plasmon resonance was proposed as a refractometer for liquid samples application. The tapered optical fiber (TOF) was fabricated by using a flame–brush method and achieved a waist diameter of 16.5 μm over the length about 20 mm. 3-mercatoptopropyltrimethoxysilane (MPTMS) with a mercapto group (-SH) was utilized as a silane agent to immobilized the nanoparticles by strong covalent bonding between sulphur and gold. The sensing mechanism relies on the transmission shifting in the output spectra due to the evanescent field absorption by immobilized gold nanoparticles on TOF surface. The performance of TOF has been evaluated with the surrounding refractive index varied over the range from 1.3330 to 1.4229. The sensitivity of the sensor was recorded at 18 nm/RIU with good repeatability and stability for 70 minutes.

1. Introduction

Fiber optics have attracted a great deal of attention in the field of sensors for detecting various physical and chemical parameters. Of the wide variety of available fiber optics sensors, fiber optics refractometer is promising and attractive in both chemical and biotechnological applications. This particular type of sensor offers many advantages such as simplicity, low-cost system, high sensitivity, and high responsivity towards measurand. On the other hand, sensing activity in hostile environments is made possible due to electromagnetic interference immunity, and compact in size [1, 2].

Among many, tapered optical fibers (TOF) is considered as one of the common types of fiber-based sensors. The evanescent waves propagating on the tapered region of a TOF can interact directly with the surrounding, thus altering the output light signal with any changes introduced to the surrounding medium. In many cases, the evanescent waves interact with an intermediate thin film transducer, which was fabricated to be permeable to particular measurands. Such devices were designed to be highly sensitive towards the surrounding's refractive index RI and or color. As a consequence, one or more wavelengths of the guided light may be selectively absorbed or attenuated. A simple detection scheme could be further used to easily monitor the absorption or attenuation suffered by the guiding light.

In this work, a single-mode optical fiber was proposed as a refractometer to detect the refractive index of surrounding liquid medium based on localized surface plasmon resonance (LSPR) effect. The structure of the optical fiber was modified by tapering it down to smaller diameter, in order to produce evanescent field propagating along with the tapered structure. A thin layer of gold nanoparticles was immobilized on the tapered region as the sensitive coating. The performance of the sensor in term of its sensitivity, stability, repeatability, as well as reversibility, was examined.

2. Experimental Setup

2.1. Synthesis of gold nanocolloid

Gold nanoparticles (AuNPs) employed in this experiment were prepared by Turkevich method. This method is based on the reduction reaction between gold chloride ($HAuCl_4$) and sodium citrate ($Na_3C_6H_5O_7$) respectively. 20 mL of 1.0 mM of $HAuCl_4$ was stirred and heated until boiled. Upon reached boiling state, a known amount of sodium citrate was added quickly to the solution. As a result, the yellowish color of gold chloride turned colorless before gradually changed to grey-purplish color. The solution was kept heated for several minutes until red-wine color appeared. After that, the heat source was removed and the solution was left at room temperature with further stirred for 10 minutes. The final solution was kept in the refrigerator to ensure the particles size was stable up to a month.

2.2. Tapered fiber LSPR sensor preparation

Standard telecommunication single-mode fiber (G652D) with core and cladding diameters of 10 μm and 125μm was used for the sensor preparation. The tapered structure was fabricated by the flame-brushing method. During the tapering process, one side of the fiber was kept static and the other side was pulled while the fiber was heated back and forth horizontally by using butane torch (max temperature at 1430 °C). Both sides of fiber were fixed by stage holder as shown in Fig. 1. Approximately 30 mm of the buffer layer was removed from the middle section of the prepared sensor by a mechanical stripper. The uncovered section of the fiber was exposed to the flame for softening purpose, while the fiber being pulled by a linear translational stage.

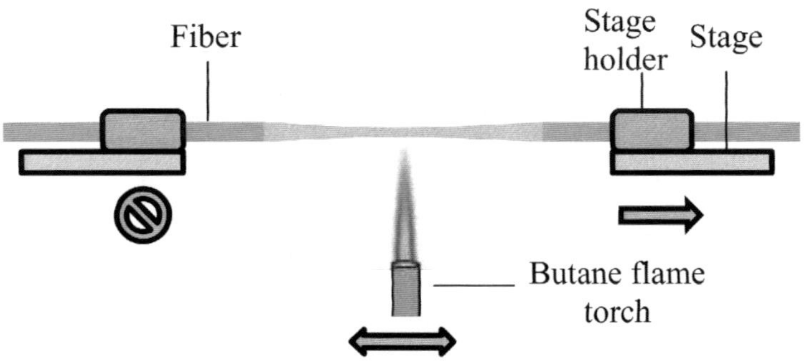

Fig. 1. Schematic diagram of tapered fabrication

The tapered fiber region was cleaned by methanol solution to remove unwanted debris from the fiber surface and let dry. The tapered fiber was then immersed in 1.0 M of sodium hydroxide, NaOH for an hour, to activate the fiber surface with -OH functional group. The tapered fiber was rinsed with distilled water thoroughly and blow-dried. Then, the prepared fiber was thiol-functionalized with 1% of MPTMS in ethanol for 12 hours. After this process, the fiber was again rinsed several times by ethanol to remove unbound monomer from the fiber surface and blow-dried. Afterwards, the tapered fiber was sent for coating with AuNPs for 48 hours through a dip coating process to form a metal nanoparticles layer on the tapered fiber surface. Finally, the coated fiber was rinsed with ethanol and dried. Fig. 2 shows the schematic diagram of self-assembled monolayer (SAM) of MPTMS on tapered fiber. Ruby-color appearance was visible on the tapered fiber surface by naked eyes.

Fig. 2. Schematic diagram of SAM of MPTMS on the tapered fiber surface

2.3. LSPR sensor setup

Fig. 3 illustrates the schematic diagram of LSPR sensor system used in this work. One end of tapered fiber sensor was connected to a halogen-tungsten broadband light source and the other end was connected to C175 Thorlabs spectrometer. The spectrometer connected to the computer and the corresponding spectra were recorded in real-time. The sensitive tapered region of the fiber was immersed in solutions with different refractive indices ranges from 1.3330 to 1.4229. Table 1 indicates the refractive indices of the 5 prepared solutions represented in refractive index unit, RIU respectively.

Table 1. Refractive index of samples test

Samples	n1	n2	n3	n4	n5	n6
Refractive index unit, RIU	1.3324	1.3523	1.3728	1.3931	1.4119	1.4254

Fig. 3. Schematic diagram of LSPR sensing setup

3. Result and Discussion

3.1. Gold nanoparticles immobilized on TOFs

Immobilization of the prepared nanoparticles on the surface of the TOF is examined by scanning electron microscope. As shown, immobilized AuNPs on the surface of TOF is evidenced in Fig. 4(a) & (b) with different resolution. The random distribution and uniformity of AuNPs layer are obtained based

Journal of Physics: Conference Series **1371** (2019) 012017 doi:10.1088/1742-6596/1371/1/012017

on the results. The mean diameter of deposited AuNPs was measured and shown in Fig. 4(c) which estimated to be 15 ± 2nm.

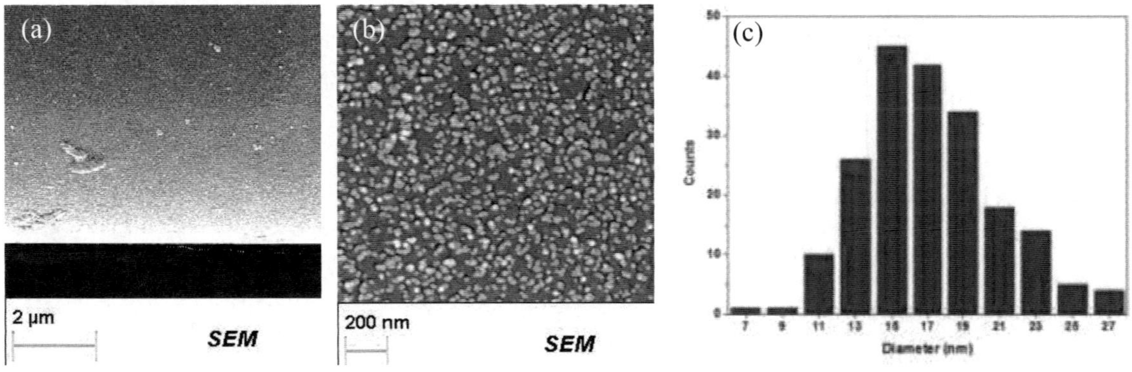

Fig. 4. (a) & (b) SEM images of Au immobilized on TOF surface; (c) Histogram analysis of Au average size based on SEM image

3.2. The sensitivity of TOF based LSPR sensor

Transmission spectra of the sensor were recorded against different refractive indices values of the controlled surrounding media. Fig. 5(a) indicates the output signals obtained from LSPR-based TOF. Based on the figure, it is shown that TOF has transmittance trends with the highest reading in n1 which refer to the lowest refractive index followed by n2 until n6. The transmission of light decreases with the increment of refractive index unit (RIU) of liquid samples due to the attenuation loss. On top of that, a significant shifting has been observed in the output spectrum as the surrounding refractive index varied from low to high due to the LSPR effect. Immobilized AuNPs on the surface of T0F acts as a new cladding and the excitation of surface plasmons on AuNPs occurs due to the presence of evanescent field at the metal-dielectric sensing layer interface. Resonance state was achieved when the evanescent field produced from incident light matched with the surface plasmons oscillation of the AuNPs, resulting in a sharp dip in the transmitted light intensity at a particular wavelength [3]. Applying the different refractive index at the surrounding surface may vary the resonance wavelength at each RIU based on the coupled light that exhibits the modulated transmittance spectrum.

In order to analyse the performance of TOF LSPR-based sensor, the spectral shift in the transmitted signal was recorded in response to refractive index changes. The result was plotted and shown in Fig. 5(b). It is shown that the proposed LSPR sensor exhibit a good linear dependency. A correlation coefficient, R^2 of 0.95490 was calculated with the corresponding sensitivity recorded at 18.4931 nm/RIU.

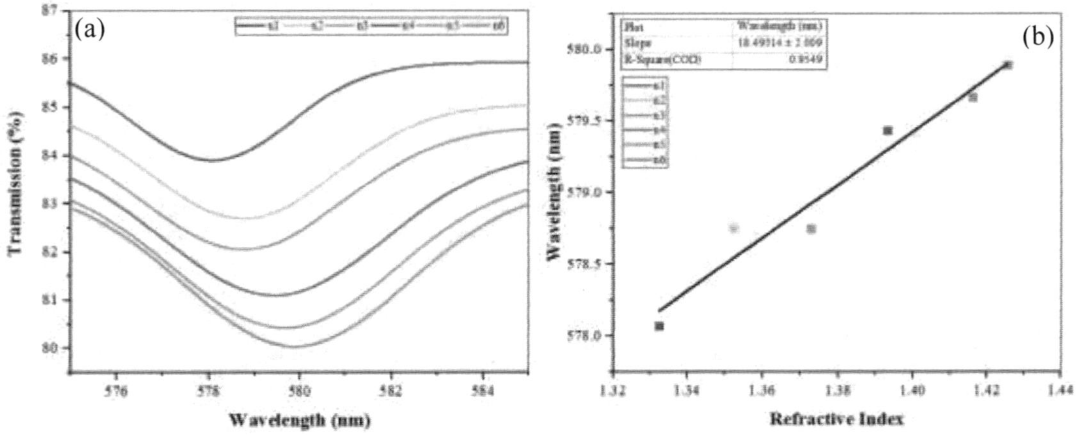

Fig. 5. (a) Transmission spectra for TSF; (b) The sensitivity analysis of TSF

3.3. Stability, Repeatability and Reversibility

The stability of the sensor was determined by continuously monitoring its output spectra for 70 minutes in 3 different solutions with RIU of 1.3324, 1.3523 and 1.4254. The controlled media was purposely selected at low, medium and high indices in order to test the actual stability of the sensor across the whole range of RIU. The peak intensities values for every reading were recorded and illustrated in Fig. 6. Result suggests that the output response of proposed LSPR-sensor is constant with very little variation over the period of 70 minutes.

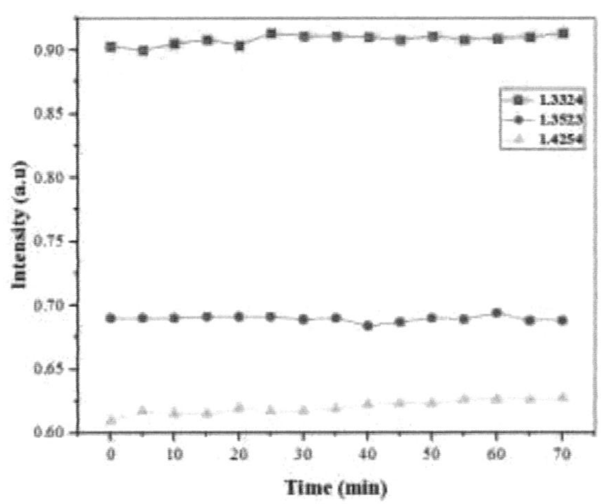

Fig. 6. Peak intensity of TSF LSPR-based sensor for 70 minutes

In order to evaluate the repeatability of the sensor, the output spectra were recorded for a complete 3 cycle periods. During the test, the sensor was subjected to the controlled refractive index solution ranges from 1.3324 to 1.4254. Fig. 7(a) shows the variations of intensities between 3 repeated cycles recorded against corresponding refractive index values. Fairly good repeatability is shown regarding the low degree of variation among those 3 cycles of measurements. To confirm the reversible performance of the sensor, measurements were made by both ascending and descending fashion. Refractive index of the surrounding medium was varied between 1.3324 to 1.4254 RIU and the corresponding transmission were recorded at 545.56 nm wavelength. Results are shown in Fig. 7(b) indicates that the sensor repeated the same transmitted intensity with little variation. This confirms the reversible and stable behaviour of the device in detecting refractive index variation of the surrounding liquid medium.

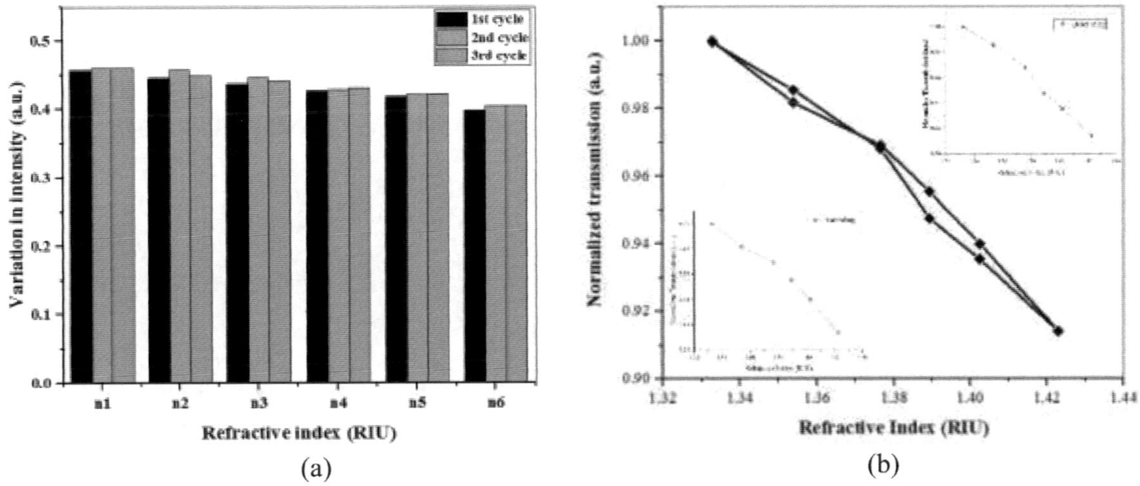

Fig. 7. (a) Histogram analysis of repeatability test for LSPR-based sensor; (b) Reversibility trends of LSPR-based sensor

4. Conclusion

In this work, Au nanoparticles were successfully synthesized by using turkevich method. The immobilization of Au on the TOF surface by SAM method was successfully employed which verified by SEM analysis. The proposed optical fiber sensor shows a relatively high sensitivity of 18.4931 nm/RIU, on top of that, it also shows good repeatability in 3 cycle periods, better stability for 70 minutes and great reversibility with a little variation. Further optimization of fiber geometry or coating parameter may result in more compact refractometric sensing with improved performance.

Acknowledgement

Authors would like to acknowledge the financial support under GUP Universiti Teknologi Malaysia through the Teir 1 fund with vote 18H67.

References

[1] Krishnan, G., Bidin, N., Abdullah, M., Ahmad, M. F. S., Bakar, M. A. A., & Yasin, M. (2016). Liquid refractometer based mirrorless fiber optic displacement sensor. Sensors and Actuators A: Physical, 247, 227-233.
[2] Fernandez-Vallejo, M., & Lopez-Amo, M. (2012). Optical fiber networks for remote fiber optic sensors. Sensors, 12(4), 3929-3951.
[3] Gupta, B. D., & Verma, R. K. (2009). Surface plasmon resonance-based fiber optic sensors: principle, probe designs, and some applications. Journal of sensors, 2009.

Indium tin oxide thin film based saturable absorber for Q-switching in C-band region

N U H H Zalkepali[1], N A Awang[1], Y R Yuzaile[1], Z Zakaria[1], A A Latif[2], A H Ali[1] and N N H E N Mahmud[1]

[1] Optical Fiber Laser Technology (OpFLAT) Focus group, Department of Physics and Chemistry, Faculty of Applied Sciences and Technology, Universiti Tun Hussein Onn Malaysia, Pagoh, Johor, Malaysia
[2] Department of Physics, Faculty of Science, Universiti Putra Malaysia, Selangor, Malaysia

norazura@uthm.edu.my

Abstract. An indium tin oxide (ITO) thin film based saturable absorber (SA) is proposed and demonstrated for the operation of Q-switched pulse within the C-band region. The ITO was deposited through DC magnetron sputtering method. The thickness of ITO was 86.40 nm and it was measured using F20 Filmetrics. The deposited fiber ferrule was incorporated in an erbium-doped fiber (EDF) laser cavity for generating a stable Q-switching pulse. The generated output pulses displayed a repetition rate that ranged between 16.74 kHz and 38.03 kHz. The shortest pulse width retrieved was 5.78 μs at the maximum pump power of 165.5mW, while the maximum pulse energy recorded was 12.30 nJ. These results show that ITO has the potential to be used for pulsed laser applications.

1. Introduction

The wide range of applications in Q-switched pulse erbium-doped fiber (EDF) lasers such as optical sensing, telecommunications and material processing [1-3] are attracted much attention due to its own characteristics such as high energy pulses. Q-switched fiber laser presents large pulse energy with low repetition rates and pulse width in the range of microsecond to nanosecond different with mode-locking fiber laser that gives low single energy pulse, but it had high repetition rates and shortest pulse width. In order to modulate the Q-factor for the generation of Q-switched pulse fiber laser, an optical modulator is involved where the saturable absorber (SA) is utilized as efficient optical modulator in passive technique, while in actively Q-switched fiber laser that requires external modulator such as acousto-optics modulator [4] and electro-optic modulator [5]. Usually, passively Q-switched fiber laser has simple in structure and more compact in geometry compared with actively Q-switched that bulky in size and expensive in maintenance.

Semiconductor saturable absorber mirror (SESAMs) are known as commercial SA, but the major challenge in this SA has narrow in bandwidth and expensive in fabrication [6]. Therefore, many researchers are still exploring to find suitable SA that has a few characteristics like low-cost, wideband and high performance in generating pulse fiber laser. Carbon nanotubes (CNTs) and graphene had found as novel SAs and widely investigated in the past few years because of their broad absorption range and good compatibility with optical fiber [7-9]. Yet, they revealed the low modulation depth and limited in applications that needed high output pulse fiber laser [10]. Topological insulators (TIs) (Bi_2Se_3, Sb_2Te_3)

[11,12], transition metal dichalcogenide (TMD) (MoS_2, WS_2) [13,14], black phosphorus [15], metal nanoparticles [16,17] and platinum [18] are other examples of SAs which are generally due to their advantages such as ultrafast carrier dynamics, easy fabrication, low cost and broadband in operation.

A few reports from the researcher have revealed that indium tin oxide (ITO) can be a good candidate to be selected as SA. This is because ITO has a large energy bandgap about 3.7 eV with unique optical and electrical properties where high transmittance in the visible region and good conductivity. Plus, its exhibit large optical nonlinearity [19], a wide region of saturable absorption [20] and lower carrier density and the plasmon frequencies are placed near-infrared region [21]. There are a few techniques for fabricating ITO thin films such as thermal evaporation [22], chemical vapour deposition [23], sol-gel process [24], pulsed laser deposition [25] and magnetron sputtering [26]. However, the magnetron sputtering technique is more reliable compared to other methods because of low cost, simple and high flexibility.

Therefore, ITO thin film-based SA is proposed and established to generate Q-switched pulse in C-band region. The ITO and DC magnetron sputtering were incorporated for the fabrication of SA by inserting SA into fiber cavity design to induce self-forming Q-switched. Self-started Q-switching at central wavelength 1561.0 nm in the EDF laser pumped by a 980 nm laser diode was obtained for a threshold pump power of 65.60 mW. The generated output pulses showed a repetition rate that ranged between 16.74 kHz and 38.03 kHz. The shortest pulse width retrieved was 5.78 μs at the maximum pump power of 165.5mW, while the maximum pulse energy recorded was 12.30 nJ. From the results, this SA has the potential to be applied in medical, optical sensing and micromachining.

2. ITO as a saturable absorber

DC magnetron sputtering method was utilized for deposition of ITO thin film onto fiber ferrule and glass substrates. A schematic diagram of the sputtering apparatus (The Quorum- Q 150R S) used in the present study is shown in figure 1. The ITO target was inserted at the negative cathode to be sputtered under conditions of glow discharge using DC power supply. The ITO was sputtered onto glass substrates and fiber ferrule on anode with (9×10^{-2} mBar) plasma vacuum and pressure of the environmental argon gas. Table 1 shows the sputtering parameters setting for deposition of ITO. The thickness of sputtered ITO thin film was measured using F20 Filmetrics which resulted in 86.40 nm.

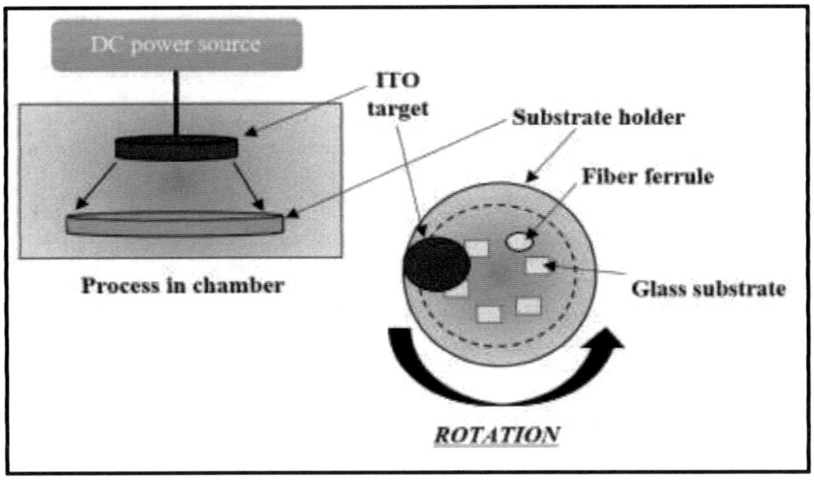

Figure 1. Schematic diagram of the sputtering apparatus

Table 1: Sputtering parameters for ITO deposition

Process parameter	**Values**
Sputtering gas	Argon
Power supply	DC
Current (mA)	80
Sputtering time (s)	600
Temperature	Room temperature
Target	ITO
Plasma vacuum (mBar)	9×10^{-2}

The EDS analysis and surface morphology of fiber ferrule of the ITO thin film using Hitachi SU1510 scanning electron microscope together with Horiba EMAX energy-dispersive spectrometer (SEM-EDS) are displayed in figure 2. The analysis shows the presence of indium (In) and tin (Sn) element in the area of fiber core. The EDS also detect the present of silicon (Si) and oxygen (O) due to fiber silica made up of silica (Si_2O). There are no elements present means no impurities are affected in this area. Figure 3 portrays the transmission loss of ITO measured using an amplified spontaneous emission (ASE) source. The transmission loss was inclusive of the insertion loss between the fiber ferrules and the ITO itself. Within the wavelength range of 1460 nm to 1600 nm, the ITO has a transmission loss of approximately 1.7 dB. Figure 4 shows the UV–VIS NIR spectra of the ITO thin film in (a) absorption (b) transmittance and (c) reflectance in the wavelength range 300 nm -1700 nm for thickness 86.40 nm of ITO thin film. The ITO thin film exhibits a stronger linear absorption in the wavelength approximately 600 nm. The absorbance of ITO thin film is almost zero in the wavelength range of 900 nm -1700 nm. The trends of transmittance and reflectance spectra were almost the same, where almost 90% of ITO thin film was transmitted and reflected. The AFM images with area 3.0 μm × 3.0 μm of ITO thin film is shown in figure 5. It is observed that the root-mean-square surface roughness is 4.10 nm.

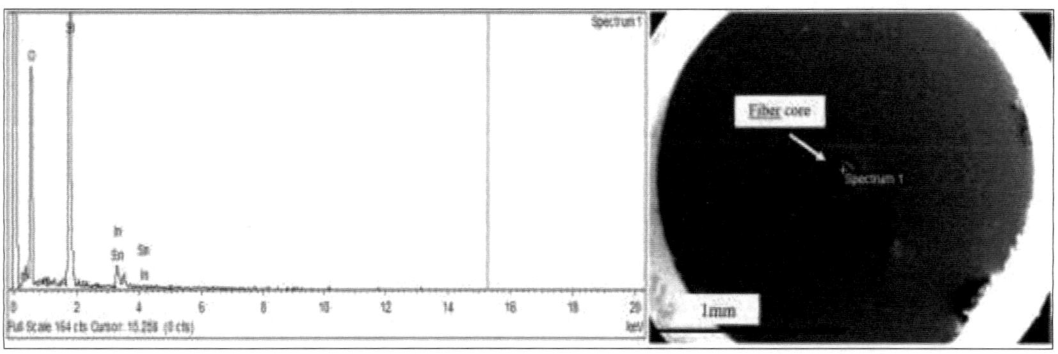

Figure 2. EDS and SEM analysis of ITO thin film.

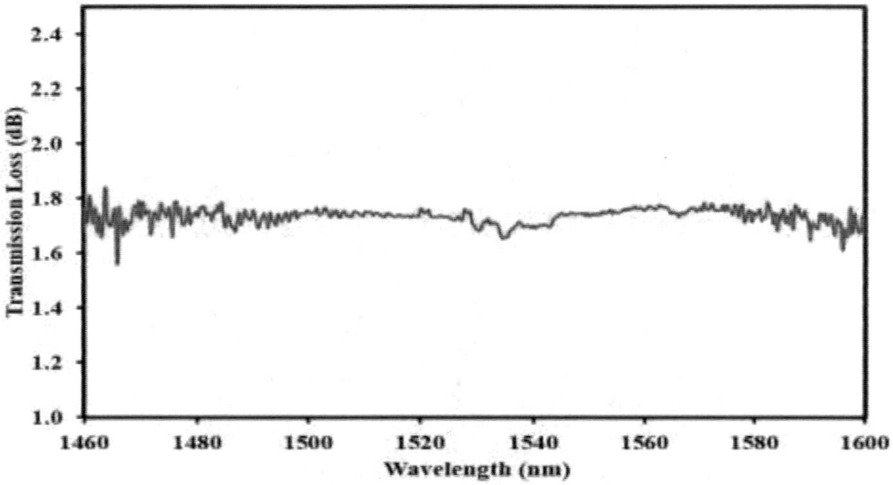

Figure 3.The transmission loss of ITO.

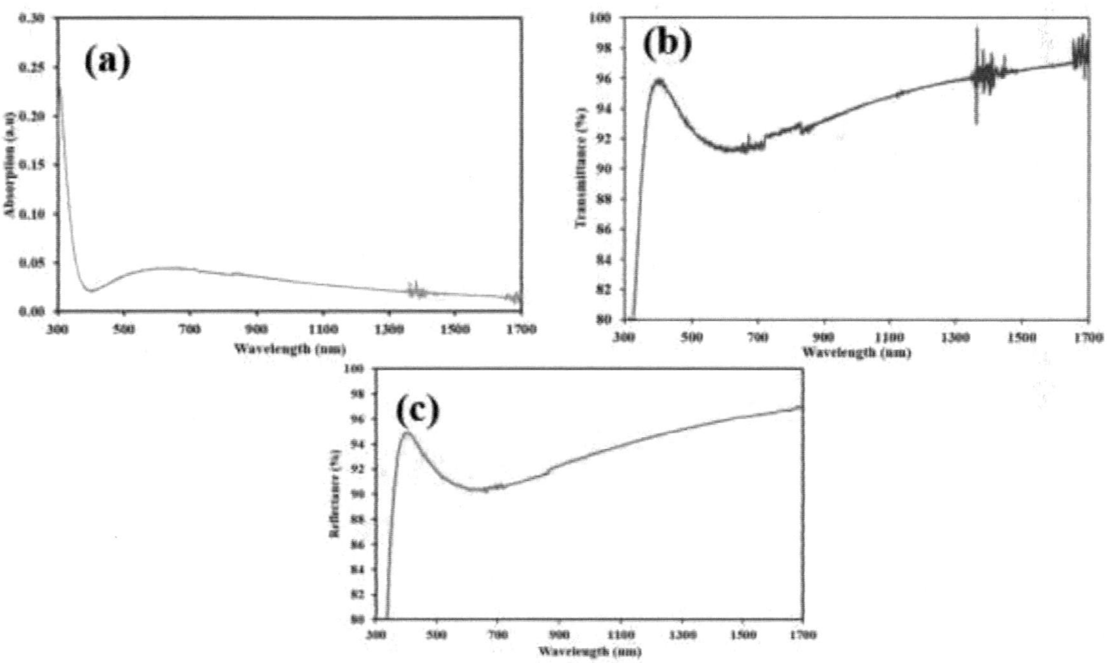

Figure 4. (a) Absorption, (b) transmittance and (c) reflectance of ITO thin film using UV-VIS-NIR spectrophotometer with thickness of 86.40 nm.

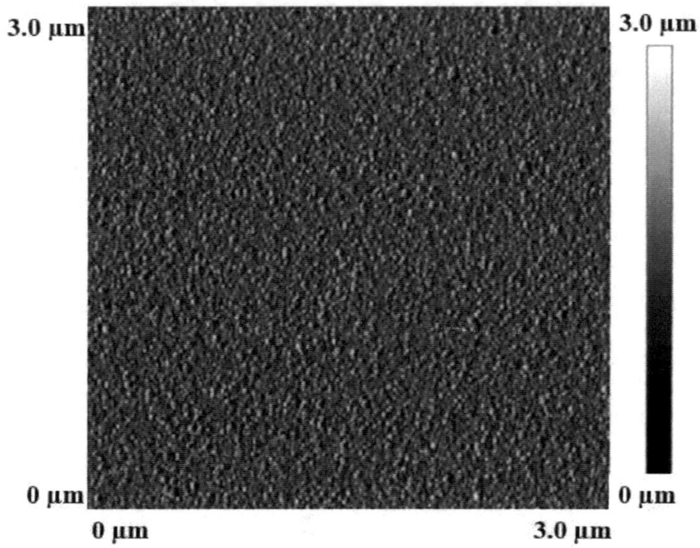

Figure 5. AFM images of ITO thin film.

3. Experimental Setup

The experimental setup of Q-switched EDF laser using ITO thin film as SA was illustrated in figure 6. The total length of fiber cavity is 11.5 m. A 5 m long of EDF has an absorption coefficient of 6.43 dB/m peak absorption at 1530 nm and 5.09 dB/m absorption at 979 nm as stated in date sheet was utilized as the gain medium. The EDF was pumped via a 980 nm laser diode. To ensure the unidirectional laser operation, an isolator was incorporated in the laser cavity, meanwhile, to vary the polarization state of intracavity light, polarization controllers (PCs) were utilized. The ITO-based SA was inserted in between PCs and isolator. For signal analysis, a 90:10 optical coupler was applied where the 10% port was tapped out from the coupler while 90% port was connected with 1550 nm port of WDM to complete the ring cavity. An Anritsu MS9740A optical spectrum analyser with a 1.0 nm resolution was used to measure the output spectrum of the generated Q-switched pulse EDF laser. Analysing the pulse train properties and radio frequency spectrum of Q-switched pulsed EDF laser make use of the Tektronix MDO3104 oscilloscope together with 5 GHz InGaAs Biased Detector, DET08CFC/M photodetector.

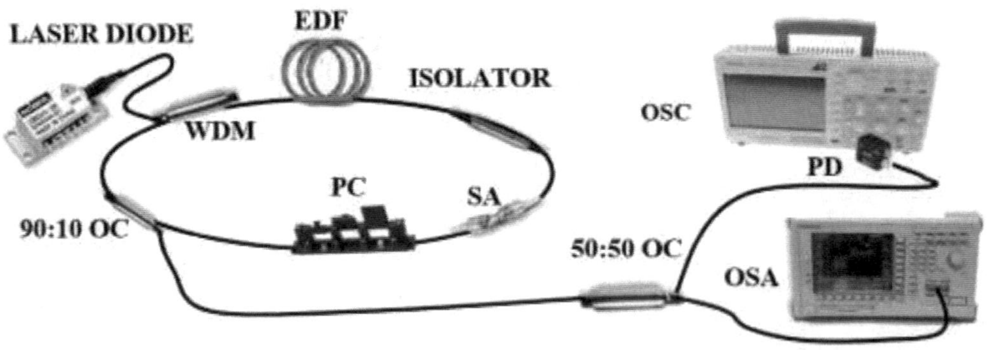

Figure 6. Schematic diagram of Q-switched pulse EDF laser using ITO thin film as SA

4. Results and Discussions

Firstly, the fiber laser cavity was investigated without incorporated the ITO thin film-based SA to observe the possibility of self-Q-switching in EDF laser. However, only the continuous wave (CW) emission was observed even the pump power was increased from the oscillation threshold until the maximum of pump power. The angle of polarizations of PCs also was adjusted from 0 to 360 degree to detect the generation of Q-switched. Then, the ITO thin film was inserted into fiber cavity in between PCs and isolator. Q-switching of fiber laser occurred as pump power at 65.60 mW. The initial repetition rate was 16.74 kHz with an average output power of 0.12 mW. The Q-switched spectrum was analysed using optical spectrum analyser by increasing the pump power until 165.60 mW as shown in figure 7. The Q-switched spectrum has a central wavelength of 1561.0 nm with peak power of -3.37 dBm. The unstable Q-switched fiber laser occurred if further increasing the pump power above the 165.60 mW. The performance of Q-switched was investigated at fix wavelength and changing the pump power.

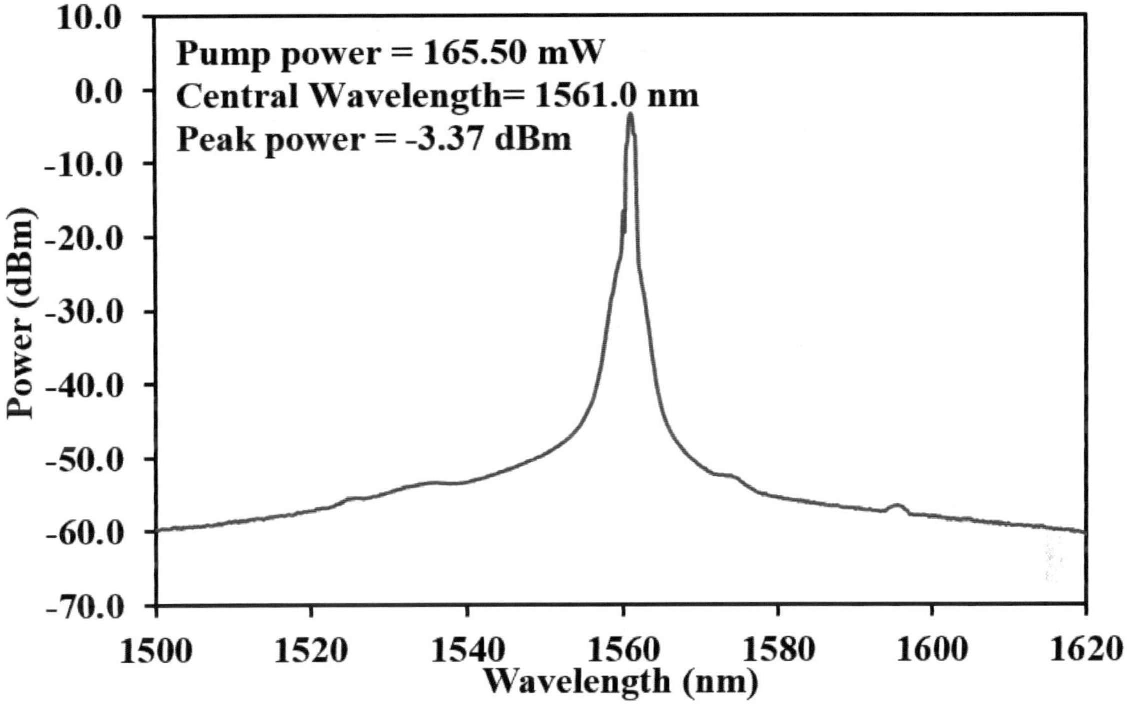

Figure 7. The optical spectrum of the Q-switched fiber laser

Figure 8 illustrated the pulse train of Q-switched EDF laser in time domain region at a maximum pump power 165.60 mW by using an oscilloscope with a coupled photodetector. The pulse train gives a repetition rate of 38.03 kHz which corresponding to the time interval of 26.29 μs between the two pulses whereby each pulse measured pulse width of 5.78 μs. Hence, the pulse train adequately stable, where the peak of intensity was constant at around 0.020 V and seemed to maintain uniform in shape. This indicates that no significant amplitudes display in this pulse train. The characteristics of Q-switched EDF laser was further investigated by measuring the RF spectrum at a fixed pump power of 165.60 mW. The fundamental frequency was depicted in figure 9 gives the repetition rate of 38.03 kHz with a signal to noise ratio (SNR) of 56.61 dB. The RF spectrum was measured in a 200 Hz resolution bandwidth (RBW) with a span of 200 kHz. These results show the high stability of the Q-switched EDF laser in this study.

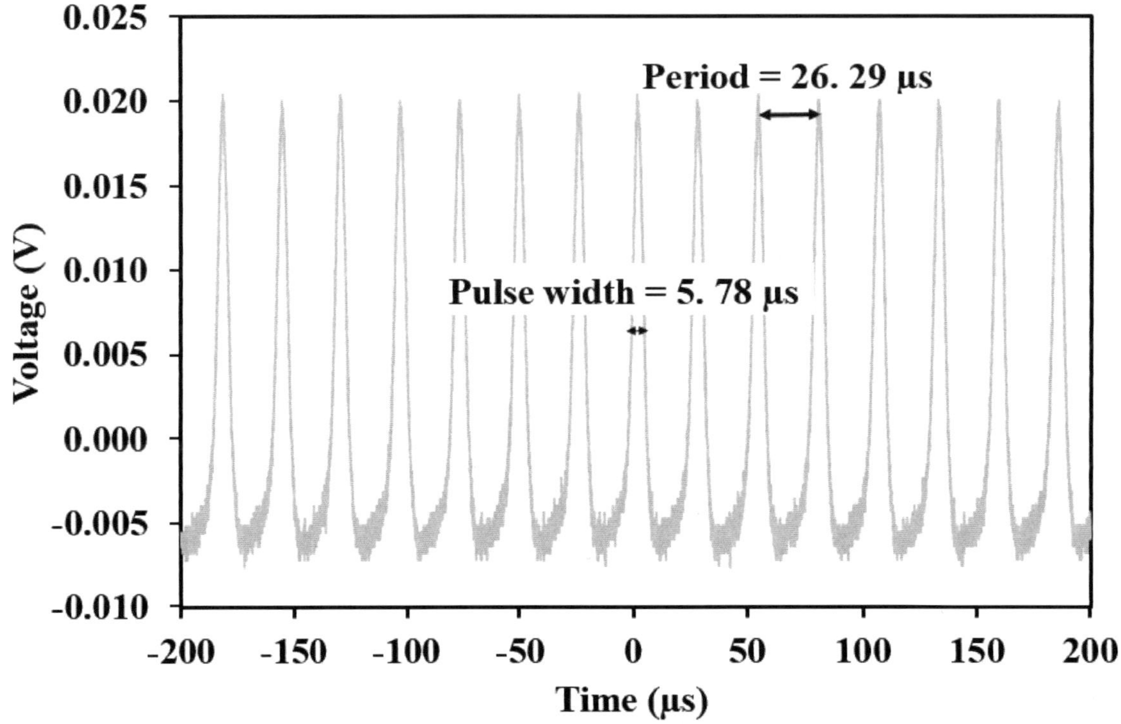

Figure 8.A typical pulse train at the pump power 165.50 mW.

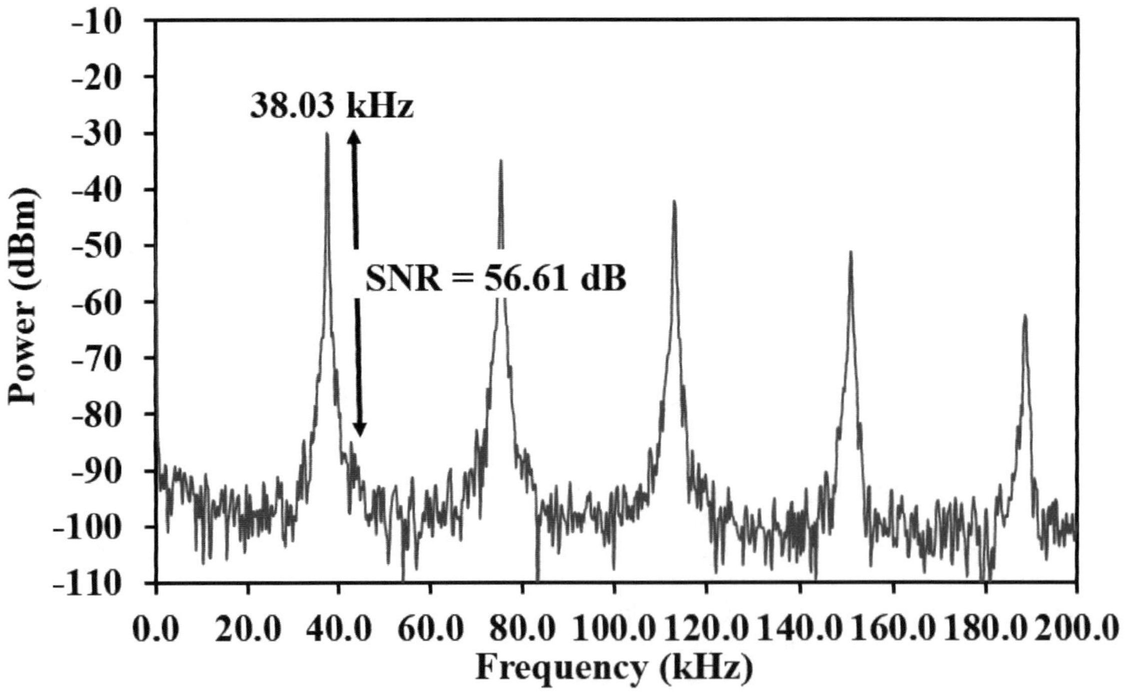

Figure 9. RF spectrum with $f = 38.03$ kHz.

The relationship between pulse width and pulse repetition of Q-switched EDF laser with respect to incident pump power portrayed in figure 10. As predictable, the repetition rate displays a rising trend as the pump power increased from 16.74 kHz at a pump power of 65.60 mW to 38.03 kHz at a maximum pump power of 165.50 mW. Meanwhile, the pulse width shows a descending order trend with responding to increasing pump power as expected. Initially, the pulse width is longer about 7.76 μs at the lower pump power of 65.60 mW and reduces to 5.78 μs at the maximum pump power of 165.60 mW. The shorter pulse width can be obtained by reducing the length of fiber in the cavity laser [27].

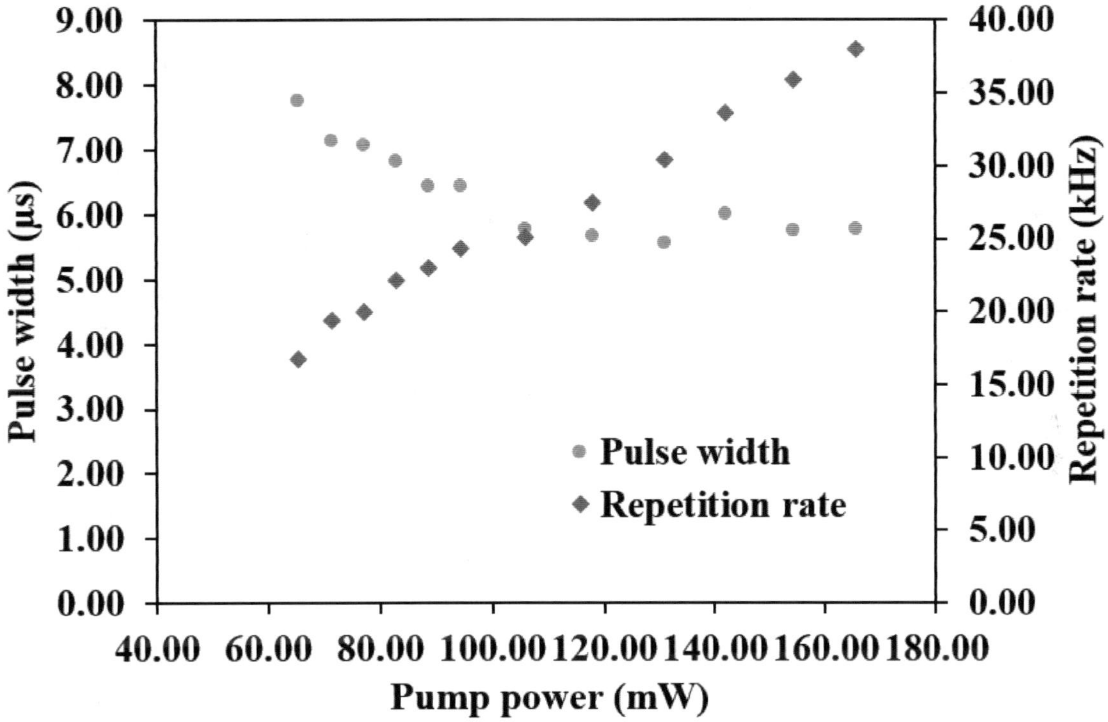

Figure 10. The pulse repetition rate and pulse width versus pump power

The pulse energy and average output power are summarized in figure 11 with respect to incident pump power. Both the pulse energy average and output power increase almost linearly with pump power, as expected. The pulse energy changed from 7.10 nJ to 12.30 nJ, while the average output power rose from 0.12 mW to 0.47 mW.

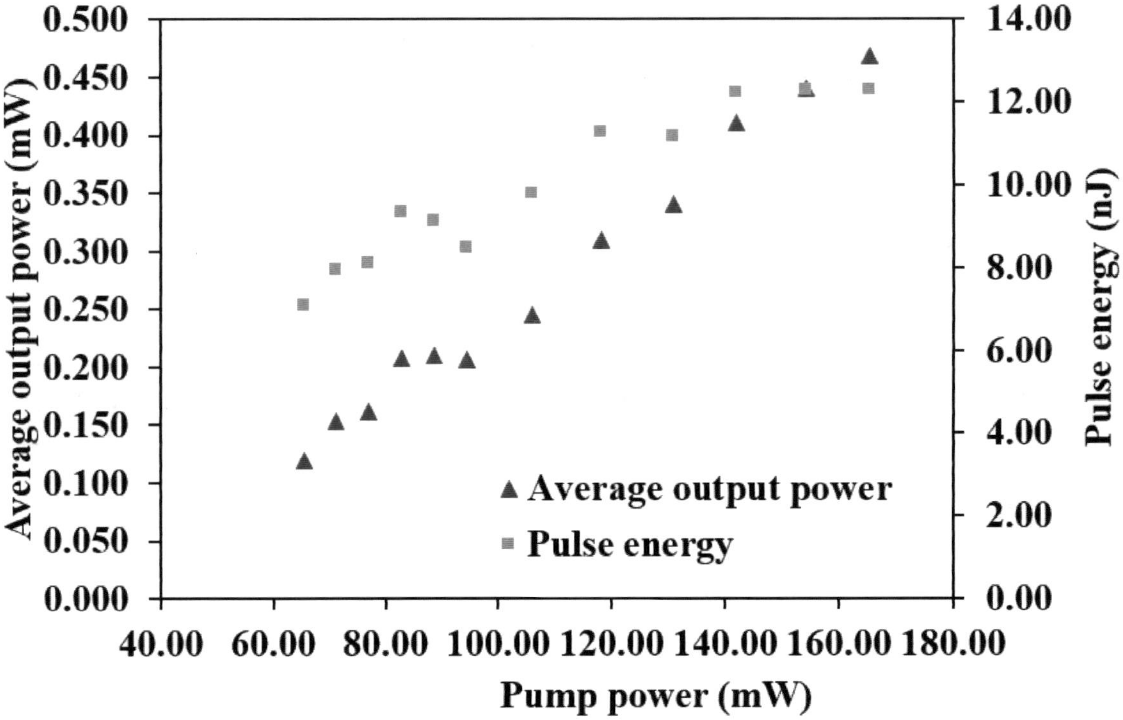

Figure 11. The average output power and pulse energy versus pump power

5. Conclusion

As conclusions, the generation of passively Q-switched pulse within the C-band region was successfully demonstrated and proposed by using ITO thin film as saturable absorber. The DC magnetron sputtering method was employed for the fabrication of saturable absorber due to the simple and easy method. The deposited fiber ferrule was inserted in fiber laser cavity for generation Q-switched pulse. The Q-switched pulse was worked at a central wavelength of 1561.0 nm with peak power of -3.37 dBm, the highest repetition rate at 38.03 kHz, the shortest pulse width at 5.78 μs and SNR at 56.61 dB. The output pulse energy was attained at 165.60 mW pump power with 12.30 nJ. Based on the outcomes from this experiment, the design can be optimized to generate ultrashort pulse fiber laser to be applied in material processing and spectroscopy applications.

Acknowledgement

We would like to thank the Universiti Tun Hussein Onn Malaysia (Pagoh Campus) for providing the instruments and probe facility to complete this work. We are also thankful to Research Management Centre (RMC) of Universiti Tun Hussein Onn Malaysia (UTHM) for funding under Postgraduate Research Grant GPPS: H051 and TIER 1 Grant: H133.

References

[1] Chernysheva M, Mou C, Arif R, AlAraimi M, Rümmeli M, Turitsyn S and Rozhin A 2016 *Sci. rep.* **6** 24220
[2] Li H, Xia H, Lan C, Li C, Zhang X, Li J and Liu Y 2015 *IEEE Pho. Techno. Lette.* **27** 69-72
[3] Goldberg L, Cole B, McIntosh C, King V, Hays A D and Chinn S R 2016 *Opt. Exp.* **24** 17397-17405
[4] Li W, Liu H, Zhang J, Long H, Feng S and Mao Q 2016 *App. Opt.* **55** 4584-4588
[5] Cong Z H, Liu Z J, Zhang X Y, Qin Z G, Men S J, Fu Q and Rao H 2016 *In 2016 I. Con. Las.*

Opt. (LO) IEEE R1-65
[6] Haiml M, Grange R and Keller U 2004 *App. Phys. B*, **79** 331-339.
[7] Zhou D P, Wei L, Dong B and Liu W K 2010 *IEEE Photo. Techno. Lette.* **22** 9-11
[8] Sheng Q, Feng M, Xin W, Han T, Liu Y, Liu Z and Tian J 2013 *Opt. Exp.* **21** 14859-14866
[9] Dong B, Hu J, Liaw C Y, Hao J and Yu C 2011 *App. Opt.* **50**1442-1445
[10] Zhao C, Zou Y, Chen Y, Wang Z, Lu S, Zhang H … and Tang D 2012 *Opt. Exp.***20** 27888-27895
[11] Bogusławski J, Soboń G J, Tarnowski K, Zybała R, Mars K, Mikuła A ... and Sotor J Z 2016 *Opt. Eng.* **55** 081316
[12] Haris H, Harun S W, Muhammad A R, Anyi C L, Tan S J, Ahmad F, ... and Arof H 2017 *Opt. & Las. Techno.* **88** 121-127
[13] Kong L C, Xie G Q, Yuan P, Qian L J, Wang S X, Yu H H and Zhang H J 2015 *Photo. Res.* **3** A47-A50
[14] Zhang M, Hu G, Hu G, Howe R C T, Chen L, Zheng Z and Hasan T 2015 *Sci. rep.* **5** 17482
[15] Yu H, Zheng X, Yin K and Jiang T 2016 *Opt. Mat. Exp.* **6** 603-609
[16] Guo H, Feng M, Song F, Li H, Ren A, Wei X, ... and Tian J 2016 *IEEE Photo. Techno. Lette.* **28** 135-138
[17] Wu D, Peng J, Cai Z, Weng J, Luo Z, Chen N and Xu H 2015 *Opt. Exp.* **23** 24071-24076
[18] Yuzaile Y R, Awang N A, Zalkepali N U H H, Zakaria Z, Latif A A, Azmi A N and Hadi F A 2019 *Optik* **179** 977-985
[19] Alam M Z, De Leon I and Boyd R W 2016 *Sci.* aae0330
[20] Kanehara M, Koike H, Yoshinaga T and Teranishi T 2009 *J. of the American Chem. Soc.* **131** 17736-17737
[21] Naik G V, Shalaev V M and Boltasseva A 2013 *Adv. Mat.* **25** 3264-3294
[22] Basak A, Hati A, Mondal A, Singh U P and Taheruddin S K 2018 *Thin Solid Films* **645** 97-101
[23] Banszerus L, Schmitz M, Engels S, Dauber J, Oellers M, Haupt F, ... and Stampfer C 2015 *Sci. adv.* **1** e1500222
[24] Hsu H P, Lin D Y, Lu C Y, Ko T S and Chen H Z 2018 *Crystals* **8** 228
[25] Chen A, Long H, Li X, Li Y, Yang G and Lu P 2009 *Vacuum* **83** 927-930
[26] Gonçalves R S, Barrozo P, Brito G L, Viana B C and Cunha F 2018 *Thin Solid Films* **661** 40-45
[27] Spühler G J, Paschotta R, Fluck R, Braun B, Moser M, Zhang G, ... and Keller U 1999 *JOSA B*, **16** 376-388

Journal of Physics: Conference Series **1371** (2019) 012019 doi:10.1088/1742-6596/1371/1/012019

Tunable passively Q-switched fiber laser of graphite flakes saturable absorber

Nik N H E N Mahmud, Fahmiruddin Esa*, Noor A Awang, Zahariah Zakaria, Yushazlina R Yuzaile and Noor U H H Zalkepali

Optical Fiber Laser Technology (OpFLAT) Focus Group, Department of Physics and Chemistry, Faculty of Applied Sciences and Technology, Universiti Tun Hussein Onn Malaysia, 84600, Pagoh, Johor, Malaysia.

fahmir@uthm.edu.my

Abstract This study demonstrated the performance of a passively Q-switched fiber laser by using graphite flakes at different EDF absorption which are EDF 1 and EDF 2 at 5.09 dB/m and 7.28 dB/m, respectively. The system is designed as a single ring cavity by employing a graphite flakes saturable absorber (SA) with 23.82% of saturable absorption. The experimental result shows that a wide-band tunable cover from 1544 nm to 1560 nm and 1552 nm to 1570 nm for EDF 1 and EDF 2 respectively. The EDF 2 is the best gain medium due to the conventional wavelength band at C band and L band region.

Introduction

Passive Q-switched is a technique to produce short pulses with a high peak. This technique is simple in setup, more compact compared to active Q-switched. Generally, a passive technique used saturable absorber to generate a pulse in the microsecond to nanosecond in laser cavities. However, in active Q-switched semiconductor SA also used to generate Q-switched whereby this technique no longer investigated due to limited operating wavelength and high cost in fabrication. In passive Q-switched, SA requires longer recovery time than the cavity round trip time because the pulse duration depends on the time to deplete the gain after the SA saturates [1].

The generation of a pulsed laser is demonstrated via Q-switched fiber laser with saturable absorber (SA) such as graphene [2-4], graphite [5], single wall carbon nanotube (SWNTs) [6] and topological insulator (TIs) [7]. This is due to the advantages of SA which has a broadband operation, easy fabrication and low cost. However, the SWNT is good in term of saturable absorption and ultrafast recovery time but it suffers intrinsic drawback. The intrinsic drawback of SWNT can be prevented if graphene SA is used. On the other hand, the gain media in single ring cavity involve a dopant either erbium, thulium and ytterbium ions in order to generate Q-switched fiber laser. Erbium-doped fiber (EDF) act as a gain medium lies in the C-band region which is important for application in communication. Nevertherless, in order to transmit more data in communication system enhance researcher to investigate other bands of EDF such as S-band [8-10].

An SA has been attracting great interest in graphene, carbon nanotube (CNT) and graphene oxide to produce Q-switched pulse due to the low-cost system. According to Popa et. Al [11], graphene Q-switched fiber laser has been proposed wide band from 1552.5 nm to 1555 nm range with pulse duration of ~2 μs with a repetition rate at 70 kHz meanwhile, CNT SA [12] can tuned wavelength from 1555 nm to 15560 nm with pulse duration of ~7μs and repetition rate at 12 kHz. In general, the random phase relationship comes up from oscillation of different modes can be determined by SA resulting in the pulse Q-switched in microseconds. Based on previous work, the pulse duration of Q-switched and repetition rate can be considered in microsecond (μ) region and kilohertz (kHz) respectively which is depending on pump power [13].

In this paper, the tunable Q-switched fiber laser based on graphite flakes is demonstrated. The laser cavity of this setup is 18.17 m with two different EDF absorption coefficient which are 5.06 dB/m: EDF 1 and 7.28 dB/m: EDF2 are used. The broadband absorption of graphite flakes can covers range from 1544 nm to 1560 nm and 1552 nm to 1570 nm for EDF 1 and EDF 2, respectively. The modulation depth of graphite flakes is 23.82 % with a saturation intensity of 0.031 MW/cm^2. The wavelength tunability of graphite Q-switched from shorter wavelength to longer wavelength created decreasing in pulse width but increasing in repetition rate.

Characterization of Erbium doped fiber

There are two types of EDF (EDF 1 & EDF 2) with different absorption coefficient used in this work; 5.09 dB/m and 7.28 dB/m. The other parameters of the two EDF can be shown in Table 1. The setup of the characterization of the Erbium-doped fiber amplifier (EDFA) is illustrated in Figure 1. The pump laser is connected to the wavelength division multiplexing (WDM) to produce 1550 nm so that the wavelength lies in the C or L band region, which is suited with EDF absorption characteristics. The output of EDF is connected to the isolator, which is known as an optical component that can ensure one direction operation in the fiber cavity. The presence of isolator can reduce noise peak in the ASE spectrum caused by oscillation reflection [14-15]. Optical spectrum analyzer (OSA) was used to observe the output signal spectrum. The wavelength of EDF depends on the EDF itself which lie at C band region or L band region as refer in Table 2.

Table 1: EDF specifications

Fiber parameters		EDF 1	EDF 2
Peak absorption (dB/m)		6.43	18.93
Absorption coefficient (dB/m)		**5.09**	**7.28**
Atenuation (dB/km)		2.8	8.23
Fiber diameter (μ)		124.9	125
Numerical Aperture		0.23	0.25
Operating Wavelength (nm)		1550	1550
Material dispersion	Dλ (ps/nm.km)	-91.6	-91.6
	β2(ps^2/km)	117.1	117.1

Journal of Physics: Conference Series **1371** (2019) 012019 doi:10.1088/1742-6596/1371/1/012019

Table 2: Optical wavelength bands [16]

Band	Wavelength range
O	1260-1360
E	1360-1460
S	1460-1530
C	1530-1565
L	1565-1625
U/XL	1625-1675

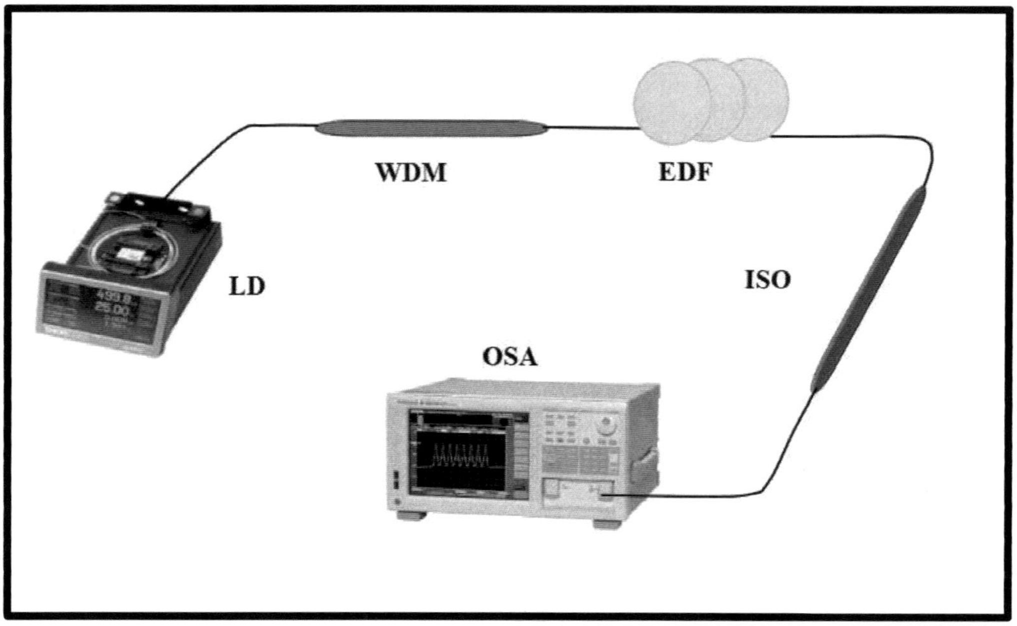

Figure 1: Setup of Erbium-doped fiber

Figure 2(a) and (b) show the spectral response of amplified spontaneous emission (ASE) for EDF 1 and EDF 2 by varying the pump power from 11.2 mW to 325.5 mW. The peak power of EDF 1 and EDF 2 are -2.39 dBm and 2.22 dBm respectively at the pump power of 325.5 mW. It is proven that the higher absorption coefficient of EDF gave a higher peak power because of more excitation of erbium ion occurs during the spontaneous emission process. The ASE spectra for EDF 1 lies in the C-band region whereas EDF 2 lies at C and L band regions [17]. This is due to the excitation of the erbium ions which provide amplification of signal in the region without introducing any effects of gain narrowing [18-20].

Journal of Physics: Conference Series **1371** (2019) 012019 doi:10.1088/1742-6596/1371/1/012019

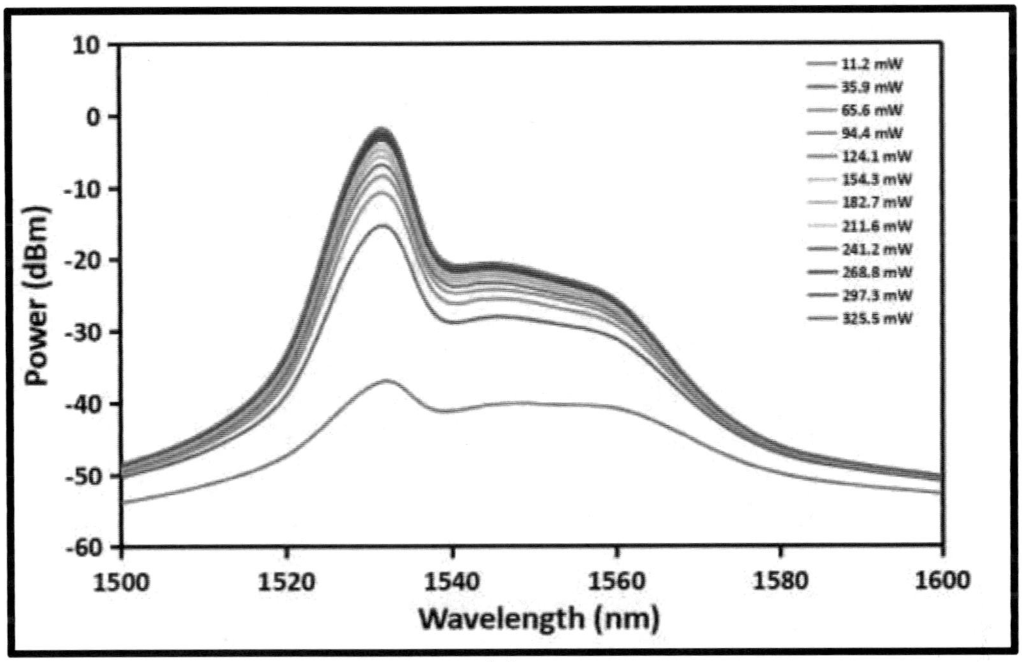

(a)

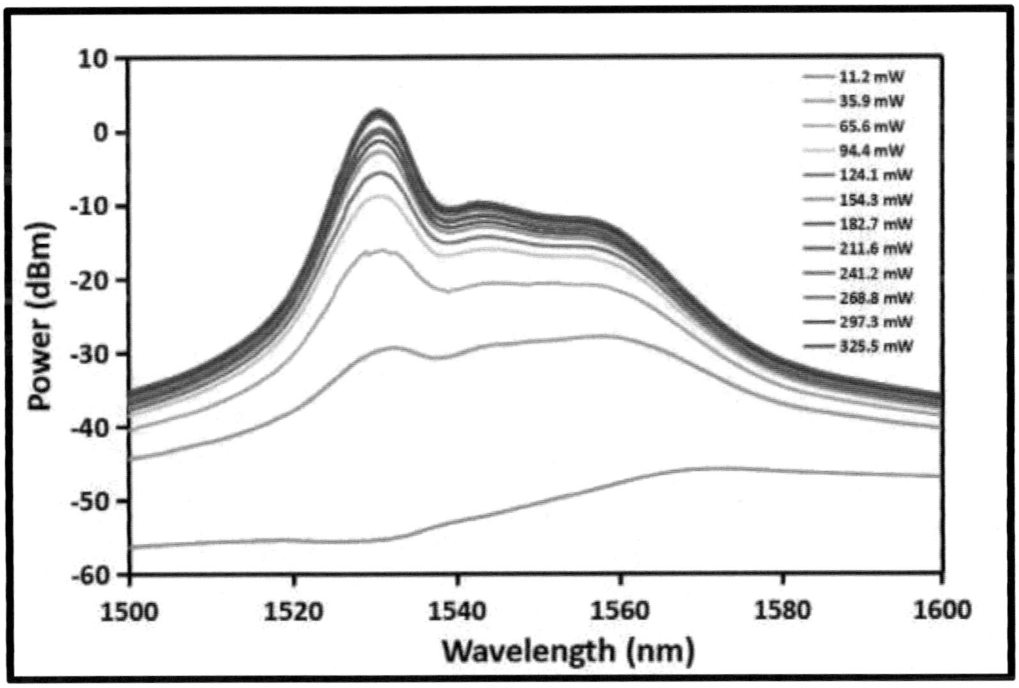

(b)

Figure 2: Spectrum ASE of (a) EDF 1 (b) EDF 2

Journal of Physics: Conference Series **1371** (2019) 012019 doi:10.1088/1742-6596/1371/1/012019

Preparation and characterization of graphite flakes SA

Graphite flakes were deposited on the surface of fiber ferrules. Before the deposition process fiber ferrules which made up of borosilicate glass was cleaned by isopropyl alcohol. The cleaning technique should be performed in the figure of eight wiping action to avoid contamination on surface fiber ferrules. Then, the surface of fiber ferrules was checked for contamination using fiberscope, FS201 Thorlabs. The graphite flakes were transferred to scotch tape for the exfoliation process as shown in Figure 3. The core and cladding of fiber ferrules is shown in Figure 4 after the exfoliation process.

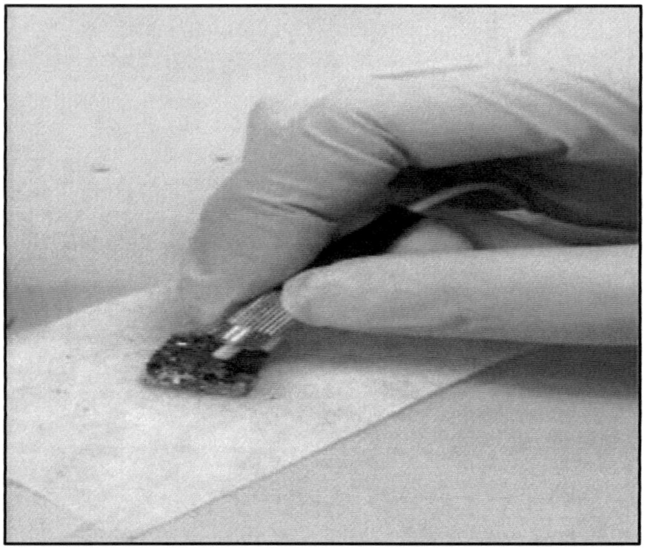

Figure 3: Exfoliation process of graphite flakes

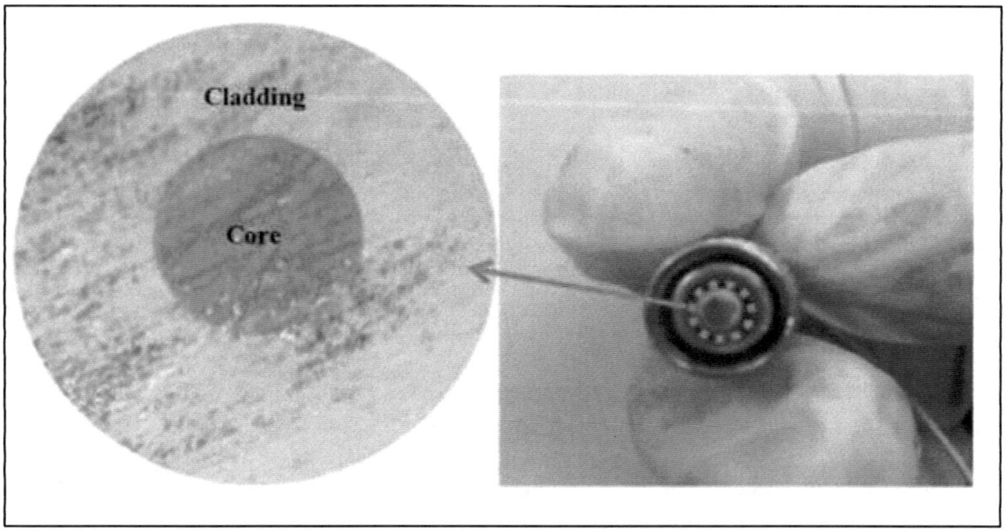

Figure 4: Zoom in of fiber core and cladding

Graphite flakes are characterized by twin detector method to measure the non-linear saturable absorption such as modulation depth, saturation intensity and non-saturable absorption of graphite flakes as shown in Figure 5. The central wavelength of mode-locked pulse seed operates at 1558nm with a repetition rate of 16.8 MHz. Then, the output port of WDM is connected to 3m EDF. The isolator is

Journal of Physics: Conference Series **1371** (2019) 012019 doi:10.1088/1742-6596/1371/1/012019

inserted in between 3m EDF and 50:50 coupler to ensure a unidirectional light propagation [21]. A graphite flakes SA is attached in between of 50:50 OC and OPM 2. Another port of OC is coupled to the OPM 1 which is designated as reference signal while OPM 2 as power absorption measurement.

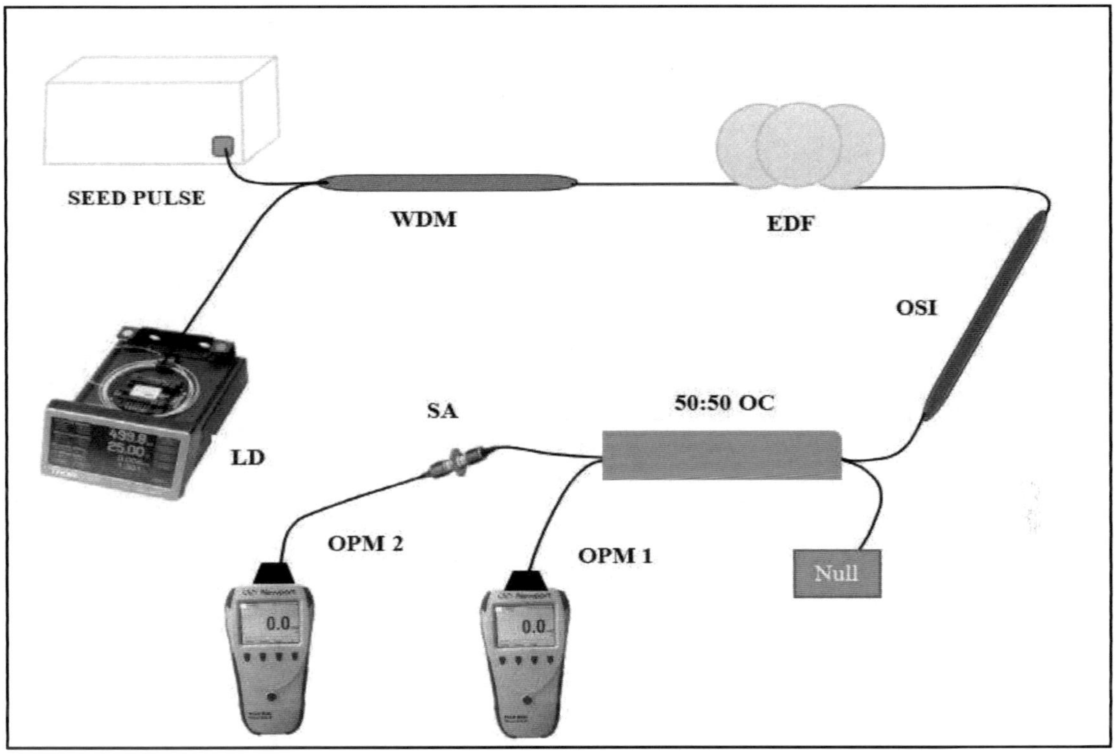

Figure 5: Setup of the twin detector method

Figure 6 shows the modulation depth and saturation intensity of the graphite flakes are 23.85 % and 0.031 MW/cm^2 respectively, which could be obtained from power dependent absorption data. The equation for saturable absorption of graphite flakes can be expressed as

$$\alpha(I) = \frac{\alpha_s}{1+\frac{I}{I_S}} + \alpha_{ns} \quad (1)$$

The non-linear saturable absorption can be observed from 28.2 % to 52.05 % at saturation. This shows that the optical absorption increases by 23.85 % when graphite flakes saturated at an intensity of 0.014 MW/cm^2, giving a modulation depth of 23.85 %. In general, a stable operation of Q-switched has a high modulation depth of more than 10% [12,22]. As the higher modulation depth, the larger non-linear saturable absorption is obtained whereby it will increase the insertion loss and reduce the performance of the laser. Thus, the absorption of the saturable absorber (SA) should be incautiously controlled [23-25]. However, a larger modulation depth of SA may indicate better pulse shaping ability and stabilization of the single pulse operation [4,26]. The pump power increase, the more gain signal produces due to the saturation of graphite SA which indicate of passively Q-switched pulses dependent on the saturation level of SA [27].

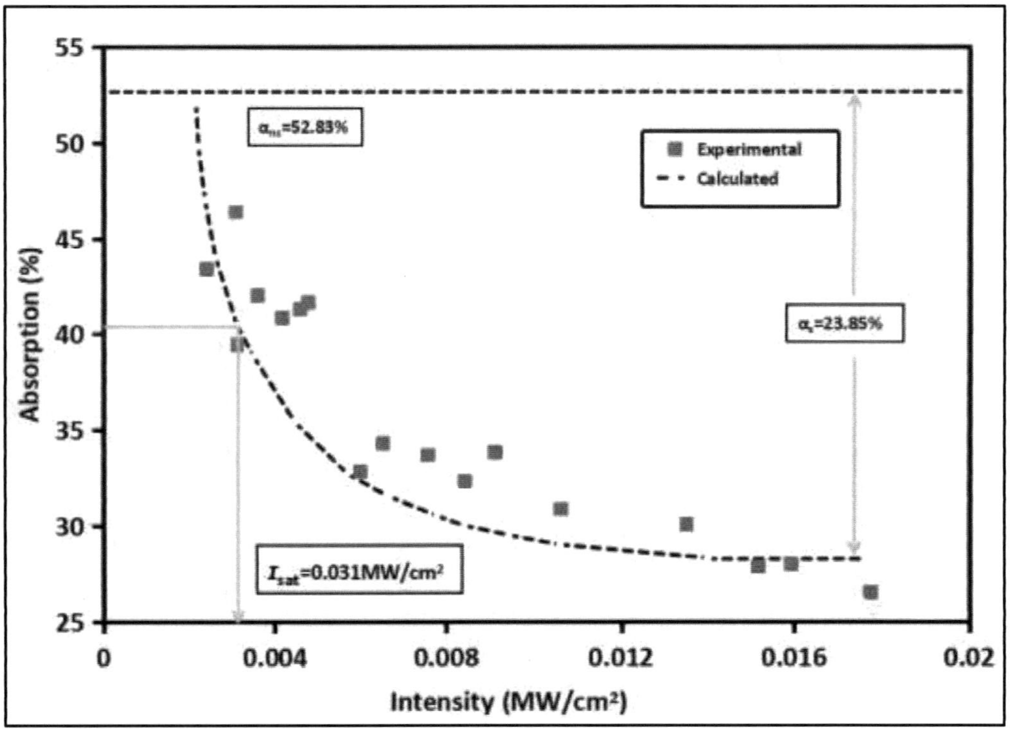

Figure 6: Modulation depth of graphite flakes

Tunable

Q-switched fiber laser was tuned at several wavelengths in C band region. The tunable bandpass filter (TBF) passively Q-switched configuration as shown in Figure 7 has a 90% output coupler that is connected to the input of the TBF. While the output of TBF is connected to 1550 nm WDM. In the generation of Q-switched fiber laser, TBF is inserted into the cavity as the tuning mechanism.

Passively Q-switched optical fiber laser was tuned using TBF in order to determine the performance range of graphite flake SA. The tuning range of the laser spectrum was measured at a fixed pump power of 118.1 mW after graphite flakes SA was inserted inside the resonator as shown in Figure 8(a) and (b). In overall, the range of wavelength was tuned from 1544 nm to 1570 nm for both EDF. Laser tunability was obtained by adjusting the TBF to the desired wavelength with tune step of 2 nm. The tuned wavelength range covers partially C-band for EDF 1 and C and L band for EDF 2. The average power of EDF 1 and EDF are less than -20 dBm and less than 0 dBm respectively. This was due to the higher concentration of Erbium ions contained in the laser cavity. Thus, the stability of the Q-switched pulse could be attained within the wavelength range of 1544 nm to 1570 nm for both EDF. However, the Q-switched pulse disappeared when the tuned wavelength exceeded the range 1560 nm and 1570 nm for EDF 1 and EDF 2 respectively.

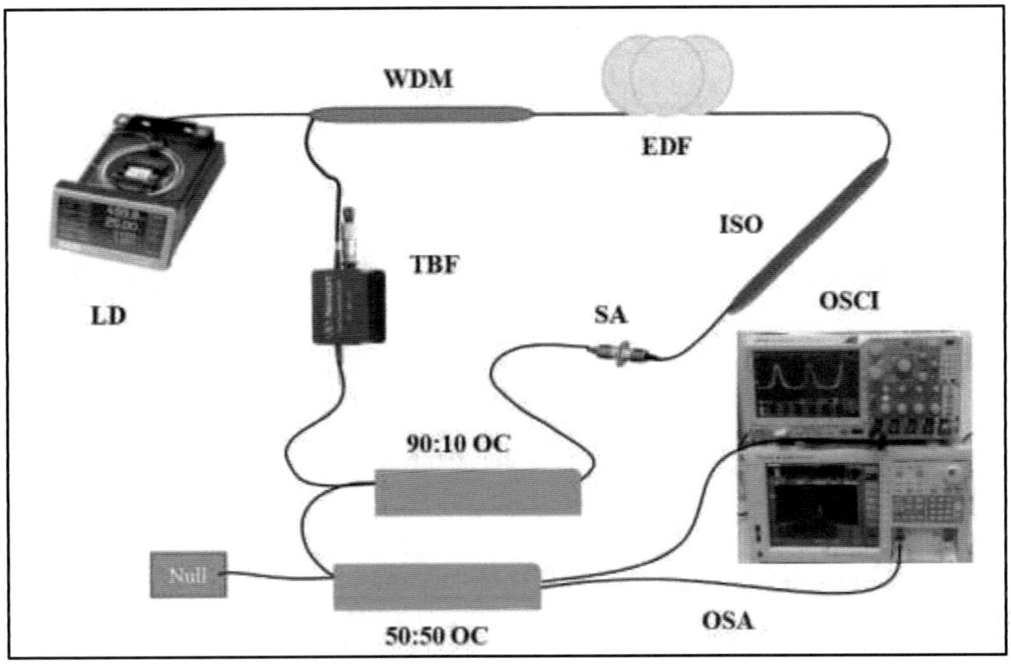

Figure 7: The setup of tunable passively Q-switched fiber laser employing TBF.

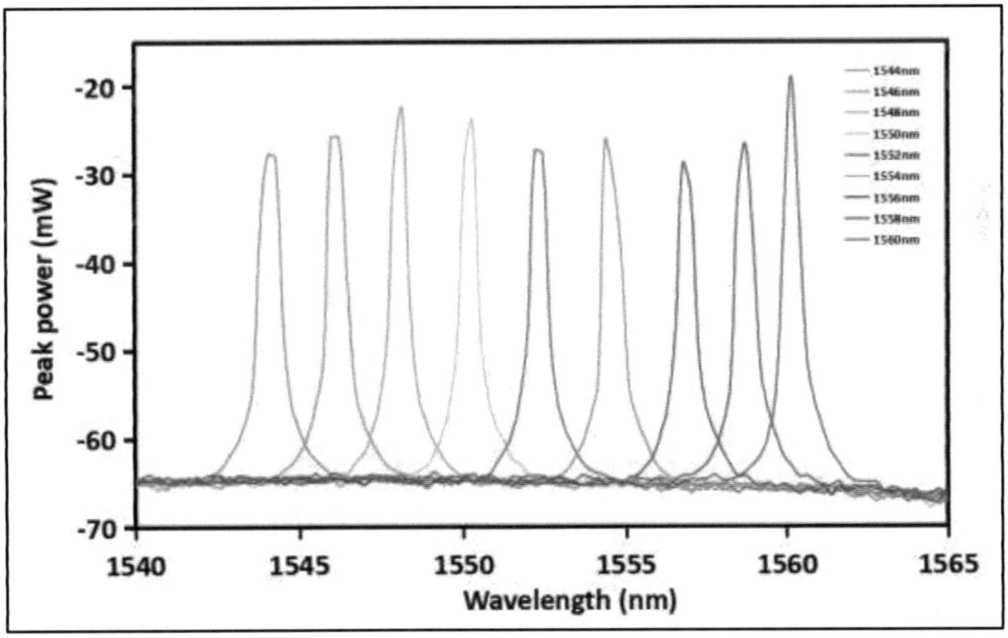

(a)

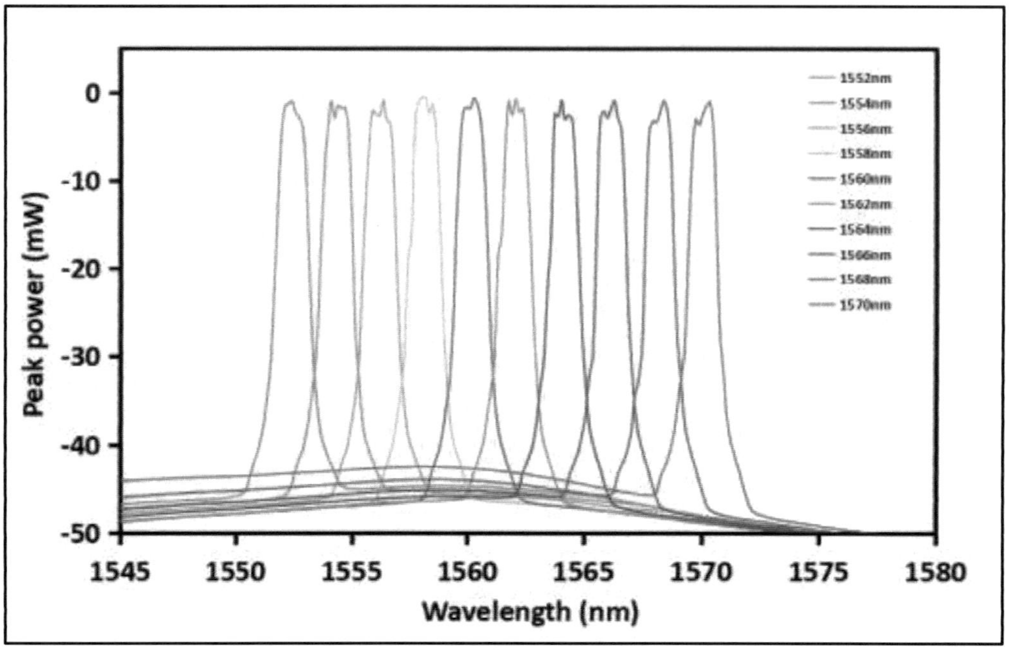

(b)

Figure 8: Tunable wavelength of passively Q-switched fiber laser at a pump power of 118.1 mW for (a) EDF 1 (b) EDF 2

The pump power for EDF 1 and EDF 2 are fixed at 118.1 mW which showed the tuning range from 1544 nm and 1570 nm. The pulse duration for EDF 1 and EDF 2 are 119.6 μs and 27.97 μs at1544 nm and 1552 nm wavelength respectively, as shown in Figure 9(a) and (b). It can be concluded that pulse duration in passively Q-switched fiber laser dependent on the pump power. As the higher the pump power applied, the more gain is needed to saturate the SA which give smaller pulse duration [28]. On the other hand, the pulse width for EDF 1 and EDF 2 are 7.38 μs and 4.17 μs. In theory, when pulse to pulse spacing is equal or longer than the upper state lifetime, most of the obtainable inversion will be lost via spontaneous decays. Therefore, the repetition rate of passively Q-switched fiber laser happens in a few kHz [29-30].

The repetition rate for both EDF 1 and EDF 2 are 8.33 kHz and 35.74 kHz as shown in Figure 10(a) and (b). From the figure below, the signal-to-noise ratio (SNR) of EDF 1 and EDF 2 can be calculated by deduction of the highest peak power to the lowest peak power [31]. The repetition rate increase with pump power as well as the SNR value. In this paper, the highest SNR value is the best compared to the others due to the more signals can be received in order to generate Q-switched pulses.

Figure 11 shows the passively Q-switched fiber laser pulse in terms of pulse repetition rate and pulse width at a wider range of wavelength. The wavelength point for Q-switching is observed at 1544 nm to 1560 nm and 1552 nm to 1570 nm for EDF 1 and EDF 2, where the pump power is fixed at 118.1 mW. The wavelength tuning for both EDF 1 and EDF 2 are obtained due to the result of the limitation of the ASE spectrum of the EDF itself lies in C-band and L- band region [48]. Figure 11(a) and (b) illustrate the relationship between repetition rate and pulse width with wavelength. As the wavelength point increases from 1544 nm to 1560 nm and 1552 nm to 1570 nm for both EDF, the pulse width shows a decreasing line for EDF 1 and increasing line for EDF 2. On the other hand, for repetition rate, the repetition rate increased in EDF 1 but decreased in EDF 2. This is because the trend of the graph of the repetition rate and pulse width are different as shown in the figure below due to the pattern of the ASE spectrum of the EDF itself.

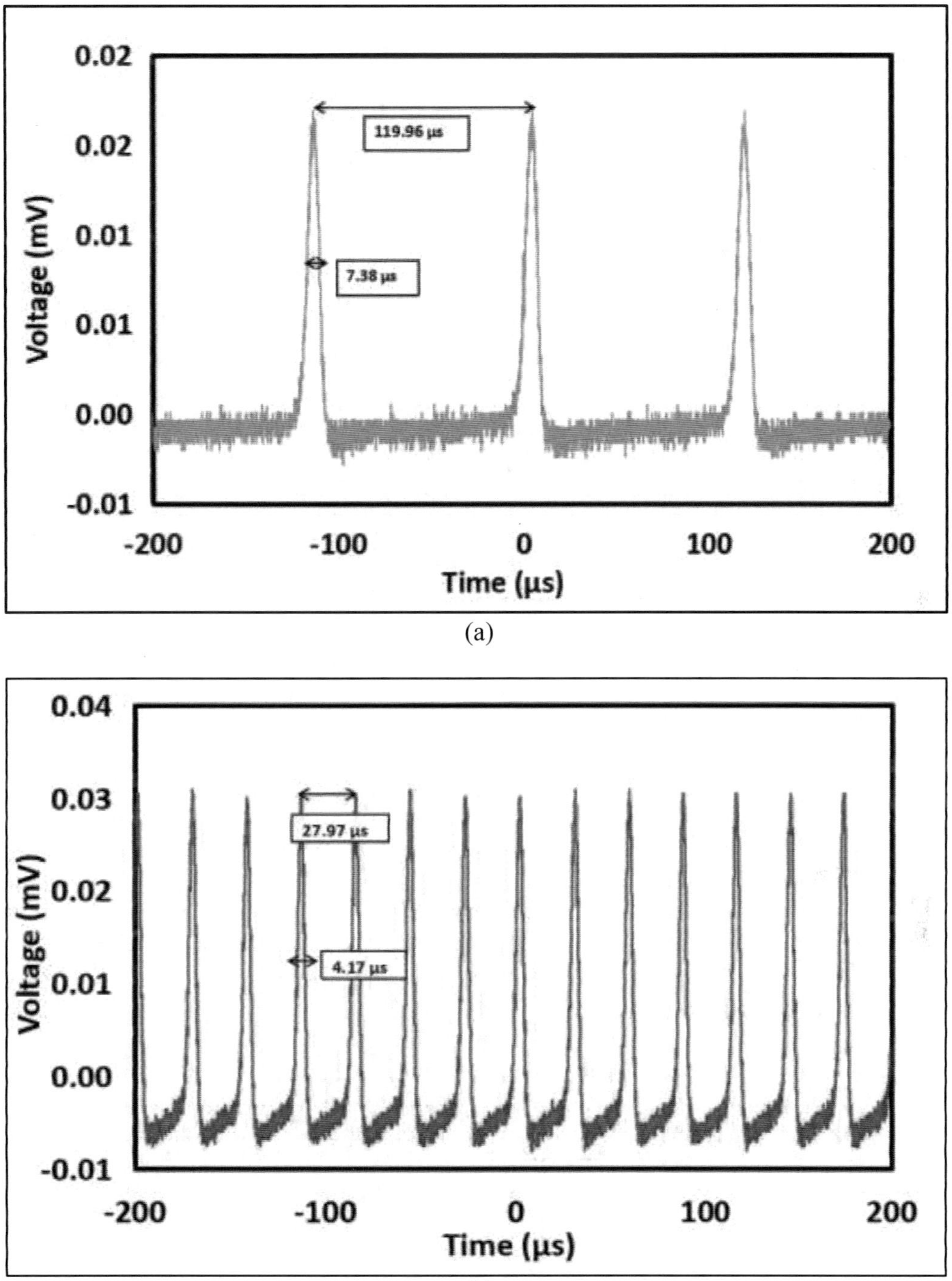

Figure 9: Pulse train of passively Q-switched fiber laser at a pump power of 118.1 mW for (a) EDF 1 (b) EDF

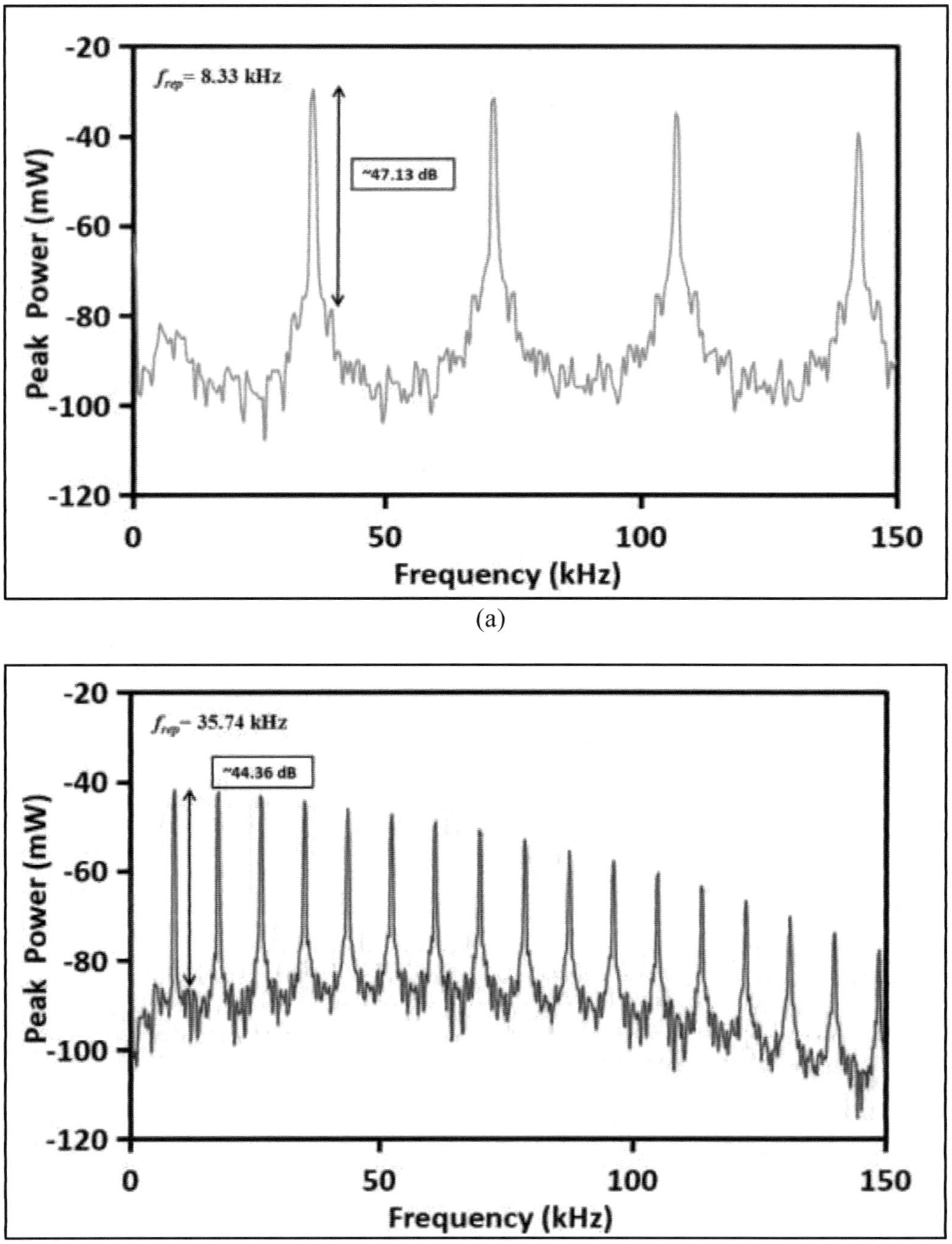

(a)

(b)

Figure 10: Repetition rate of passively Q-switched fiber laser at a pump power of 118.1 mW for (a) EDF 1 (b) EDF

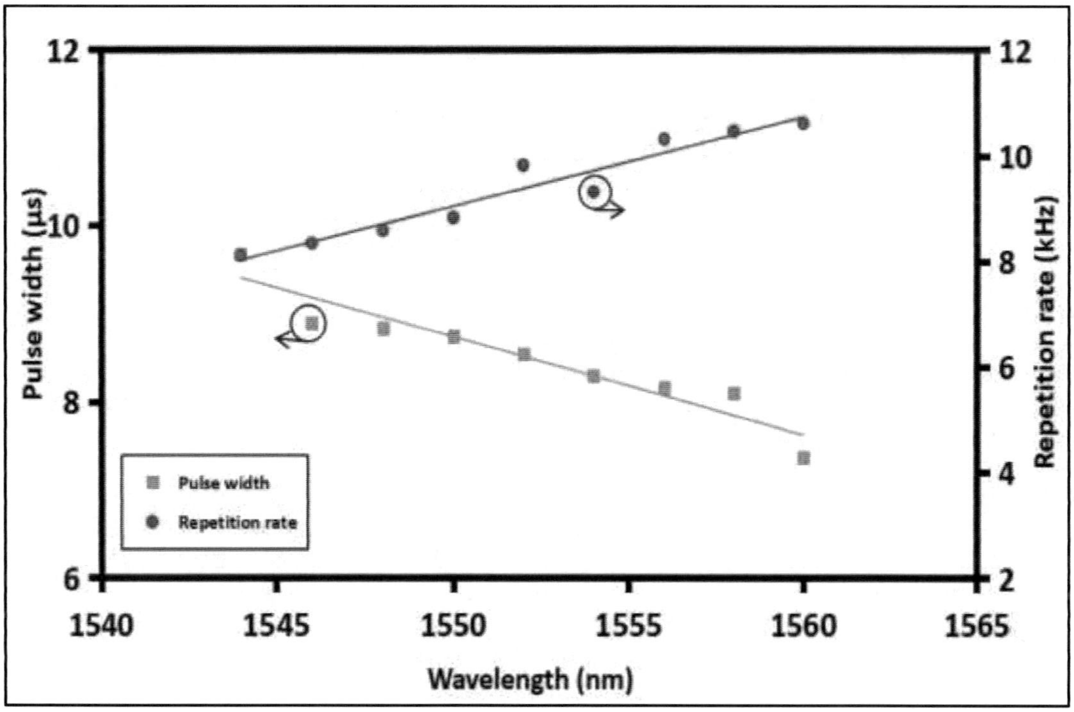

Figure 11: Pulse width and pulse repetition rate as a function of wavelength for (a) EDF 1 (b) EDF 2

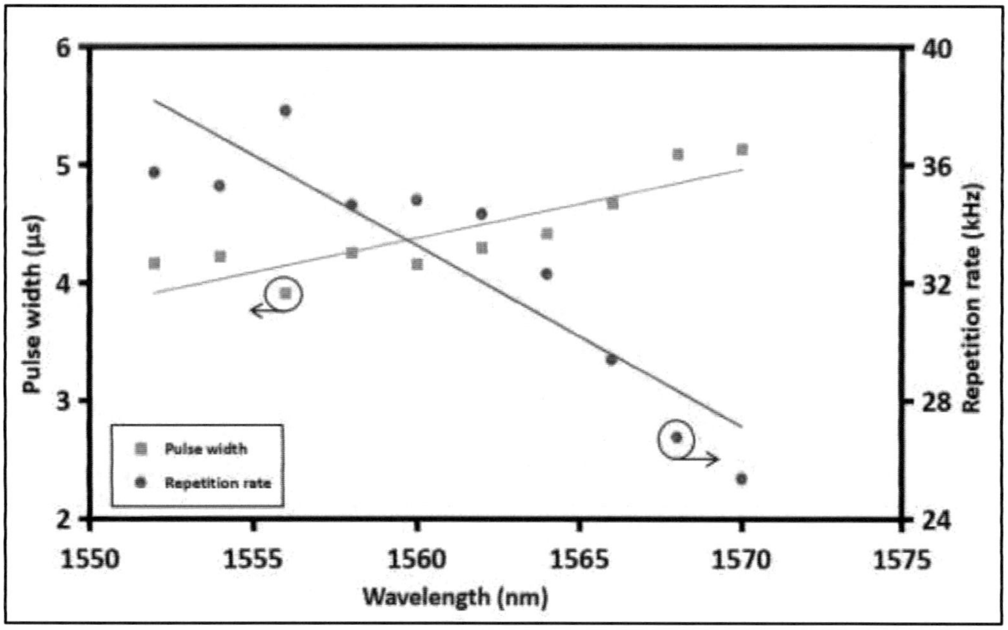

Figure 12: Pulse width and pulse repetition rate as a function of wavelength for (a) EDF 1 (b) EDF 2

Conclusion

The performance of a passively Q-switched fiber laser by using graphite flakes at different EDF absorption which are EDF 1 and EDF 2 at 5.09 dB/m and 7.28 dB/m, respectively, have been demonstrated. The system was designed based on a single ring cavity by employing a graphite flakes saturable absorber (SA) with 23.82% of saturable absorption. The experimental results showed that a wide-band tunable covering from 1544 nm to 1560 nm and 1552 nm to 1570 nm for EDF 1 and EDF 2, respectively, could be achieved. The EDF 2 was observed to be the best gain medium due to the conventional wavelength band at C band and L band region.

Acknowledgement

Acknowledgements Authors would like to thank Research Management Centre (RMC) of Universiti Tun Hussein Onn Malaysia for supporting this work under TIER 1 Grant: H133 and H139. Authors would also thank Universiti Tun Hussein Onn Malaysia (Pagoh Campus) for providing the facilities and instruments to complete this work.

References

[1] Svelto O & Hanna DC (1998), *Principles of lasers* (Vol. 4). New York: Plenum press.

[2] Zhang H, Tang DY, Zhao LM, Bao QL & Loh KP (2009), Large energy mode locking of an erbium-doped fiber laser with atomic layer graphene. *Optics Express*, *17*(20), 17630-17635.

[3] Liu J, Xu J & Wang P (2012), Graphene-based passively Q-switched 2 μm thulium-doped fiber laser. *Optics Communications*, *285*(24), 5319-5322.

[4] Ahmed MHM, Al-Masoodi AHH, Yasin M, ArofH & Harun SW (2017), Stretched and soliton femtosecond pulse generation with graphene saturable absorber by manipulating cavity dispersion. *Optik*, *138*, 250-255.

[5] Lee J, Lee J, Koo J & Lee JH (2016), Graphite saturable absorber based on the pencil-sketching method for Q-switching of an erbium fiber laser. *Applied optics*, *55*(2), 303-309.

[6] Wang Y, Wang Y, Zhang X & Wen Q (2014), Passively Q-switched Nd: YAG laser with single-walled carbon nanotube in heavy water. *Optics Communications*, *321*, 172-175.

[7] Lin YH, Yang CY, Lin SF, Tseng WH, Bao Q, Wu CI & Lin GR (2014), Soliton compression of the erbium-doped fiber laser weakly started mode-locking by nanoscale p-type Bi2Te3 topological insulator particles. *Laser Physics Letters*, *11*(5), 055107.

[8] Harun SW, Abd Rahman F, Dimyati K & Ahmad H (2006), An efficient gain-flattened C-band Erbium-doped fiber amplifier. *Laser Physics Letters*, *3*(11), 536-538.

[9] Harun SW, Cheng XS, Saat NK & Ahmad H (2005), S-band Brillouin erbium fibre laser. *Electronics Letters*, *41*(4), 174-176.

[10] Harun SW, Zulkifli MZ & Ahmad H (2006), A linear cavity S-band Brillouin/erbium fiber laser. *Laser Physics Letters*, *3*(7), 369.

[11] Popa D, Sun Z, Hasan T, Torrisi F, Wang F & Ferrari AC (2011), Graphene Q-switched, tunable fiber laser. *Applied Physics Letters*, *98*(7), 073106.

[12] Zhou DP, Wei L, Dong B & Liu WK (2010), Tunable passively Q-switched erbium-doped fiber laser with carbon nanotubes as a saturable absorber. *IEEE Photonics Technology Letters*, *22*(1), 9-11.

[13] Ahmad H, Reduan SA, Ruslan NE, Lee CSJ, Zulkifli MZ & Thambiratnam K (2019), Tunable Q-switched erbium-doped fiber laser in the C-band region using nanoparticles (TiO2). *Optics Communications*, *435*, 283-288.

[14] Jarabo S, Sola IJ & Saez-Landete J (2003), Spectral hole burning induced by reflected amplified spontaneous emission in erbium-doped silica optical fiber pumped at 980 nm. *JOSA B*, *20*(6), 1204-1211.

[15] Shi H, Song Y, Li R, Li Y, Cao H, Tian H & Hu M (2018), Review of low timing jitter mode-locked fiber lasers and applications in dual-comb absolute distance measurement. *Nanotechnology and Precision Engineering*, *1*(4), 205-217

[16] Bands O, Laurent B & Draka G (2008), From O to L: The Future of optical wavelength-bands. *Broadband properties*, 83-85.

[17] Altuncu A, Basgumus A, Uzunca B & Haznedaroglu E (2005), Design and Characterization of High Performance C and L Band Erbium Doped Fiber Amplifiers (C, L-EDFAs). In *4th International Conference on Electrical and Electronics Engineering* (Vol. 2, pp. 279-283).

[18] Basak A, Hossain MM & Islam MR (2013), Performance Analysis of Erbium-Doped Fiber Amplifier in Fiber Optic Communication Technique. *Global Journal of Research In Engineering*.

[19] Mahad FD, Supa'at ASBM & Sahmah A (2009), EDFA gain optimization for WDM system.

[20] Chiba, Y., Takada, H., Torizuka, K., & Misawa, K. (2015). 65-fs Yb-doped fiber laser system with gain-narrowing compensation. *Optics express*, *23*(5), 6809-6814.

[21] Lau KY, Muhammad FD, Latif AA, Bakar MA, Yusoff Z & Mahdi MA (2017), Passively mode-locked soliton femtosecond pulses employing graphene saturable absorber. *Optics & Laser Technology*, *94*, 221-227.

[22] Lau KY, Abidin NZ, Bakar MA, Latif AA, Muhammad FD, Huang NM & Mahdi MA (2018), Passively mode-locked ultrashort pulse fiber laser incorporating multi-layered graphene nanoplatelets saturable absorber. *Journal of Physics Communications*, *2*(7), 075005.

[23] Bao Q, Zhang H, Yang JX, Wang S, Tang DY, Jose R & Loh KP (2010), Graphene–polymer nanofiber membrane for ultrafast photonics. *Advanced functional materials*, *20*(5), 782-791.

[24] Luo Z, Zhou M, Weng J, Huang G, Xu H, Ye C & Cai Z (2010), Graphene-based passively Q-switched dual-wavelength erbium-doped fiber laser. *Optics letters*, *35*(21), 3709-3711.

[25] Ma J, Xie GQ, Lv P, Gao WL, Yuan P, Qian LJ & Tang DY (2012), Graphene mode-locked femtosecond laser at 2 μm wavelength. *Optics letters*, *37*(11), 2085-2087.

[26] Rozhin AG, Sakakibara Y, Namiki S, Tokumoto M, Kataura H & Achiba Y (2006), Sub-200-fs pulsed erbium-doped fiber laser using a carbon nanotube-polyvinylalcohol mode locker. *Applied physics letters*, *88*(5), 051118..

[27] Mao D, Cui X, Zhang W, Li M, Feng T, Du B & Zhao J (2017), Q-switched fiber laser based on saturable absorption of ferroferric-oxide nanoparticles. *Photonics Research*, *5*(1), 52-56.

[28] Ismail MA, Harun SW, Ahmad H & Paul MC (2016), Passive Q-switched and Mode-locked Fiber Lasers Using Carbon-based Saturable Absorbers. In *Fiber Laser*. IntechOpen.

[29] Nady A, Ahmed MHM, Numan A, Ramesh S, Latiff AA, Ooi CHR & Harun SW (2017), Passively Q-switched erbium-doped fibre laser using cobalt oxide nanocubes as a saturable absorber. *Journal of Modern Optics*, *64*(13), 1315-1320.

[30] Nady A, Ahmed MHM, Latiff AA, Ooi CR & Harun SW (2017), Femtoseconds soliton mode-locked erbium-doped fiber laser based on nickel oxide nanoparticle saturable absorber. *Chinese Optics Letters*, *15*(10), 100602.

[31] Bosworth BT, Bernecky WR, Nickila JD, Adal B & Carter GC (2008), Estimating signal-to-noise ratio (snr). *IEEE Journal of Oceanic Engineering*, *33*(4), 414-418.

Uric acid detection by using contactless intensity modulation based displacement sensor

N N G M A Gan[1], N A M Taib[1*], M Abdullah[2], W M Mukhtar[1] and A R A Rashid[1]

[1]Faculty of Science and Technology, Universiti Sains Islam Malaysia
[2] Institute of Nano Optoelectronics Research and Technology, Universiti Sains Malaysia

athirahtaib@usim.edu.my

Abstract. A simple and contactless intensity modulated displacement sensor is proposed and demonstrated for sensing uric acid concentration. For a concentration change of uric acid from 0 ppm to 500 ppm, two peak voltages are obtained from the displacement curve correspond to the highest reflectivity of each concentration. Those peak light intensities increase linearly with the concentration due to the increase of the refractive index of the uric acid solution. This implies that the higher concentration of uric acid tends to detect a stronger signal. The measured sensitivities are obtained at 0.0015 V/ppm and 0.0016 V/ppm for the first peak voltage and second peak voltage respectively. The result showed the percentage of similarity for the first peak to the second peak is almost 94% and the linearity more than 97% for both peak voltages is obtained suggested the consistency of the sensor system. The stability and simplicity of the contactless sensor offer a good and valuable opportunity for many applications especially in the hazardous chemical, pharmaceutical, process control and diagnostic sectors.

1. Introduction

Uric acid is a heterocyclic organic compound with C5H4N4O3 formula (7,9-dihydro-1H-purine-2,6,8(3H)-trione). It is the final oxidation product of purine metabolism as shown in Figure 1. The elevated uric acid level causes the deposition of monosodium urate crystals. As urate concentration increasest in the blood, uric acid crystal formation increases. The normal reference interval of uric acid in human blood is 1.5 to 6.0 mg/dL in women and 2.5 to 7.0 mg/dL in men [1]. Meanwhile, Chauhan and Pundir cited the normal range of uric acid should fall in between 0.15-0.45 mmol/L for blood and 1.19-2.98 mmol/day in urine excretion [2].

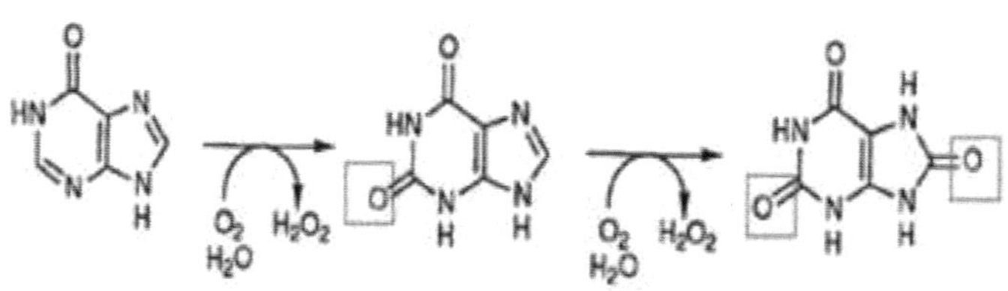

Figure 1. The final product of purine metabolism (uric acid)

Overproduction of uric acid by the metabolism of purines has been proven to play emerging roles in human disease. The high uric acid content is regarded as a biomarker of gout that often linked with obesity, diabetes, muscle stiffness and heart diseases which can be diagnosed and treated based on the level of attack [3]. Similarly, uric acid may be potentially a sign for kidney disease [4] and there are some works have suggested that uric acid contributest to the development of cardiovascular disease [5] and respiratory disease [6]. In order to avoid those diseases from proliferating, the monitoring of uric acid level is pressing. Therefore, it is necessary to develop a simple and reliable method for the detection of uric acid level. Monitoring uric acid is important because it can be used as a powerful indicator for early sign of diseases.

There have been several kinds of research done on the detection of the uric acid level. UV-VIS spectrometer is the simplest high technology method capable of differentiating uric acid based on their respective absorbance [7]. Apart from that, the performance response of spectrophotometric by [8] is excellent and close to clinical method however required to deal wthe ith the sensitive enzymatic reaction which is expensive and prone to metabolite.an In the electrochemical method, the modified electrode surface by nanomaterials such as reduced graphene oxide-zinc oxide nanorods composite [9] and gold nanoparticles [10] showed a very good result of low detection limit and large linear range capabilities for selective determination in uric acid. Another way to determine the uric acid level is by fabrication of fiber-coated sensitive medium [11–13]. These coating sensitive medium based fiber optic operate based on intensity modulation of guided light and evanescent wave concept. It is indeed undeniable that the incorporation of nanoparticles and nanomaterials will enhance the performance such as the sensitivity, limit of detection as well as the linear range, however, some of them are very costly, complicated and required a highly skilled to handle with during the modification and fabrication processes. Some part of the probes needs to be removed and immersed in the sample solution in other to obtain the output which tends to increase the chance of contamination and inconsistency area of removal parts.

In this paper, a simple, cost-effective and reliable contactless fiber-optic sensor based on displacement method is proposed and demonstrated for the detection of uric acid concentration. The sensor uses a bundled fiber of having a transmitting fiber and 16 receiving fiber sa as a probe and the probe is non-contact to the uric acid solution. The detection is based on contactless sensing of uric acid concentration using intensity modulation technique via displacement method that is the analysis of light reflectivity on the surface of uric acid solution.

2. Methodology

The schematic diagram of the experimental setup for a contactless displacement sensor which is used to measure the uric acid concentration by using intensity modulation is shown in Figure 2. The setup consists of a light source, bundled fiber probe, automated displacement controller, silicon detector, oscilloscope and computer. A bundled fiber probe comprised of a transmitting fiber and 16 receiving fibers is connected to a 650 nm red laser diode source. The cross-sectional view for the bundle probe is also depicted in Figure 2. When the light from diode laser with output power of 9.3 mW is coupled into a transmitting fiber and illuminated the liquid surface which is meniscus due to surface tension effect, the light is reflected back assuming that the liquid surface is a virtual reflector [14], then the light signals are received by the receiving fibers which cascading them directly to the silicon detector. The silicon detector (DET100A2, Thorlabs) has an optical response of 320 nm to 1100 nm, making it compatible with a wide range of visible light for detection of light signals including 650 nm visible red laser used for optimum transmission. The working principle of the sensor is verified by the displacement as the main parameter. The position of the automated displacement controller is monitoredy by software provided by Cytron Technologies installed in a computer. The probe displacement is accomplished by attaching the displacement controller rigidly to a vibration free optical table. A precise displacement of 40 mm is taken for every sample by shifting the bundled probe away from liquid meniscus in a vertical axis. The range of the displacement is taken between 0 to 40 mm as a fingerprint of the sensor works effectively within this range.

The position of the fiber probe must be as close as possible to the liquid surface and located in the middle on the main axis of the liquid meniscus for this position is acting as a reference point of the probe shift. A single shot signal is captured by an oscilloscope (TDS1002B; 2 CH, 60 MHz, 1 Gs/s for data analysis and evaluation subsequently yielding an informative response graph which is important in sensor characterization. The uric acid concentration ranging from 0 ppm to 500 ppm is used as a sample and prepared by dissolving in distilled water. The different uric acid concentration has proven to have different refractive index [11].

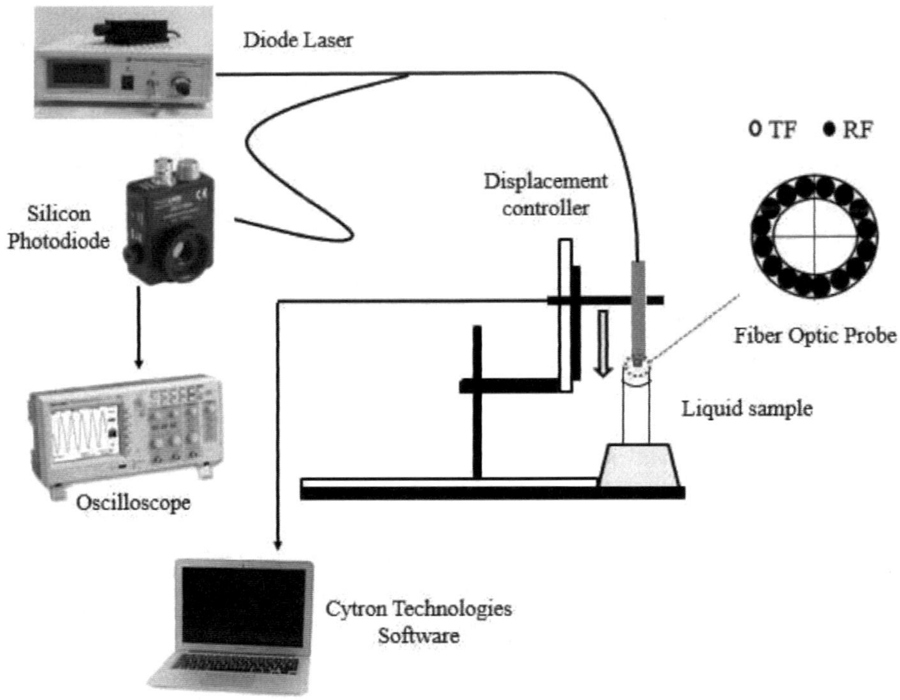

Figure 2. Schematic experimental setup of uric acid detection

3. Result and discussion

The relationship between concentration and refractive index of uric acid is plotted in Figure 3. The uric acid refractive index was found to be linearly increased from 1.3330 RIU to 1.3336 RIU with respect to its concentration. There is a high potential that different refractive index of uric acid in the solution influence the amount of light being reflected by the surface.

Figure 4 depicts the sensor fingerprint of output voltage against the displacement for different uric acid concentration. In general, all displacement curves have a similar trend; there are two maximum output voltages with a slightly broad distance between each curve. The positive front slope of the first peak curve rises up steeply and the negative back slope of the second peak drop gradually along with the displacement. The shape of displacement response occurs as a result of the meniscus formation of the liquid sample while the maximum voltages correspond to the difference in the refractive index value.

At small displacement, that means the sensor probe is nearly touched with the meniscus, there is a little overlapping between the core of receiving fiber and emitted a light cone of transmitting fiber leads to a quite low intensity has been detected by the photodiode. As the displacement increases, the intensity increases as the size of overlapping area becomes larger and eventually reaches the first maximum point when the core of the receiving fiber is totally immersed by the cone of the emitted light. The intensity reduces with further increase in displacement because of inverse square law. Interestingly, as the sensor probe approaches a distance of roughly around 9 mm, the displacement response comes to local minima.

The output voltage once again rises up attaining the second maximum point which is located at nearly 10 times of the first peak distance and decreases again following the inverse square law. That is, the power density of light from transmitting fiber now is weak and far enough from the meniscus and eventually, the output voltage drop as a result of less power is collected by the receiving fiber.

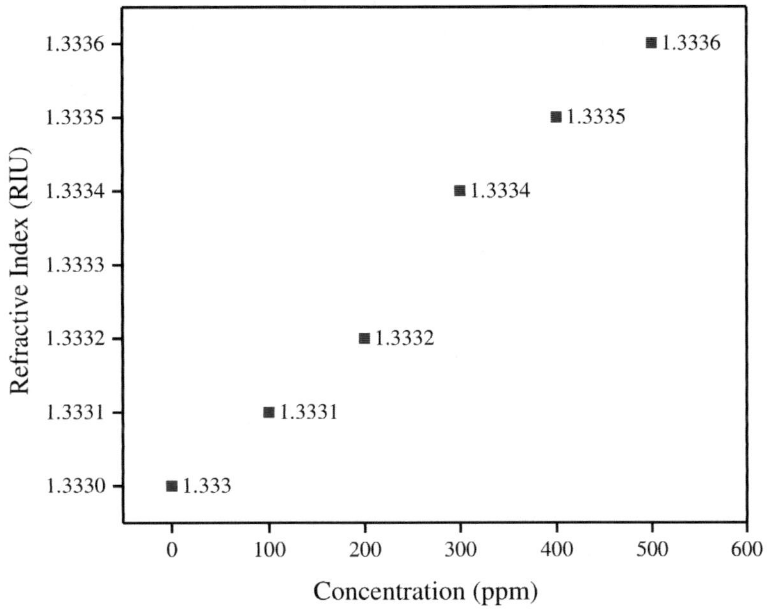

Figure 3. Refractive index and concentration relationship

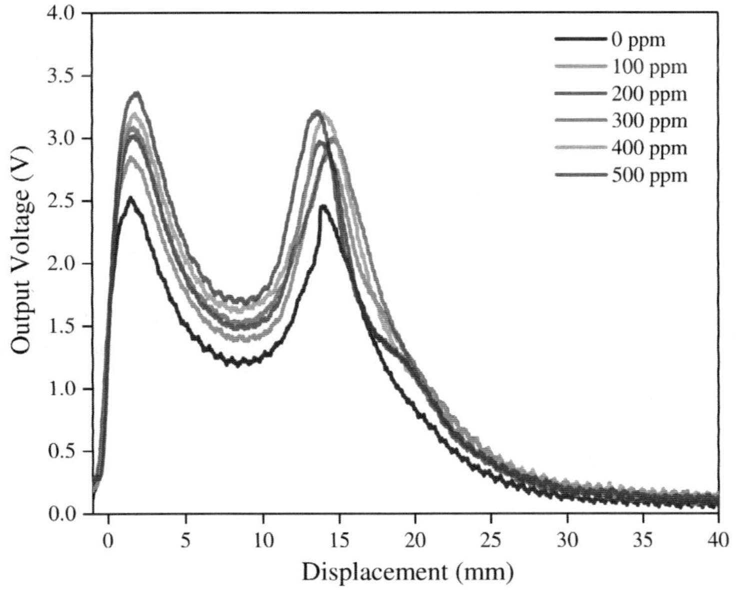

Figure 4. Sensor fingerprint as a function of displacement

The peak voltage from each response curve of the uric acid solution is plotted against its respective concentration as shown in Figure 5. It can be observed that the peak voltage increases when the concentration increases for both peak voltages. The slope is used to determine the sensitivity of the sensor. In the figure, the sensor achieves approximately similar sensitivity up to 94% for both first and second peak voltage which are 0.0015 V/ppm and 0.0016 V/ppm respectively. This implies the consistency of the sensor able to detect the same output even the probe is displaced far away from the liquid surface. Moreover, the sensor exhibits good linearity of more than 97% with a low detection limit of 16.8 ppm and 15.8 ppm indicating that the sensor could work efficiently in detecting uric acid level in the human body and applicable to be a potential sensor for medical application concerning the issue of the uric acid-related diseases. The sensor performance of the uric acid detection is tabulated in Table 2.

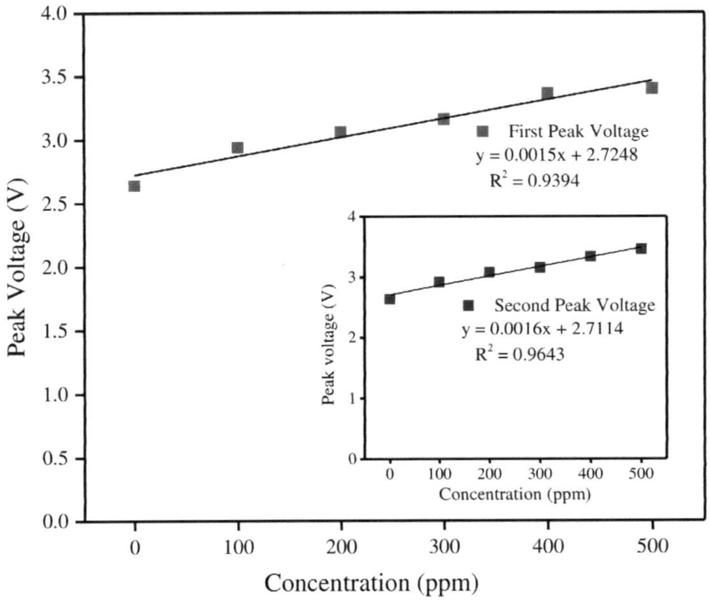

Figure 5. Peak voltage (V) as a function of concentration (ppm)

Table 1. Sensor performance of uric acid detection

Parameter	**Peak Voltage (V)**	
	First Peak Voltage	**Second Peak Voltage**
Sensitivity (V/ppm)	0.0015	0.0016
Measurement range (ppm)	0 -500	
Linearity (%)	> 97	> 98
Resolution (ppm)	16.8	15.8

4. Conclusion

An uric acid sensor has been introduced using an intensity modulation technique as a device to detect uric acid concentration. The sensor operates based on the contactless displacement sensor. In the experiment, the probe is non-contact with the liquid surface but must be as close as possible in order to attain uniform displacement pattern. Two peak voltages extracted from the displacement curves show excellent characteristics in determining the concentration of uric acid in the range of 0 ppm to 500 ppm. The peak voltages are proportional to the concentration due to the difference in the refractive index value. The sensitivities obtained are 0.0015 V/ppm and 0.0016 V/ppm respectively with slope linearity of more than 97% and detection limit of 16.8 ppm and 15.8 ppm which are differed in slightest value suggesting that the sensor is capable of producing output voltage consistently. This finding has great potential and useful for many applications especially in a diagnostic sector such as for liquid parameter calibration like surface tension, which has becoming a new interest in the medical area due to surface tension in human fluids are discovered varying with pathological disorders.

References

[1] Maiuolo J, Oppedisano F, Gratteri S, Muscoli C, and Mollace V 2016 Regulation of uric acid metabolism and excretion. *Int. J. Cardiol*, 213, 8–14.
[2] Chauhan N and Pundir C S 2011 An amperometric uric acid biosensor based on multiwalled carbon nanotube-gold nanoparticle composite. *Anal. Biochem*, 413(2), 97–103..
[3] Tausche A K, Jansen T L, Schröder H E, Bornstein S R, Aringer M, and Müller-Ladner U 2009 Gout--current diagnosis and treatment. *Dtsch. Ärzteblatt Int*, 106(34–35), 549–555.
[4] Giordano C, Karasik O, King-Morris K, and Asmar A 2015 Uric acid as a marker of kidney disease: Review of the current literature. *Dis. Markers*, 2015.
[5] Borghi C 2015 The role of uric acid in the development of cardiovascular disease. *Curr. Med. Res. Opin,* 31(S2), 1–2.
[6] Horsfall L J, Nazareth I, and Petersen I. 2014 Serum uric acid and the risk of respiratory disease: A population-based cohort study. *Thorax*, 69(11), 1021–1026.
[7] Norazmi N, Abdul Rasad Z R, Mohamad M, and Manap H 2017 Uric acid detection using UV-VIS spectrometer. *IOP Conf. Ser. Mater. Sci. Eng.* 257(012031), 1-6.
[8] Hamzah H H 2013 Spectrophotometric determination of uric acid in urine based-enzymatic method uricase with 4-Aminodiphenylamine diazonium sulfate (Variamine Blue RT Salt)," *J. Anal. Bioanal. Tech.* S7, 1-6.
[9] Fu L, Zheng Y, Wang A, Cai W, Deng B, and Zhang Z 2016 An electrochemical sensor based on reduced graphene oxide and ZnO nanorods-modified glassy carbon electrode for uric acid detection. *Arab. J. Sci. Eng*, 41(1), 135–141.
[10] Shi Y, Wang J, Li S, Yan B, Xu H, Zhang K and Du Y 2017 The enhanced photo-electrochemical detection of uric acid on Au nanoparticles modified glassy carbon electrode. *Nanoscale Research Letter,* 12(455), 1–6.
[11] Batumalay M, Harun S W, Ahmad F, Nor R M, Zulkepely N R and Ahmad H 2014 Tapered plastic optical fiber coated with graphene for uric acid detection. *IEEE Sens. J*, 14(5), 1704–1709.
[12] Batumalay M, Harith Z, Rafaie H A, Ahmad F, Khasanah M, Harun S W, Nor R M and Ahmad H 2014 Tapered plastic optical fiber coated with ZnO nanostructures for the measurement of uric acid concentrations and changes in relative humidity. *Sensors Actuators, A Phys*, 210(2014), 190–196.
[13] Batumalay M, Ahmad F, Lokman A, Jasim A A, Harun S W, and Ahmad H 2014 Tapered plastic optical fiber coated with single wall carbon nanotubes polyethylene oxide composite for measurement of uric acid concentration. *Sens. Rev.*, 34(1), 75–79.
[14] Krishnan G, Bidin N, Abdullah M, Ahmad M F S, Bakar M A A, and Yasin M 2016 Liquid refractometer based mirrorless fiber optic displacement sensor. *Sensors Actuators, A Phys.*, 247(2016), 227–233.

Journal of Physics: Conference Series **1371** (2019) 012021 doi:10.1088/1742-6596/1371/1/012021

The pH sensor based optical fiber coated with PAH/PAA

Khairiah Mohd Yatim[1], Ganesan Krishnan[1,2*], Hazri Bakhtiar[1,2], Suzairi Daud[1,2] and Sulaiman Wadi Harun[3]

[1] Laser Center, Ibnu Sina Institute for Scientific and Industrial Research (ISI-SIR), Universiti Teknologi Malaysia, 81310 Johor Bahru, Johor, Malaysia
[2] Department of Physics, Faculty of Science, Universiti Teknologi Malaysia, 81310 Johor Bahru, Johor, Malaysia.
[3] Department of Electrical Engineering, University of Malaya, Malaysia.

k.ganesan@utm.my

Abstract. pH-sensitive optical fiber sensors were fabricated with layer-by-layer (LbL) method. Multiple layers of poly(allylamine hydrochloride) (PAH) and poly(acrylic acid) (PAA) bilayers were coated on an unclad fiber to create the pH-sensitive region. The coating process was done using a coating system developed in this work. 10 to 40 bilayers of PAH/PAA were coated on the fiber core to evaluate its performance as pH sensor. The existences of coating on the fiber were verified with field emission scanning electron microscopy (FESEM) and energy-dispersive x-ray spectroscopy (EDX) analysis. It was shown that the fabricated sensor is pH sensitive and independent of variation in the refractive index of the surrounding. The best performance was obtained with 30 bilayers fiber sensor. The sensitivity and resolution of the sensor were 0.453 a.u/pH and 0.0022 pH units respectively. With the developed coating system, various physical and chemical sensors can be realized by varying the coating materials.

1. Introduction

Optical fiber sensor has undergone tremendous development in the past decades. Its advancement has attracted the attention of researchers due to its capabilities as a sensor and becomes an active research area due to its electromagnetic immunity, isolation to electrical parameter variation, ruggedness and remote sensing [1]. Physical and chemical parameters also could be detected using fiber sensors where it propagates the light from a remote source to a detector.

The development of optical fiber as pH sensors based on the immobilization of thin films and other waveguide materials has become an increasingly active area of research. One of the methods used for deposition of thin-film is layer-by-layer (LbL) due to the good uniformity and convenient. This method is based on the electrostatic attraction of the sequential adsorption of oppositely charged molecules [2]. For pH sensing, poly(allylamine hydrochloride) (PAH) and poly(acrylic acid) (PAA) layers were coated on an unclad fiber via LbL method in the previous study [1]. At the sensing region, an evanescent wave (attenuated total internal reflection) spectrum of PAH/PAA was used to monitor the effect of pH [1]. Most of the investigation focused on optical fiber sensor based on fluorescence and UV-Visible spectroscopy [1-4]. However, the recent publication reported the use of an infrared region for sensing and it shows a promising performance [5]. The work based on the generation of lossy mode resonance (LMR) using PAH/PAA polymeric pH-sensitive showing an average sensitivity of 0.027 pH units/nm with resolution varied from 0.00054 pH and 0.2268 pH units [5]. LbL technique was used to fabricate the sensors, yet the coating system was not discussed in detailed [5]. Therefore, in the current paper, the development of an automated coating system and the performance of the fabricated pH sensors are discussed.

2. Materials and methods

2.1. Materials

The polymers poly(allylamine hydrochloride) (PAH) (Mw ~ 19000) was obtained from Huichem and poly(acrylic acid) (PAA) (Mw ~ 1800) was obtained from Sigma Aldrich and used without further purification. Aqueous solutions of the polymers (10 mM) were prepared using distilled water and adjusted at pH 4.4. Potassium hydroxide (KOH) 1M and pH 3, 4, 5, 6 and 7 buffer solutions were obtained from Quirec. The pH solutions were verified using pH-meter by Hanna Instruments and a few drops of NaOH or HCl was added to adjust the pH solution. Optical fiber with 200/225 μm core/cladding diameter (FT200EMT) produce by Thorlabs was used.

2.2. Development of Coating System

The coating system was set up using Arduino driver, 3 stepper motor with a driver, power supply, breadboard and 3D printer axis as shown in figure 1. 2 stepper motor were set to control the vertical movement of a platform containing the coating materials and the stepper motors were placed parallel to each other. Another stepper motor was set to control the horizontal movement of the platform and the stepper motor was placed perpendicular to 2 stepper motors. All stepper motors were connected to its respective stepper motor drivers and the drivers were connected to a DC power supply as shown in figure 1. The power supply supplied a voltage of 4.2 V and a current of 0.84 A to the motors. An Arduino Uno microcontroller board was used to control the movement of the stepper motors. The movements were coded with Arduino IDE software according to the dipping time required during the coating process.

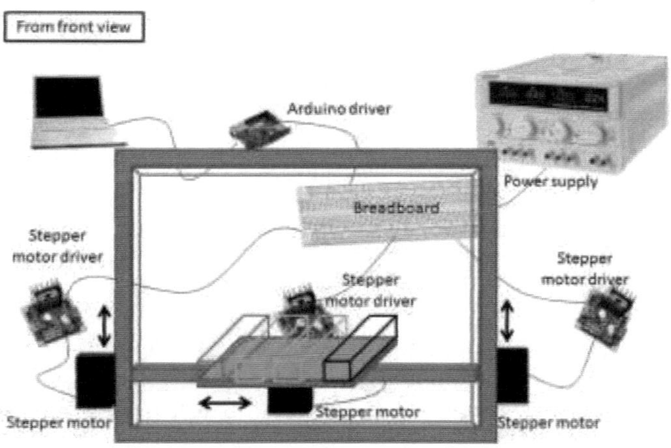

Figure 1. The schematic setup for the coating system.

2.3. Coating Process

The 4 cm long cladding was removed by heating the fiber and then followed by wiping and cleaning the remaining optical fiber with acetone. The optical fiber was attached to a holder as represented in figure 2.

The PAH/PAA coating was coated on the fiber core via LbL technique and the coating process was carried out using the developed coating system. Firstly, the unclad region was immersed in 1M of KOH solution for 10 minutes. This was to acquire the negative charge on the surface of fiber core and then was cleaned up in distilled water for 1 minute. Next, the unclad region was immersed into the 10 mM of PAH solution for 2 minutes to absorb the positive charged. This step was followed by rinsing in distilled water for 1 minute to remove the excess of the material adsorbed and dried for 30 seconds. Then, the unclad region was immersed into10mM of PAA solution for 2 minutes to absorb the negative charged material and was rinsed again in distilled water for 1 minute and dried for 30 seconds. As a result, a single bilayer of [PAH/PAA] coating had been coated on the core. This process was repeated until the desired number of bilayers coated on the core. Fiber sensors with 10 to 40 bilayers were fabricated in this work to evaluate its performance as a pH sensor.

2.4. Device Characterization

Figure 2 shows the setup for pH sensing using the fabricated fiber sensors system. One end of the optical fiber was connected to white halogen lamp (Avalight) in order to couple light into the optical fiber. The light propagates through the sensitive region located in the transmission path and the other end of the fiber was connected to a spectrometer (USB4000 Ocean Optics). The sensing region was attached to a holder as shown in figure 2. The absorbance response was recorded during the immersion of the sensing region into the pH buffer solution. The measurements were performed at room temperature (25°C).

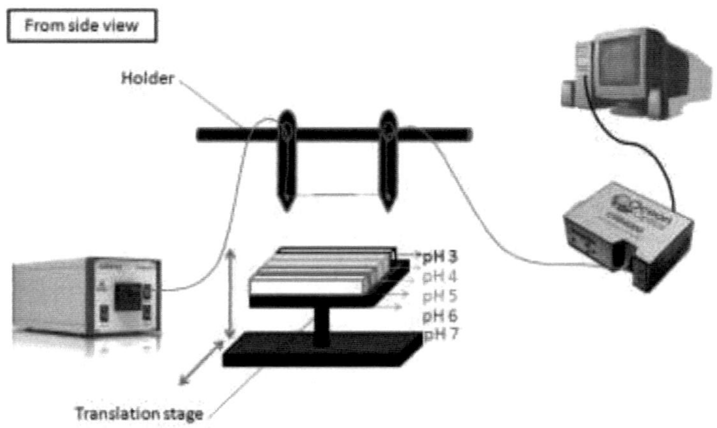

Figure 2. Schematic image of the experimental setup for device characterization.

3. Results and Discussion

3.1. Coating Characterization

Figure 3 shows the field emission scanning electron microscopy (FESEM) and energy-dispersive x-ray spectroscopy (EDX) results of the cross-section of the coated fiber. The result shows that PAH/ PAA layer was successfully coated on optical fiber core as shown in figure 3. From figure 3, it can be seen the core of fiber was 200.4 μm. Also, a layer of coating was seen on the circumference of the core. The thickness coating ranges between 201.0 nm and 240.1 nm for 25 bilayers. The thickness of the coating obtained in this study is comparable to the one obtained by Zamarenno et al. for 25 bilayers [5].

The EDX results display all the elements exist on the surface of the coated fiber. The highest percentage is gold (Au). This is due to the sputtered gold element, which is to avoid charging effect during image scanned. Other than that, there is no impurity element exist. The carbon (C) and oxygen (O) are the constituents of PAH/PAA with the molecular formula for PAH is $(C_3H_8Cl)n$ and PAA is $(C_3H_4O_2)n$. On the other hand, Si is an element in the fiber core which is glass made from silica.

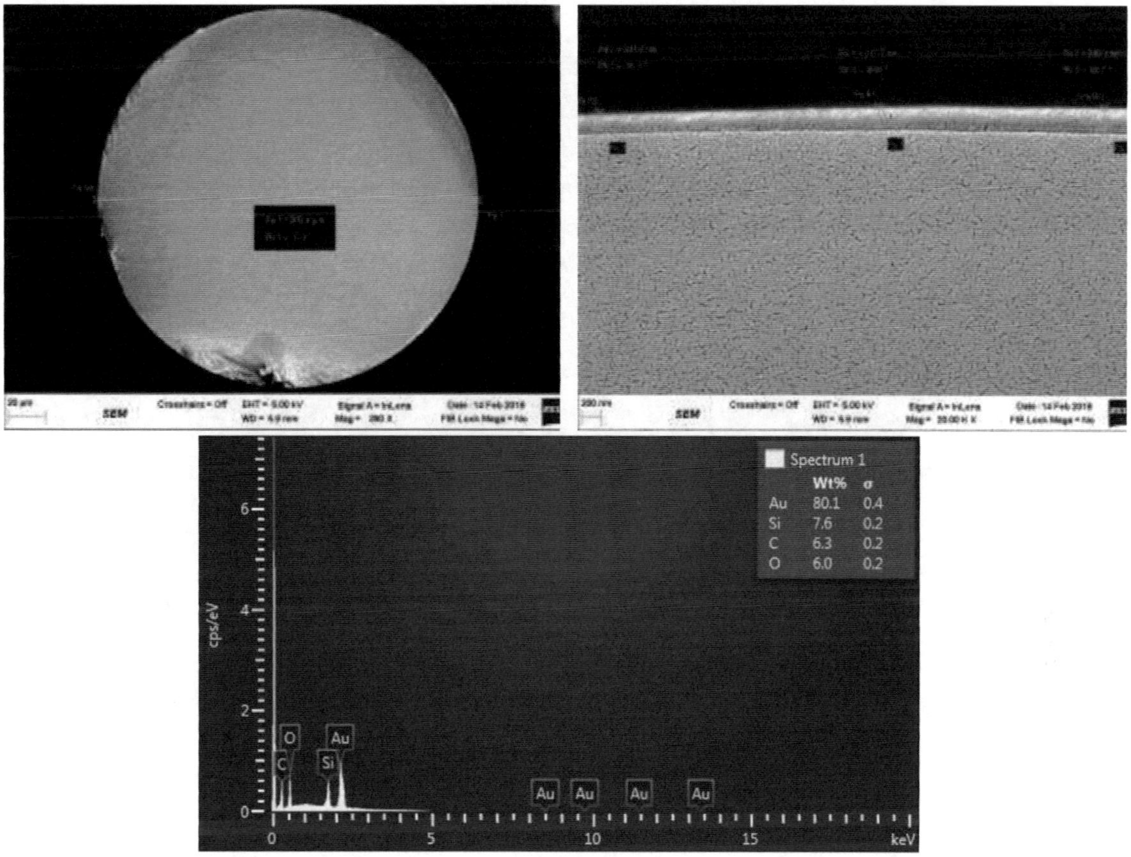

Figure 3. Cross-section and EDX analysis of optical fiber coated with 25 bilayers of PAH/PAA .

3.2. Device Fabrication Response

The fabricated device showed a spectral response when immersed in different pH buffer. Figure 4 shows the absorbance spectrum for 10 bilayers and 0 bilayer (uncoated fiber). It could be observed that for 0 bilayer the changes in absorbance value is negligible compared to 10 bilayers. For the 0 bilayer fiber, the change in absorbance is 0.001 a.u/pH was shown in figure 5. This change may cause by the variation in the refractive index of pH buffer solution and not due to the change in pH. The change in absorbance for 10 bilayers is 0.120 a.u/pH was showed in figure 5. It shows that the PAH/PAA coating layer on optical fiber core is pH sensitive and it is not an influence of the surrounding refractive index. The utilization of PAH/PAA as polyelectrolyte materials pair is one due to its intrinsic swelling pH behaviour of the resultant films. The structural changes in the bilayer form from PAH/PAA can be induced by changes in pH [6]. The changing in absorbance intensity is due to the swelling/deswelling behaviour of the coating that contributes to the thickness alteration during the immersion of fabricated fiber into different pH buffer.

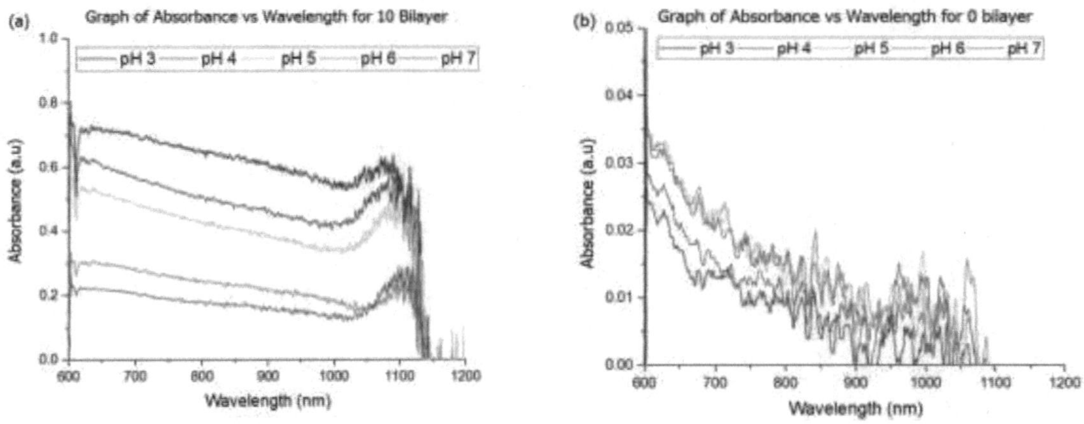

Figure 4. Absorbance spectrum of fabricated fiber (a) 10 bilayer and (b) 0 bilayer during immersed in different pH buffer respectively.

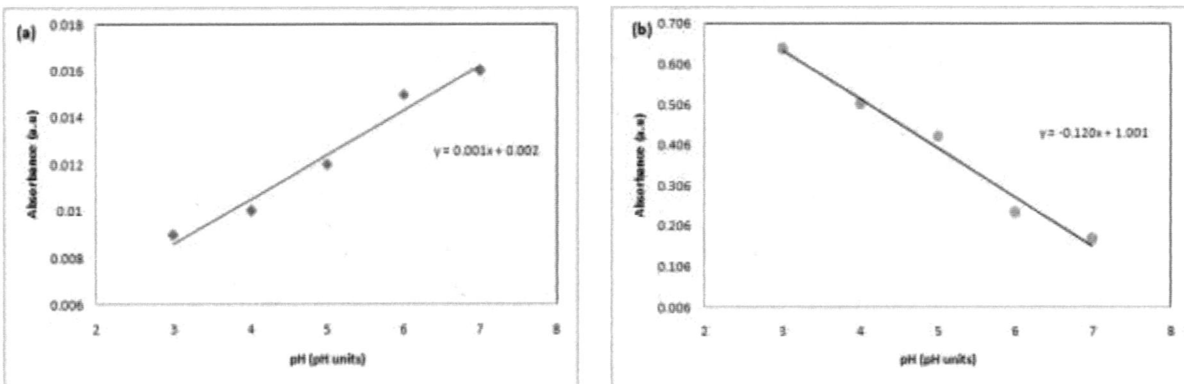

Figure 5. Absorbance value at 800 nm of (a) 0 bilayer and (b) 10 bilayers fiber in various pH buffer solutions.

The comparison of performance fiber sensor coated with different bilayers is listed in table 1. From the table, the best sensitivity and resolution of the fabricated device is showed by 30 bilayers of PAH/PAA coating with 0.453 a.u/pH for sensitivity and 0.0022 pH units for resolution. From 0 to 30 bilayers, the sensitivity value is increasing, however, the sensitivity value is decreased for 40 bilayers which is 0.058 a.u/pH. This result shows that the performance of sensors is coating thickness dependent and there exists an optimum coating thickness. The linearity of sensors is decreasing with the increase in the number of bilayers due to nonlinear swelling behaviour of the coating with pH.

Table 1. The list of performance comparison for each bilayer for pH sensor.

Bilayer	Linearity, (R^2)	Sensitivity, (a.u/pH)	Resolution, (pH units)
0	0.970	0.001	1
10	0.981	0.120	0.0083
20	0.968	0.241	0.0042
30	0.959	0.453	0.0022
40	0.953	0.058	0.0172

4. Conclusion

pH sensors based optical fiber were successfully fabricated with the developed coating system. In term of the number of bilayers, 30 bilayers have better sensitivity and resolution compared with other numbers of bilayers. The linearity of 30 bilayers can be improved in future study and other sensor performance parameters such as repeatability will be reported. With the development of this coating system, various physical and chemical sensors can be realized by varying the coating materials.

Acknowledgement

Authors are thankful to the Malaysian Ministry of Education and Universiti Teknologi Malaysia for the financial assistance through UTM-TDR 06G83. Also, the authors would like to extend their gratitude to the research management centre (RMC) of Universiti Teknologi Malaysia.

References

[1] Rivero, P. J., Goicoechea, J., & Arregui, F. J. (2019). Layer-by-Layer Nano-assembly: A Powerful Tool for Optical Fiber Sensing Applications. Sensors, 19(3), 683.

[2] Del Villar, I., Hernaez, M., Zamarreño, C. R., Sánchez, P., Fernández-Valdivielso, C., Arregui, F. J., & Matias, I. R. (2012). Design rules for lossy mode resonance based sensors. Applied optics, 51(19), 4298-4307.

[3] Zamarreño, C. R., Bravo, J., Goicoechea, J., Matias, I. R., & Arregui, F. J. (2007). Response time enhancement of pH sensing films by means of hydrophilic nanostructured coatings. Sensors and Actuators B: Chemical, 128(1), 138-144.

[4] Goicoechea, J., Zamarreño, C. R., Matias, I. R., & Arregui, F. J. (2007). Minimizing the photobleaching of self-assembled multilayers for sensor applications. Sensors and Actuators B: Chemical, 126(1), 41-47.

[5] Zamarreño, C. R., Hernáez, M., Del Villar, I., Matías, I. R., & Arregui, F. J. (2011). Optical fiber pH sensor based on lossy-mode resonances by means of thin polymeric coatings. Sensors and Actuators B: Chemical, 155(1), 290-297.

[6] Fery, A., Schöler, B., Cassagneau, T., & Caruso, F. (2001). Nanoporous thin films formed by salt-induced structural changes in multilayers of poly (acrylic acid) and poly (allylamine). Langmuir, 17(13), 3779-3783.

Development and preliminary data of integrated temperature-Insensitive lateral force sensor based on linear chirp Fiber Bragg Grating

Abdulfatah A G Abushagur, Ahmad Ashrif A Bakar and Norhana Arsad

Center of Advanced Electronic and Communication Engineering, Faculty of Engineering and Built Environment, Universiti Kebangsaan Malaysia, 43600 UKM Bangi, Selangor, Malaysia

abushagur@ukm.edu.my; a.abushagur@gmail.com

Abstract. Lateral force measurement using a cantilever ruler-beam (CRB) sensor based on linear chirp fiber Bragg grating (LCFBG) is proposed and experimentally demonstrated. The lateral force is determined by altering the reflection optical power of LCFBG of which an axial-strain gradient along its sensing is induced. The reflection spectrum intensity responds linearly and monotonically with lateral force increasing from 0 to ≈ 0.6 N. Experimental results demonstrate that the CRB force sensor can successfully achieve lateral forces prediction with less than 0.0015 N resolution using simply Thorlabs PM100USB optical power meter. And can accomplish an adequate accuracy with RMS error < 0.04N of full scale (0-0.749N). The CRB is chosen in one way or another to resemble steerable catheter flexibility as our intention is to developing catheter force sensor based on temperature-insensitive contact force measurement.

1. Introduction

Fiber Bragg gratings (FBG) has emerged widely in many fields especially in an application wherein the general strain gauges and encoders are not applicable due to the environment, cable networks complexity, and electromagnetic interference [1]. Features like flexibility, nontoxic, chemical inert, and instant response furthered the FBG-based sensor as the best candidate for several medicine disciplines [2].

Extensive studies have proposed FBG-based transducer in miniaturized medicine instruments to measure three-dimensional (3-D) forces in a variety of fields such as soft shapes/assisted-robotics sensors [3], [4], minimally invasive surgery as well as microsurgery [3]–[5]. High sensitivity wavelength shift method (WSM) which is commonly used to monitor the changes of the physical measurand, however, besides the influence of both strain and temperature on it, it is also an error maker when 3-axis force components measurement are desired. Axial strain along device-incorporated fiber houses the FBGs can be induced by both axial and lateral forces, leads to cross-talk error of the wavelength shift response in *z-direction* under exerted force in *x or y-direction.*

Xingchi He et.al. [3], [6] developed 3-D vitreoretinal microsurgery force sensor using 4-FBGs. Authors relied on mechanical means and nonlinear mathematical model to mitigate the WSM-induced cross-talk error between axial/lateral. An analogous presentation can be found in [7] for the development of 3-D catheter distal contact forces for cardiac ablation. Although the inner FBG was mounted in the neutral axis of the developed flexure of the catheter-tip, due to bending deformation under lateral forces yet can introduce significant cross-talk error to interfere with the axial sensing. Recently bandwidth of the reflection spectrum of the FBG is utilized to monitor force-induced strain instead. Bandwidth can only be tuned if nonuniform strain induced along with the fiber housing the FBG. Thus, temperature and forces in *z-direction* unlikely to be sensed by the bandwidth. Our previous work presented a novel method utilizing the bandwidth method and tapered FBG to eliminate WSM-induced axial/lateral cross-talk noise [8]. Temperature-insensitive force sensor for catheters using FBG is proposed in [9]. They have presented that the contact force applied on the catheter-tip can be determined directly by measuring optical power meter. This approach proves temperature-insensitive force measurement in which optical power meter is sensitive only to contact forces. However, the results show a nonlinear relationship between the reflection power and forces applied, authors, didn't provide a clear reason.

However, in this paper, we demonstrate a flexible-ruler beam incorporated linear chirp FBG (LCFBG) for lateral force measurement. The main contribution of this paper; (i) we show the linear relationship between forces applied and power measured, (ii) to the best of our knowledge, this the first time using LCFBG and bandwidth tuning, a real-time force measurement compared with actual forces is presented experimentally.

The paper is organized as follow; section 2 briefly introduces the bandwidth and corresponding total reflection power sensing principle, experimental setup clearly described in section 3, section 4 discuss the results obtained, and finally, work concluded in section 5.

2. Sensing Principle

When no force is applied, the original bandwidth of the LCFBG can be expressed as;

$$BW_{chirp} = \lambda_B(L) - \lambda_B(0) = 2n_{eff}[\Lambda(L) - \Lambda(0)] \tag{1}$$

Here $\lambda_B(L)$ and $\lambda_B(0)$ denotes the longer and shorter wavelengths of LCFBG sensor (see Figure 1(right)), the Λ is the grating period and n_{eff} is the effective refraction index.

2.1. Bandwidth Response

The LCFBG is a 10 mm long and is bonded as shown in Figure 1(right) such that the position of the longer grating period (longer wavelengths λ_L) is located proximal to the fixed end of the CRB, in contrary the shorter wavelengths λ_0 are located further up towards the free end. This arrangement theoretically leads to broadening in the bandwidth of the sensor. As the calibration weight is applied transversely, the CRB bends causing the top surface to stretch and as a result, a nonuniform strain induced along the CRB. The nonuniform strain changes with respect of the length of CRB starting from zero-strain at the free-end a and reaching its maximum value at the fixed-end b.

Consequently, when the CRB bends under the lateral forces a local grating period elongation occurred which could tune the bandwidth of the sensor, both resonance wavelengths denoted in Equation (1) shift accordingly and the difference between these shifts determine the fraction bandwidth variation; it can be expressed as follows;

$$\Delta\left(BW_{chirp}\right) = 2n_{eff}(1 + \wp_e).[\Delta\Lambda(L) - \Delta\Lambda(0)] \tag{2}$$

The expression above ignores the small differences between the elasto-optic effects $\wp_e$ at both ends of the sensor when compared to the pitch elongation. Here the elongation of each period can be found as follow;

$$\Delta\Lambda(0) = \Lambda(0).\varepsilon(0) \ and \ \Delta\Lambda(L) = \Lambda(L).\varepsilon(L) \tag{3}$$

Where $\varepsilon(0)$ and $\varepsilon(L)$ are the local strain at the starting and ending point of the LCFBG sensor. In our case the $\varepsilon(L) > \varepsilon(0)$ from which a positive or broaden bandwidth will be observed. From a cantilever beam theory, strain induced by lateral forces can be found as;

$$\varepsilon(0) = \frac{F*Z_0*R}{E*I}, \varepsilon(L) = \frac{F*(Z_0+L)*R}{E*I} \tag{4}$$

Here the Z_0 and $(Z_0 + L)$ are the distance from the free end to the beginning and ending points of the LCFBG sensor, $E \ and \ I$ are Young's module and area moment of inertia, respectively. While R denotes the thickness distance of the CRB from the neutral axis to the surface at which the LCFBG placed. Thus, the reading signal of the effective bandwidth change is calculated as the summation of the original bandwidth and the fraction variation which is tuned by forces applied as follow;

$$\Delta BW_{eff}(\varepsilon_{ave}) = BW_{chirp} + \Delta\left(BW_{chirp}\right)$$
$$= BW_{chirp} + \frac{2n_{eff}(1+\wp_e) * R * F}{E * I}.\left[\Lambda(L) * (Z_0 + L) - \Lambda(0) * Z_0\right] \quad (5)$$
$$= BW_{chirp} + k_\varepsilon * F$$

Here the coefficient k_ε defines the scale factor or responsivity ratio constant of the relationship between bandwidth response and forces applied.

2.2. Integral Power/Energy Response

The energy conservation implies that whenever the effective bandwidth is altered, the total area under the curve proportionally varies. If the bandwidth has broadened, it means more frequencies carry the energy will be in the signal. The total reflection power, therefore, would increase as a result, the expression below describes the variation in total reflected optical power calculation as a function of average strain;

$$\Delta P = \Delta BW_{eff}(\varepsilon_{ave}) * \rho_{LS} \quad (6)$$

Where the ρ_{LS} is the spectral density of the laser source in mW/nm.

3. Experimental Setup

The CRB force sensor is calibrated with an automated system using LabVIEW software. Figure 1 illustrates the actual calibration setup. A calibration weight (100 g) is attached to the tip of the CRB through a stiff wire. A very precise scale (1 mg resolution) is used to measure the lateral load magnitude applied which is converted later to a force magnitude (gravity magnitude is assumed 9.81 N/Kg) using the program. A 3-axis precision motorized stage is utilized to control the position of the calibration weight above the precision scale. Starting with the calibration weight is fully resting on the scale ($F = 0\ N$), later by rising up the CRB with respect to the scale, the portion of the weight applied on the RB's tip is determined. The LCFBG sensor response is sampled at 3 kHz rate using I-MON-256USB (Ibsen) interrogator.

Figure 1. Automated calibration setup system (left), Close-up view of the LCFBG position within the RB (right)

Figure 2 shows the schematic of the experiment setup including the 50/50 an optical splitter and a power meter (PM100USB-THORLAB).

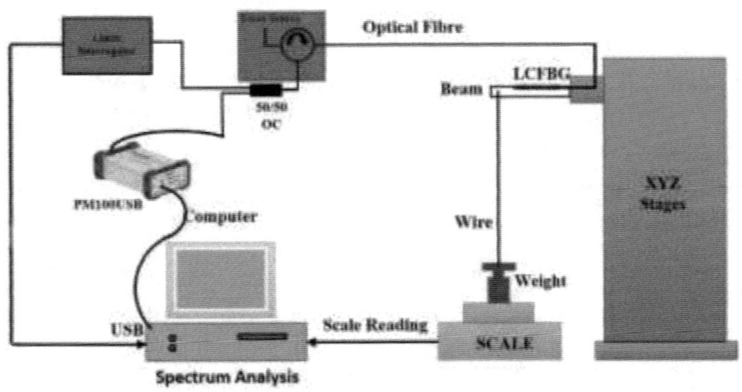

Figure 2 The schematic of the complete experiment setup to measure optical power as loads applied

The optical power meter is used to measure the real-time total reflection power of LCFBG as loads applied. By leveraging bandwidth interrogation method, the temperature variation will not change the bandwidth (while only shift the reflection spectrum), hence the total reflection power is only sensitive to the forces applied. A temperature controller, heat sink, and a relay are used to control the environmental temperature in the vicinity of the LCFBG sensor as shown in Figure 3. The LCFBG sensor is tested under different temperature values (25°C-35°C). National instrument DAQ is used to acquire the temperature values through a thermocouple attached directly on the RB.

The CRB sensor is calibrated twice, when force is increased by lifting up CRB or the calibration weight (Inc. force) and when force decreased by lowering down the calibration weight (Dec. force), both with 1 mm incremental. In the following sections, the calibration data is used to determine the relationship between total reflection power of the LCFBG and force magnitude applied. Further up, the calibration result is then used accordingly to map the sensor reading to the real-time lateral forces measurement.

Figure 3. The temperature control of the sensor' environment

4. Results and Discussion

Referring to Equation (1) and Equation (2), the non-uniform stretching of each grating pitch induced due to the lateral forces will govern the Bragg wavelength resonances $\lambda(z) = 2n_{eff}\Lambda(z)$. Bending downward the CRB with LCFBG bonded above the neutral axis of CRB leads to stretching every grating pitch according to the strain distribution. To this end, the wavelength shift to the longer wavelength can be observed. Figure 4 illustrates the spectrum change due to an incremental increase in the applied forces which has been observed during the calibration procedure by I-MON256USB interrogator. The x-axis of the graph indicates the pixel number of the Photodiode array detectors of the interrogator. It is worth mentioning that as the consequence of placement of sensor array design, the incident light of shorter wavelength impinges the higher pixel number.

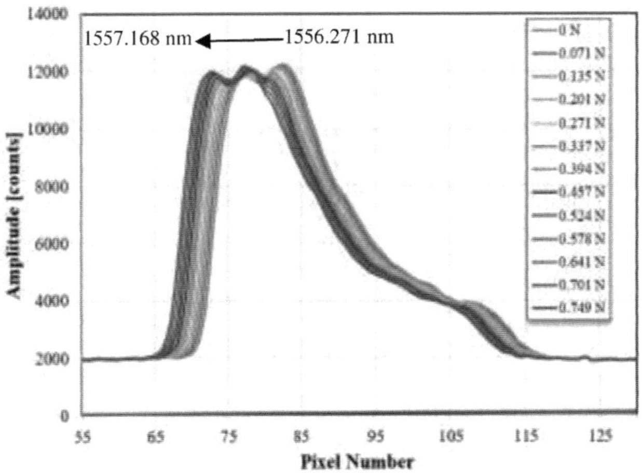

Figure 4. Measured change in the spectral profile of the LCFBG under applied forces

It is seen that a total center wavelength shift of 0.897 nm has been observed from 1556.271 nm to 1557.168 nm with an increase of applied forces range (0 N-0.749 N). As mentioned earlier, since the LCFBG is attached above the neutral axis, bending downward the CRB results in local grating-pitch elongation ($\Delta\Lambda(L) > \Delta\Lambda(0)$ which in turn results in broader bandwidth towards longer wavelength direction. This according to the Equation (6) enables the actual force applied to be encoded in the total reflection power.

As an unequal change in temperature within a short segment of the LCFBG is unlikely to occur, thus the main feature of the proposed CRB sensor is a temperature-independent strain measurement guaranteed. The temperature impact is investigated using a heat sink to which temperature controllable heat element is attached to change the environment vicinity of the sensor. Figure 5 depicts the response of the total reflection power of the sensor as the temperature varies from approximately 25 °C to 35 °C. As shown the total change in reflected optical power is smaller than 0.08dB. Even though this value is small enough to consider, however, we attribute this result to the uniformity of the applied adhesive and the accuracy of the power meter. Thus, the results confirm that the output of the system is the only function of the applied forces, independent of temperature.

Two calibration process is performed to guarantee the repeatability of our LCFBG sensor. Increased force denoting raising up the CRB incrementally while recording every step interval (vertically moving up 1 mm) of both forces applied and the total reflection power. The same thing is performed when lowering down the CRB which means decreasing the force. Figure 6 illustrates the linear relationship in both directions of the calibration process, the response of the reflected power of the LCFBG for both directions clearly agree with the mathematical model expressed in previous sections.

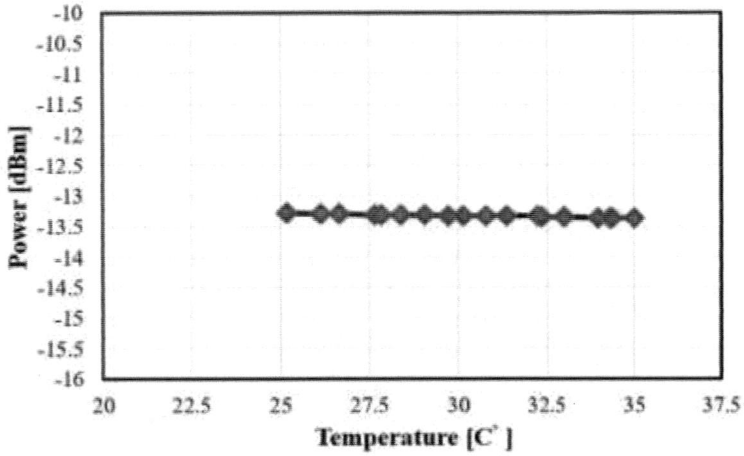

Figure 5. System output in response to the temperature at strain-free

It is obvious from the Figure 6, although the force change only ranges from (0N-to-0.749N), yet the change in reflection power is significant enough to be measured with a responsivity of 0.0785 dBm/N for increased force and -0.0825 dBm/N for decreased force. The slight inconsistency in the scale factor of both directions could be attributed to the optical power fluctuation, for which another power meter as a reference power measure capable of removing this noise. Another reason could also contribute to this issue is that the wire-ring loose at the beam tip-hole as the position of the ring was not firmly fixed and changes in its position were observed during the calibration process. The CRB sensor shows excellent temperature-insensitive as expected, in which the total reflection power varies purely only when forces applied.

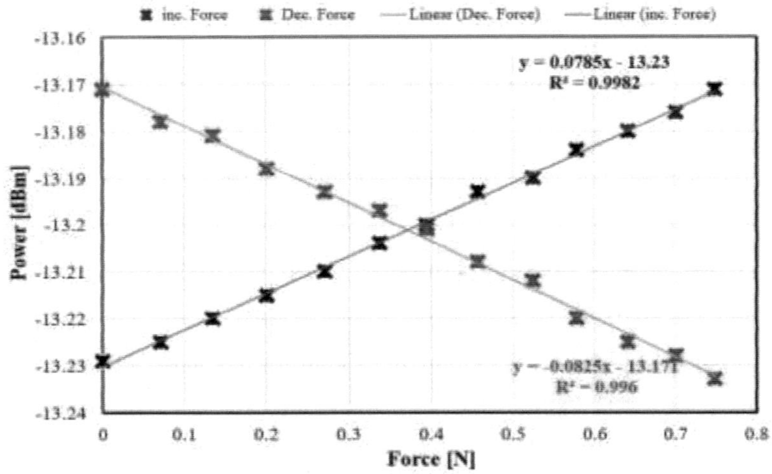

Figure 6. The total reflection power of the LCFBG sensor measured under the different magnitude of forces applied

The mapping due to the calibration results using linear fitting is illustrated in Figure 7. It shows the calculated lateral force against the actual lateral force in only one direction (one-degree-of-freedom 1-DOF).

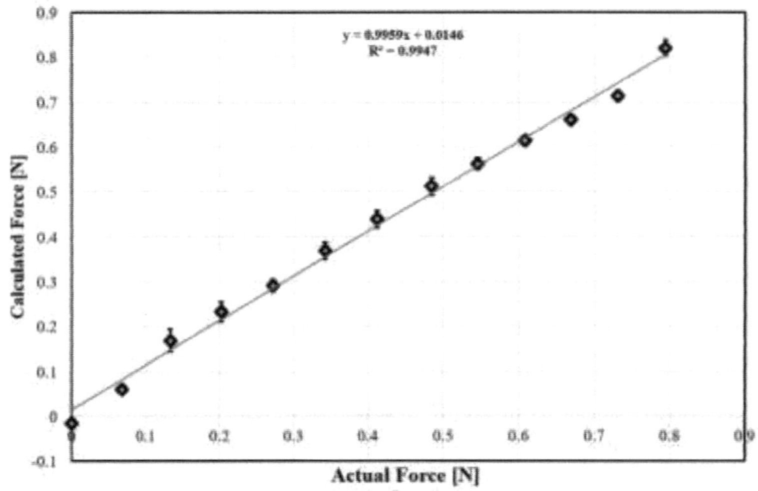

Figure 7. The calculated lateral force versus the actual force values using linear fitting

The perfect result of fitting would be a straight line through the origin with a slope of one, the calculated force values, however, are fairly consistent with their corresponding actual values. The slope is close to 1, a trivial error and deviation can be observed with (RMS) error of < 0.04N over the full scale of (0-0.749N) and the highest standard deviation of the results observed is 0.025 N.

5. Conclusion

Real-time force sensing is crucial in nowadays minimally invasive surgeries. To this end, a flexible ruler-beam with an embedded linear chirp FBG sensor to measure lateral forces is presented experimentally. Real time force monitoring is achieved directly using an optical power meter, rather than the slow, bulky optical spectrum analysis which is used in the past. The total reflection power from the LCFBG sensor has shown a linear relationship as desired against the applied lateral forces. The results show that the CRB sensor can provide lateral force measurement with RMS error less than 0.04N. The advantage of the proposed sensor is the temperature-independency, avoiding, as a result, the cross-talk noise of the conventional wavelength shift method. Furthermore, the proposed force sensor experimentally also shows accurate force measurement in real time with a slope of almost 1 when compared with the actual forces. Since the bandwidth of the LCFBG sensor exhibits accurately and linearly response with applied forces, a catheter incorporated LCFBG for 2-DOF contact force sensor mimicking ablation catheterization procedure will be considered as future work.

Acknowledgement

This work was supported by the Ministry of Energy, Science, Technology, Environment and Climate Change, under the grant scheme of Sciencefund (03-01-02-SF1317).

References

[1] A. Abushagur, N. Arsad, M. Reaz, and A. Bakar, "Advances in Bio-Tactile Sensors for Minimally Invasive Surgery Using the Fibre Bragg Grating Force Sensor Technique: A Survey," *Sensors*, vol. 14, no. 4, pp. 6633–6665, 2014.
[2] R. H. Morris and M. I. Newton, Eds., *Magnetic Resonance Sensors*. Sensors, 2014.
[3] X. He, P. Gehlbach, J. Handa, R. Taylor, and I. Iordachita, "Development of a miniaturized 3-DOF force sensing instrument for robotically assisted retinal microsurgery and preliminary results," in *5th IEEE RAS/EMBS International Conference on Biomedical Robotics and Biomechatronics*, 2014, pp. 252–258.
[4] D. Shin, H.-U. Kim, A. Kulkarni, and T. Kim, "Development and Evaluation of Tri-Axial Fiber

Bragg Grating Force Sensor for Catheter," *Proceedings*, vol. 1, no. 4, 2017.

[5] U. Kim, Y. B. Kim, J. So, D. Y. Seok, and H. R. Choi, "Sensorized surgical forceps for robotic-assisted minimally invasive surgery," *IEEE Trans. Ind. Electron.*, 2018.

[6] X. He, J. Handa, P. Gehlbach, R. Taylor, and I. Iordachita, "A submillimetric 3-DOF force sensing instrument with integrated fiber bragg grating for retinal microsurgery," *IEEE Trans. Biomed. Eng.*, 2014.

[7] T. Li, C. Shi, and H. Ren, "Three-Dimensional Catheter Distal Force Sensing for Cardiac Ablation Based on Fiber Bragg Grating," *IEEE/ASME Trans. Mechatronics*, vol. 23, no. 5, pp. 2316–2327, 2018.

[8] A. A. G. Abushagur, A. A. A. Bakar, M. S. Dzulkefly Bin Zan, and N. Arsad, "A Novel Technique Employing Tapered Fiber Bragg Grating to Solve the Axial/Transverse Forces Crosstalk in Microsurgical Instruments," *IEEE Sens. J.*, 2016.

[9] L. Xu, M. I. Miller, J. Ge, K. R. Nilsson, Z. T. H. Tse, and M. P. Fok, "Temperature-Insensitive Fiber-Optic Contact Force Sensor for Steerable Catheters," *IEEE Sens. J.*, vol. 16, no. 12, pp. 4771–4775, Jun. 2016.

Titanium dioxide-based picoseconds pulsed fiber laser performances comparison in the 1.5-micron region

M F A Rahman[1,2], P H Reddy[3,4], M C Paul[4], S Das[4], A Dhar[4], M F M Rusdi[2], A A Latiff[5], K Dimyati[2], and S W Harun[2]

[1] Fakulti Teknologi Kejuruteraan Elektrik & Elektronik, Universiti Teknikal Malaysia Melaka, 76100 Hang Tuah Jaya, Melaka, Malaysia.
[2] Photonics Engineering Laboratory, Department of Electrical Engineering, Faculty of Engineering, University of Malaya, 50603 Kuala Lumpur, Malaysia
[3] Academy of Scientific and Innovative Research (AcSIR), CSIR-CGCRI Campus, Kolkata, India.
[4] Fiber Optics and Photonics Division, CSIR-Central Glass and Ceramic Research Institute, 196, Raja S.C. Mullick Road, 700032, Kolkata, India.
[5] Faculty of Electronic and Computer Engineering, Universiti Teknikal Malaysia Melaka, 76100 Hang Tuah Jaya, Melaka, Malaysia.

mfauziar@utem.edu.my

Abstract. We demonstrated and compared picoseconds pulsed fiber lasers based on Titanium dioxide based saturable absorbers (SAs); 20 cm long Titanium dioxide-doped fiber (TiO_2DF) and Titanium dioxide PVA film (TiO_2PF) in the 1.5-micron region. The laser cavity utilized 2.4 m long Erbium-doped fiber (EDF) as the gain medium. A self-starting pulsed laser with a consistent repetition rate of ~1 MHz emerged stably with the incorporation of TiO_2 based SAs. The TiO_2DF SA produced 9.74 ps pulsed laser at a central wavelength of 1553 nm within a pump power range of 106-142 mW. The fiber SA promoted slightly higher slope efficiency and maximum pulse energy of 13.17% and 8.56 nJ, respectively in comparison with the film SA. On the other hand, the TiO_2PF SA generated stable 3.89 ps pulsed laser at an operating wavelength of 1560 nm within 86-142 mW pump power range. The film SA also produced slightly greater maximum output power of 12.17 mW and maximum peak power of 3.43 kW, respectively at the maximum pump power. The results confirmed that both TiO_2 SAs can be good alternative pulse modulator in the 1.5-micron region.

Keywords. Optical fiber laser, fiber saturable absorber, Titanium dioxide, thin film

1. Introduction

Passive pulsed fiber lasers have several advantages in comparison with the active one, which includes, simplicity, flexibility and compactness. They also have significant potentials for use in many applications, such as telecommunication, biomedical and material engraving. Numerous methods have been claimed able to facilitate passively pulsed fiber lasers such as nonlinear polarization rotation [1], semiconductor saturable mirror (SESAMs) [2], carbon nanotubes (CNT) [3], graphene [4, 5], transition metal dichalcogenides (TMD) [6] and several others [7-26] including those which have been reported by our research group lately.

Although SESAMs have been successfully developed for commercial use in the last past two decades, they suffer from a relatively narrowband operation and has to be designed for explicit wavelength. In addition to that, they are also quite bulky for fiber lasers and have a complex fabrication. NPR based lasers are environmentally unstable and have to be securely located in limited movements surroundings. On the other hand, CNT requires quite a complex bandgap engineering or diameter control for obtaining broader operation bandwidth. Additionally, graphene has a small absorption of 2.3% at 1550 nm. Thus, in this regard, new SA material that is small, easy to integrate, possessing quite a fast recovery time and can perform well in a robust environment is highly desirable.

Titanium dioxide (TiO_2), that belongs to transition metal oxides (TMO) have seized attention in recent years as alternative SAs in several regions; 1 [27], 1.55- [28, 29] and 2-micron [30]. It has a recovery time of ~1.5 ps at room temperature [31], while Z scanning performed on anatase and rutile TiO_2 proved that it has saturable absorption feature. The modulation depth (35.41%) and the saturation intensity (0.013%) of TiO_2 [29] also are comparable to the others SAs [6]. Although it has a bandgap of ~3.2 eV (~387 nm), it was claimed able to operate in the visible [32] and infrared [33] regions.

In this report, we demonstrated and compared TiO_2DF SA and TiO_2PF SA based picoseconds lasers performances in the EDFL cavity. The TiO_2PF was fabricated in-house, using polyvinyl alcohol (PVA) as the polymer film. On the other hand, the TiO_2DF was fabricated by doping TiO_2 element into silicate glass fiber core. Experimentally, we obtained about the same value of pulse repetition rates (~1 MHz) for both TiO_2 based SAs. The fiber laser performances of these TiO_2 based SAs are summarized and elaborated in the following topics to come.

2. Fabrication and Characterization of the Titanium dioxide-based SA

2.1. TiO_2DF SA (fiber) fabrication

The TiO_2DF was fabricated by doping TiO_2 element into silica glass matrix via a conventional modified chemical vapor deposition (MCVD) through bubbling of Helium (He) and Oxygen (O_2) gases into Titanium tetrachloride ($TiCl_4$) and Silicon tetrachloride ($SiCl_4$) containing bubbles. The developed TiO_2DF has core diameter and numerical aperture (NA) of ~40 μm and 0.21, respectively. The fabricated fiber also contains 4.0 mol % doping levels of TiO_2. To effectively works as fiber SA, the TiO_2DF was cut into 20 cm length. Both of its arms were then fusion spliced with 10 cm long single-mode fiber (SMF28), to form 10 cm SMF28 – 20 cm TiO_2DF – 10 cm SMF28. However, in this report, the whole piece is regarded as TiO_2DF SA only. The TiO_2DF fabrication process in details, along with its important fiber characteristics can be found in [13].

2.2. TiO_2PF SA (film) fabrication

The TiO_2 powder used in the film SA was purchased from Sigma Aldrich. The powder is 99 % pure and has a diameter of lower than 45 μm. First, the host solution was prepared by dissolving 1 g of polyvinyl alcohol (PVA) into 120 ml of deionized (DI) water. Then, the host mixture was stirred using a magnetic stirrer at 90°C temperature, until the PVA powder fully dispersed. The solution was then cooled down to room temperature. Additionally, TiO_2 solution was prepared by dissolving TiO_2 powder into DI water with the help of 1% sodium dodecyl sulphate (SDS) solvent. The mixture was stirred for 5 minutes and centrifuged at 3000 rpm for ~15 minutes, to yield supernatant containing TiO_2 suspension. The TiO_2 suspension was collected and then poured onto the prepared PVA host solution. Next, the mixed solution was centrifuged to yield a composite precursor solution. The precursor solution was then transferred onto a flat and clean petri dish and left for two days at room temperature, to form TiO_2 composite film SA with a thickness of ~30 μm. The TiO_2 film fabrication details and its important characteristics are explained in [30, 34]. To be effectively served as an effective SA, the composite film (TiO_2PF SA) was cut into a square shape of 1 mm x 1 mm and sandwiched between two clean fiber ferrules with the aid of an index matching gel. Then, a clean fiber adapter was carefully fixed at the coupling spot. The whole piece was then inserted into the EDFL cavity.

2.3. Linear absorption profile of the TiO_2 based SAs.

Figure 1 depicts linear absorption profiles of the TiO_2PF SA and the TiO_2DF SA taken at their respective mode-locked operating regime. The measurements were conducted by coupling the SA between a broadband white light source (WLS) and an optical spectrum analyzer (OSA). The net reduction of the injected WLS intensity is regarded as the linear absorption of the tested SA. As shown, the TiO_2PF SA shows relatively more consistent absorption profile than the TiO_2DF SA, due to the highly uniform

Journal of Physics: Conference Series **1371** (2019) 012023 doi:10.1088/1742-6596/1371/1/012023

structure obtained in the thin film. The TiO_2PF SA has slightly lower absorption of 1.55 dB at 1560 nm wavelength as compared to the TiO_2DF SA (that has an absorption of 2.95 dB at 1553 nm wavelength).

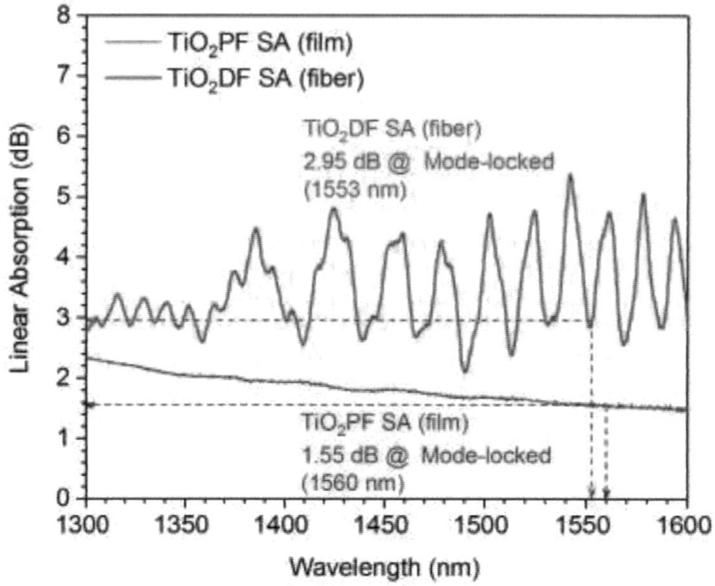

Figure 1. Linear absorption profile of TiO_2 based SAs.

3. Picoseconds pulsed fiber laser cavity design

The proposed design of the mode-locked based TiO_2 element is shown in Figure 2. The laser cavity is driven by a 980 nm laser diode pump. The light enters into a wavelength division multiplexer (WDM) and then into 2.4 meters long Erbium-doped fiber (EDF) gain medium. An isolator ensures unidirectional light propagation in the cavity. The EDF is a commercial glass fiber (IsoGain, I-25) from Fibercore and has core diameter and NA of 4 μm and 0.23, respectively. In addition to that, the EDF has core concentricity of less than 0.5 μm and absorption of 23-27 dB/m at 980 nm wavelength. As shown, 3 dB coupler 1 is used to separate the laser output into equal halves. Half of the laser output is then channeled into 3 dB coupler 2, while the remaining half, is left to be looped inside in the cavity. The 3 dB coupler 2 divides the light into further halves, providing real-time light measurements at two different ports. The TiO_2 based SAs, acts as mode-lockers; exploiting its linear and non-linear absorption characteristic to generate stable picoseconds pulsed lasers. A commercial single-mode fiber (SMF28) with a length of 195 nm, is appended before the SA to provide sufficient intracavity nonlinear and dispersion in the cavity. At the same time, it also reduces the pulse repetition rate and improves the pulse energy. This improved pulse energy would be sufficient enough to saturate the SA for a reliable self-starting mode-locked operation. A digital oscilloscope (OSC) and a radio frequency spectrum analyzer (RFSA) with a pre-connected photo-detector were employed to measure the pulse in the time and frequency domain, respectively, while an OSA records the output laser spectrum. The output power is measured via an optical power meter.

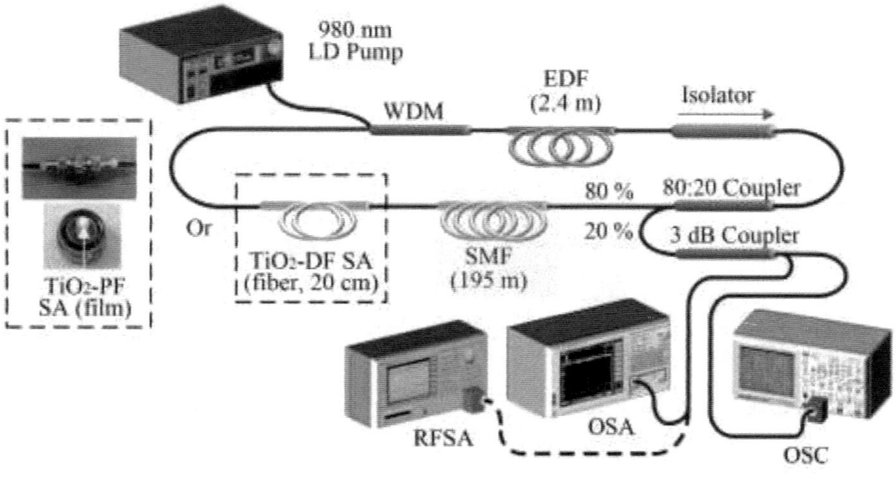

Figure 2. Mode-locked EDFL based on TiO_2 based SA cavity design.

4. Picoseconds pulsed fiber laser performances

The mode-locked EDFL based on TiO_2DF SA self-started at 106 mW threshold pump power and destructed into continuous laser (CW) laser when the pump power goes beyond 142 mW. Figure 3(a) depicts the mode-locked laser optical spectrum with a central wavelength of 1553 nm. The 3 dB spectral bandwidth (3 dB SB) is found to be 0.269 nm (33.58 GHz). The pulsed laser has a pulse repetition rate of 0.984 MHz. The autocorrelator trace as depicted in Figure 3(b), shows a single enveloped pulse with a pulse duration of ~9.74 ps (using $sech^2$ fitting analysis). Based on the obtained 3 dB SB of 33.58 GHz, the time-bandwidth product (TBP) is measured to be 0.33 which is slightly higher than the transform limit value of 0.315. This phenomenon indicates that the pulse is marginally chirped.

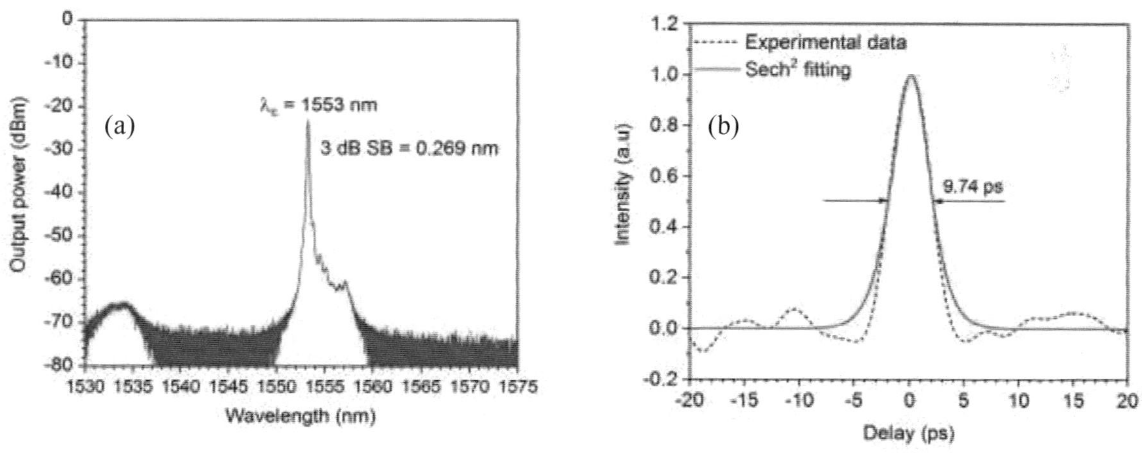

Figure 3. Mode-locked EDFL based on TiO_2DF SA (a) optical spectrum (b) autocorrelator trace

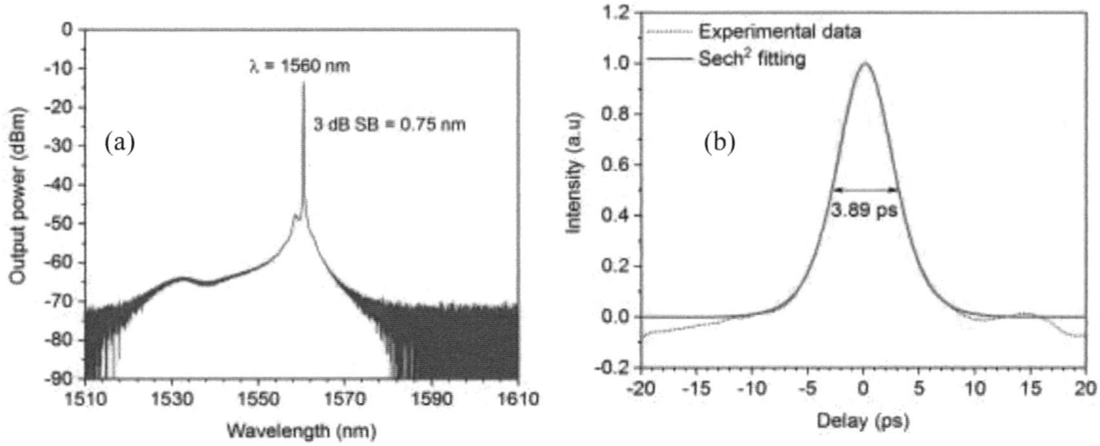

Figure 4. Mode-locked EDFL based on TiO_2PF SA (a) optical spectrum (b) autocorrelator trace

On the other hand, mode-locked EDFL based on TiO_2PF SA self-emerged at a threshold pump power of 86 mW and remained stable until 142 mW. Figure 4(a) illustrates the generated pulsed laser output spectrum, indicating a central wavelength of 1560 nm. As seen, the 3 dB SB is measured as 0.75 nm (72.57 GHz). The pulsed laser has a pulse repetition rate of 1.01 MHz. The autocorrelator trace as provided in Figure 4(b), depicts a pulse duration of ~3.89 ps (based on the $sech^2$ fitting analysis). By considering the 3 dB SB as 72.57 GHz, the time-bandwidth product (TBP) is calculated to be 0.36, slightly greater than the transform limit of 0.315, which also indicates that the pulse is chirped.

The TiO_2DF SA based pulsed laser performances as a function of pump power (106-142 mW) is demonstrated in Figure 5(a). As depicted, the output power increases linearly within a range of 3.73-8.42 mW. The pulse energy and the peak power also ascend with the increase of pump power. The maximum pulse energy and peak power are measured to be 8.56 nJ and 0.88 kW, respectively. The calculated slope efficiency is found to be 13.17 %.

On the other hand, the TiO_2PF SA based pulsed laser performances against increasing pump power (86-142 mW) is demonstrated in Figure 5(b). As illustrated, the output power ascends linearly from 7.58 mW to 12.17 mW. The pulse energy and the peak power also increases almost linearly with the rise of pump power. The maximum pulse energy and peak power are found to be 4.21 nJ and 3.43 kW, respectively. The obtained slope efficiency is measured to be 8.17 %.

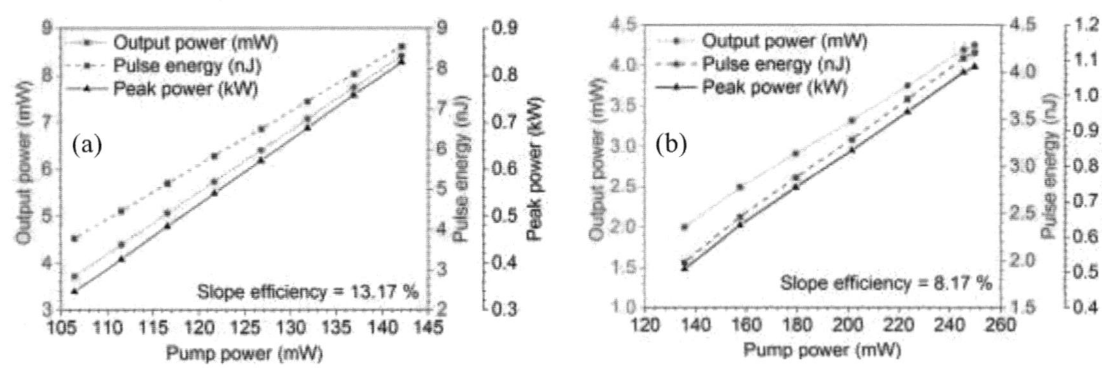

Figure 5. Pulsed laser performances based on (a) TiO_2DF SA (b) TiO_2PF SA

Table 1 summarizes and compares the TiO_2DF SA and TiO_2PF SA based pulsed laser performances and characteristics. Both TiO_2 based SAs, have about the same repetition rate of ~1 MHz, due to the same cavity length (~204 m) used in the design. The generated pulse repetition rates via these TiO_2 SAs are examined to be associated with a single cavity round trip period, suggesting that the generated pulsed lasers are mode-locked lasers. As demonstrated, the TiO_2PF SA generates higher output power and peak power as compared to the TiO_2DF SA. Conversely, the TiO_2DF SA generates better slope efficiency and higher pulse energy. Higher pulse energy generated via the film SA is attributed to a slightly broader pulse width obtained in the optical spectrum. In summary, both SAs produces almost comparable laser performances. However, in the aspect of durability, the TiO_2DF was observed to have better pulse stability when tested under several hours of laser operation at the maximum pump power operation [13].

Table 1. Mode-locked EDFL performances comparison between TiO_2DF SA (fiber) and TiO_2PF SA (film).

Optical quantity	Unit	TiO_2DF SA (fiber)	TiO_2PF SA (film)
Central wavelength	nm	1553	1560
Pump power	mW	106-142	86-142
Output power	mW	3.73-8.42	7.58-12.17
Peak power	kW	0.39-0.88	2.14-3.43
Slope efficiency	%	13.17	8.17
Pulse energy	nJ	3.79-8.56	1.98-4.21
Repetition rate	MHz	0.984	1.01
Pulse width	ps	9.74	3.89
SNR (FF)	dB	54	67
Linear absorption	dB	1.55	2.95

5. Conclusion

We have demonstrated and compared picoseconds pulsed fiber laser at 1.5-micron by utilizing TiO_2 based SAs; fiber and film. Both pulsed fiber lasers have almost the same repetition rate of ~1MHz. The TiO_2DF SA generated 9.74 ps pulsed laser at 1553 nm wavelength. In addition, it generated higher slope efficiency and maximum pulse energy of 13.15% and 8.56 nJ, respectively, in comparison to the TiO_2PF SA. On the other hand, the TiO_2PF SA produced 3.89 ps pulsed laser at 1560 wavelength. The TiO_2 film based pulsed laser had relatively higher maximum output power and maximum peak power of 12.17 mW and 3.43 kW, respectively. In overall, the TiO_2DF provides advantages in terms of thermal damage threshold, flexibility and durability. These features are important for a portable laser source design.

Acknowledgement

The authors would like to acknowledge Universiti Teknikal Malaysia Melaka (JURNAL/2019/FTKEE/Q00029) for the financial support.

References

[1] Feng X, Tam H-y and Wai P 2006 Stable and uniform multiwavelength erbium-doped fiber laser using nonlinear polarization rotation *Optics express* **14** 8205-10
[2] Hakulinen T and Okhotnikov O G 2007 8 ns fiber laser Q switched by the resonant saturable absorber mirror *Optics letters* **32** 2677-9
[3] Schibli T, Minoshima K, Kataura H, Itoga E, Minami N, Kazaoui S, Miyashita K, Tokumoto M and Sakakibara Y 2005 Ultrashort pulse-generation by saturable absorber mirrors based on polymer-embedded carbon nanotubes *Optics Express* **13** 8025-31
[4] Martinez A and Sun Z 2013 Nanotube and graphene saturable absorbers for fibre lasers *Nature Photonics* **7** 842
[5] Bao Q, Zhang H, Wang Y, Ni Z, Yan Y, Shen Z X, Loh K P and Tang D Y 2009 Atomic-layer graphene as a saturable absorber for ultrafast pulsed lasers *Advanced Functional Materials* **19** 3077-83
[6] Chen B, Zhang X, Wu K, Wang H, Wang J and Chen J 2015 Q-switched fiber laser based on transition metal dichalcogenides MoS 2, MoSe 2, WS 2, and WSe 2 *Optics express* **23** 26723-37
[7] Rahman M, Latiff A, Rusdi M, Dimyati K and Harun S 2018 Passively Q-switched fibre laser utilizing erbium-doped fibre saturable absorber for operation in C-band region *Journal of*

Modern Optics 1-5

[8] Tey K, Rahman M, Rosol A, Rusdi M, Mahyuddin M, Harun S and Latiff A 2017 Q-switched double-clad Ytterbium-doped fiber laser using MoS2 flakes saturable absorber. In: *IOP Conference Series: Materials Science and Engineering*: IOP Publishing) p 012054

[9] Rahman M, Rusdi M, Harun S and Yasin M 2018 Q-Switched Thulium-Doped Fiber Laser Using Vanadium Oxide Saturable Absorber *Nonlinear Optics, Quantum Optics: Concepts in Modern Optics* **48**

[10] Rahman M F, Latiff A A, Rusdi M F, Dimyati K and Harun S W 2018 Q-switched ytterbium-doped fiber laser via a thulium-doped fiber saturable absorber *Applied Optics* **57** 6510-5

[11] Ab Rahman M F, Latiff A A, Rosol A H A, Dimyati K, Wang P and Harun S W 2018 Ultrashort Pulse Soliton Fiber Laser Generation With Integration of Antimony Film Saturable Absorber *Journal of Lightwave Technology* **36** 3522-7

[12] Rahman M, Latiff A, Zaidi U, Rusdi M, Rosol A, Bushroa A, Dimyati K and Harun S 2018 Q-switched and mode-locked thulium-doped fiber laser with pure Antimony film Saturable absorber *Optics Communications* **421** 99-104

[13] Reddy P H, Rahman M, Paul M, Latiff A, Rosol A, Das S, Dhar A, Bhadra S K, Dimyati K and Harun S 2018 Titanium dioxide doped fiber as a new saturable absorber for generating mode-locked erbium doped fiber laser *Optik* **158** 1327-33

[14] Rahman M, Zhalilah M, Latiff A, Rosol A, Lokman M, Bushroa A, Dimyati K and Harun S 2018 Pure antimony film as saturable absorber for Q-switched erbium-doped fiber laser *Journal of Modern Optics* **65** 811-7

[15] Rahman M, Dhar A, Das S, Dutta D, Paul M, Rusdi M, Latiff A, Dimyati K and Harun S 2018 An 8 cm long holmium-doped fiber saturable absorber for Q-switched fiber laser generation at 2-μm region *Optical Fiber Technology* **43** 67-71

[16] Rahman M, Rusdi M, Lokman M, Mahyuddin M, Latiff A, Rosol A, Dimyati K and Harun S 2017 Holmium Oxide Film as a Saturable Absorber for 2 μm Q-Switched Fiber Laser *Chinese Physics Letters* **34** 054201

[17] Rusdi M, Latiff A, Ahmad M, Rahman M, Rosol A, Harun S and Ahmad H 2019 Nickel Oxide as a Q-switcher for Short Pulsed Thulium Doped Fiber Laser Generation. In: *Journal of Physics: Conference Series*: IOP Publishing) p 012029

[18] Rahman M, Rusdi M, Muhammad A, Latiff A, Ahmad A, Dimyati K and Harun S 2019 Ytterbium doped fiber saturable absorber for a stable passively Q-switched fiber laser in the 1-micron region. In: *Journal of Physics: Conference Series*: IOP Publishing) p 012008

[19] Rahman M, Rusdi M, Latiff A, Hisyam M, Dimyati K and Harun S 2019 Passively Q-switched Erbium doped fiber laser by incorporating a segment of Thulium doped fiber saturable absorber. In: *Journal of Physics: Conference Series*: IOP Publishing) p 012010

[20] Rahman M, Latiff A, Reddy P, Das S, Dhar A, Paul M and Harun S 2019 Passively Q-switched fiber laser utilizing new hafnium–bismuth–erbium co-doped fiber as saturable absorber *Indian Journal of Physics* 1-5

[21] Rahman M, Reddy P, Paul M, Das S, Dhar A, Baharom M, Latiff A, Rusdi M, Wang P and Dimyati K 2019 Titanium dioxide fiber saturable absorber for Q-switched fiber laser generation in the 1-micrometer region *Applied optics* **58** 3495-500

[22] Baharom M, Rahman M, Latiff A, Wang P and Harun S 2019 Lutetium (III) oxide film as passive mode locker device for erbium-doped fiber laser cavity *Optics Communications*

[23] Rusdi M, Rosol A, Rahman M, Mahyuddin M, Latiff A, Ahmad H, Harun S and Yasin M 2019 Q-switched and mode-locked thulium doped fiber lasers with nickel oxide film saturable absorber *Optics Communications* **447** 6-12

[24] Rahman M F A, Latiff A A, Rusdi M F M, Dimyati K and Harun S W 2019 Passively Q-switched fibre laser utilizing erbium-doped fibre saturable absorber for operation in C-band region *Journal of Modern Optics* **66** 235-9

[25] Baharom M, Rahman M, Latiff A, Wang P, Arof H and Harun S 2019 Lutetium oxide film as a passive saturable absorber for generating Q-switched fiber laser at 1570 nm wavelength *Optical Fiber Technology* **50** 82-6

[26] Rahman M, Mahyuddin M, Latiff A, Paul M, Dhar A, Das S, Yupapin P, Yasin M and Harun S 2019 Holmium based nanoseconds pulsed fibre laser generation in the 2-micron region *Optik*

163157

[27] Ahmad H, Salim M, Ali Z, Ismail M F, Thambiratnam K, Latif A, Nayan N and Harun S W 2016 Titanium dioxide-based Q-switched dual wavelength in the 1 micron region *Chinese Optics Letters* **14** 091403

[28] Reddy P H, Rahman M, Paul M, Latiff A, Rosol A, Das S, Dhar A, Bhadra S K, Dimyati K and Harun S 2018 Titanium Dioxide Doped Fiber as a New Saturable Absorber for Generating Mode-Locked Erbium Doped Fiber Laser *Optik-International Journal for Light and Electron Optics*

[29] Ahmad H, Reduan S, Ali Z, Ismail M A, Ruslan N, Lee C, Puteh R and Harun S W 2015 C-Band Q-Switched Fiber Laser Using Titanium Dioxide (TiO_2) As Saturable Absorber *IEEE Photonics Journal* **8** 1-7

[30] Rusdi M F M, Latiff A A, Paul M C, Das S, Dhar A, Ahmad H and Harun S W 2017 Titanium Dioxide (TiO_2) film as a new saturable absorber for generating mode-locked Thulium-Holmium doped all-fiber laser *Optics & Laser Technology* **89** 16-20

[31] Elim H, Ji W, Yuwono A, Xue J and Wang J 2003 Ultrafast optical nonlinearity in poly (methylmethacrylate)-TiO 2 nanocomposites *Applied physics letters* **82** 2691-3

[32] Kuznetsov V N and Serpone N 2006 Visible light absorption by various titanium dioxide specimens *The Journal of Physical Chemistry B* **110** 25203-9

[33] Luxon J T and Summitt R 1969 Interpretation of the infrared absorption spectra of stannic oxide and titanium dioxide (rutile) powders *The Journal of Chemical Physics* **50** 1366-70

[34] Ahmad H, Salim M, Ali Z, Ismail M, Thambiratnam K, Latif A, Nayan N and Harun S 2016 Titanium dioxide-based Q-switched dual wavelength in the 1 micron region *Chinese Optics Letters* **14** 091403

Journal of Physics: Conference Series **1371** (2019) 012024 doi:10.1088/1742-6596/1371/1/012024

Q-switched Thulium-doped fiber laser with Bismuth-doped fiber saturable absorber

B Musa[1], A A Latiff [2], M F M Rusdi[3], M F A Rahman [2], N F Zulkipli[3], F S M Samsamnun[3], A A A Jafry[3], S W Harun[3]

[1]Faculty of Electrical Engineering, University Teknologi Mara (Terengganu), 23000 Dungun, Terengganu, Malaysia
[2]Centre for Telecommunication Research & Innovation (CeTRI), Universiti Teknikal Malaysia Melaka (UTeM), 76100 Hang Tuah Jaya, Melaka, Malaysia
[3]Photonics Engineering Laboratory, Department of Electrical Engineering, Faculty of Engineering, University of Malaya, 50603 Kuala Lumpur, Malaysia

anasabdullatiff@utem.edu.my

Abstract. We demonstrate a passive Q-switched all-fiber ring configuration in thulium-doped fiber laser (TDFL) cavity using a bismuth-doped fiber (BDF) as a saturable absorber (SA). A piece of 20 cm long BDF was spliced and incorporated in the laser cavity for Q-switching pulse generation. By raising the pump power from 454 mW to 564 mW, the repetition rate of the laser is tunable within a range from 18.10 kHz to 25.22 kHz while the pulse width reduces from 9.04 µs to 6.98 µs. At maximum pump power, the Q-switched TDFL has output power, a slope efficiency, pulse energy and peak power of 5 mW, 3.53 %, 0.2 µJ and 28.40 mW, respectively.

1. Introduction

Q-switched fiber lasers have attracted more attention in recent years, as they provide more advantages such as reliability, simplicity and more economic, reliability and zero maintenance as compared to the conventional bulk lasers [1, 2]. The Q-switched lasers are fascinating in various applications such as material processing, telecommunications, remote sensing and medical surgery [3, 4]. Various saturable absorber (SA) have been successfully demonstrated including semiconductor saturable absorber mirrors (SESAMs), carbon nanotubes (CNTs), black phosphorus (BP), topological insulators (TIs), transition metal dichalcogenides (TMDs) and graphene, including a few more others [5-13] for passively generating Q-switched pulses. Graphene (classified as 2-dimensional (2D) materials) has made outstanding achievement in many fields especially in photonics, because of their excellent characteristics such as nearly zero energy bandgap, fast recovery time and cover broadband operation [14, 15]. With all these advantages, graphene is reported to have low saturable absorption per layer which is about 2.3% per layer [10]. Since that, many new materials are proposed such as Molybdenum disulphide (MoS_2) [16], titanium dioxide (TiO_2) [4, 17], Bismuth (III) telluride (Bi_2Te_3) [18], Lutetium (III) oxide [19, 20] and nickel oxide [10] for SA applications .

However, all-fiber configuration set-up is preferable since it can provide simpler and much easier handling, nearly zero maintenance and is capable to generate a much stable pulse generation. Rahman et al. reported a stable Q-switched Thulium-doped fiber laser (TDFL) based on all-fiber set-up using an 8 cm long holmium-doped fiber as SA. The TDFL provides the maximum pulse repetition rate and energy of 38.89 kHz and 129 nJ, respectively [21]. In this paper, we demonstrate an all-fiber Q-switched pulsed fiber laser generation by using Bismuth-doped fiber (BDF) as SA.

2. Experimental Setup

The experimental set-up of the proposed Q-switched TDFL is shown in Figure 1. It consists of a 1550 nm Erbium-Ytterbium doped fiber laser (EYDFL) as a pump source, a 1550/2000 wavelength division multiplexer (WDM), 5 m Thulium-doped fiber (TDF) gain medium and a 90/10 coupler. The 90/10 coupler is used to split the light into two ways where 90% is reverted back into the cavity, while the other output is used for various measurements. The TDF (Nuferm SM-TDF-9/125) is a commercial fiber and has a numerical aperture (NA) of 0.15, a core diameter of 9 μm and thulium ion absorption of .3 dB/m at 1180 nm and 793 nm, respectively. Meanwhile, the fiber SA, BDF has a core and cladding diameter of 10 and 100 μm, each, with a respective NA of 0.18. The output was measured by using the optical spectrum analyzer (OSA- Yokogawa, AQ6370B), a digital oscilloscope (OSC- 500 MHz LeCroy Wavejet 352A), a radio frequency spectrum analyzer (7.8 GHz Anritsu MS2683A), a digital power meter (THORLABS PMD100D) and a photodetector (EOT ET-5010F InGaAs). The total cavity length of the ring resonator is measured to be around 13.1 m.

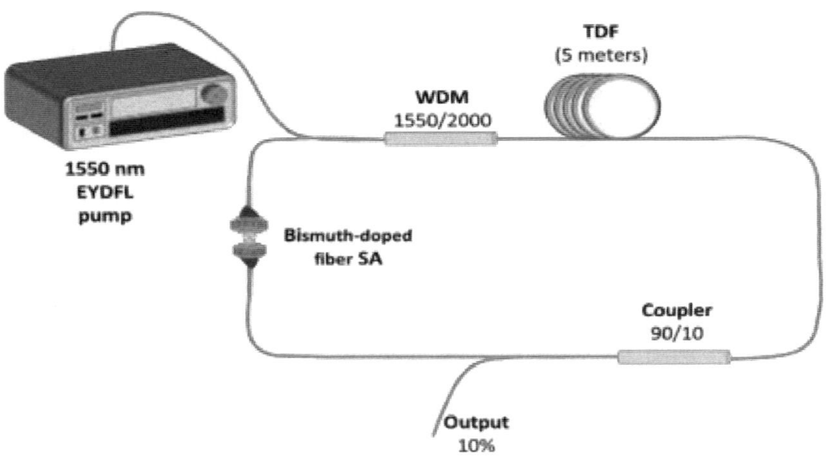

Figure 1. Schematic of the experimental set-up for the Q-switched TDFL with BDF SA.

3. Result and discussion

The performance of the Q-switched TDFL, which was configured with BDF SA is summarized in Figure 2. Figure 2(a) shows the output spectrum of the TDFL; with and without the SA. It is shown that the peak laser wavelength shifts from 1981.95 nm to 1967.91 nm as the SA applied onto the TDFL ring cavity. This is attributed to the cavity loss in the resonator which make the laser to operate at the shorter wavelength. Figure 2(b) shows the tuneable of repetition rate and the pulse width versus the pump power. The repetition rate of the BDF SA increases linearly from 18.1 kHz to 25.22 kHz with the increment of pump power level from 454 mW to 564 mW. At the same time, the pulse width decreases from 9.04 μs to 6.98 μs as the pump power increases from 454 mW to 564 mW. This is a typical trend of a Q-switched laser. Figure 2(c) shows the oscilloscope trace of the Q-switched pulse train at the maximum pump power of 564 mW. At this pump power, a passively Q-switched laser provides an average output power of 5 mW with the repetition rate of 25.2 kHz. The measured RF spectrum at the pump power of 564 mW is shown in Figure 2(d). The fundamental frequency was obtained at 25.2 kHz with a signal-to-noise ratio (SNR) of about 39.17 dB. The constant separation of peak-to-peak of RF spectrum indicates the high stability of the laser operation.

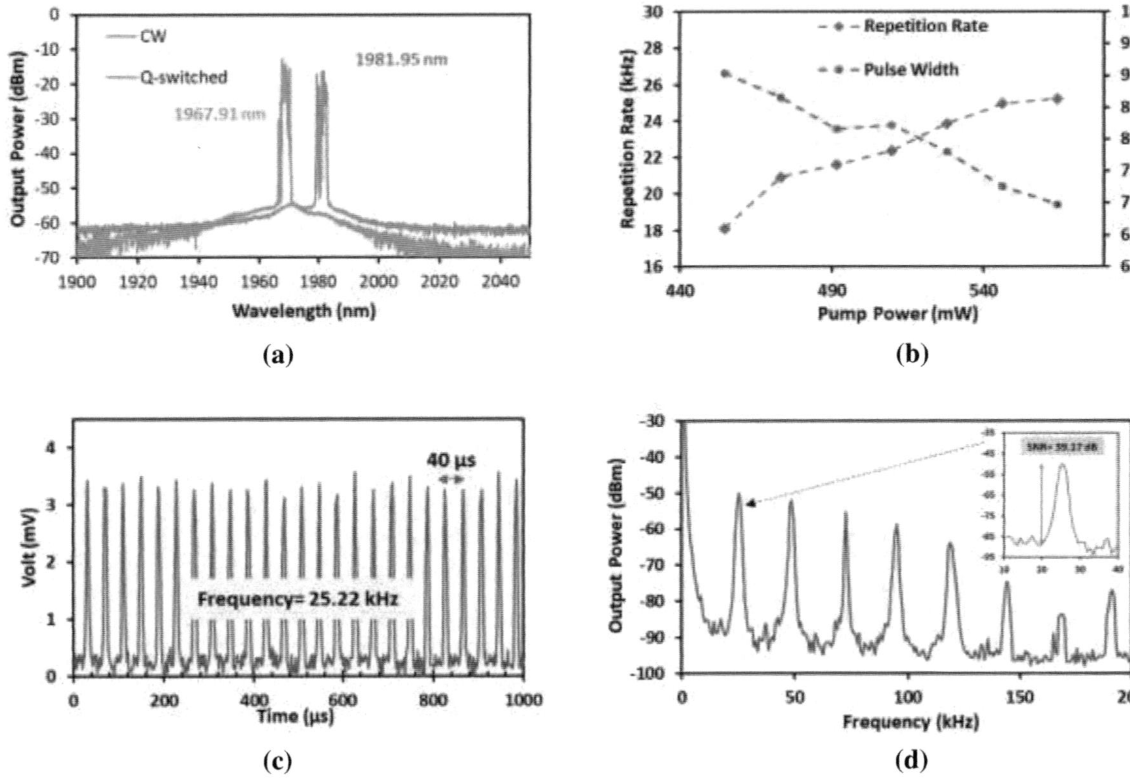

Figure 2. Q-switched laser performance. (a) Output spectrum. (b) Repetition rate and pulse width. (c) Pulse train. (d) RF spectrum.

4. Conclusion

We have successfully demonstrated a passively Q-switched fiber laser using BDF as SA with the single wavelength output that operating at 1967.91nm. The proposed laser has achieved tuneable repetition rate from 18.1 kHz to 25.22 kHz as the pump power is varied from 454 mW to 564 mW. The corresponding pulse width also reduces from 9.04 µs to 6.98 µs as the pump power is varied within the same range.

Acknowledgement

The authors would like to acknowledge the University of Malaya, Universiti Teknikal Malaysia Melaka (PJP/2018/FKEKK(1B)/S01613).

References

[1] Nishizawa N 2014 Ultrashort pulse fiber lasers and their applications *Japanese Journal of Applied Physics* **53** 090101

[2] Rahman M, Latiff A, Reddy P, Das S, Dhar A, Paul M and Harun S 2019 Passively Q-switched fiber laser utilizing new hafnium–bismuth–erbium co-doped fiber as saturable absorber *Indian Journal of Physics* 1-5

[3] Xie G, Ma J, Lv P, Gao W, Yuan P, Qian L, Yu H, Zhang H, Wang J and Tang D 2012 Graphene saturable absorber for Q-switching and mode locking at 2 µm wavelength *Optical Materials Express* **2** 878-83

[4] Rusdi M F M, Latiff A A, Paul M C, Das S, Dhar A, Ahmad H and Harun S W 2017 Titanium dioxide (TiO2) film as a new saturable absorber for generating mode-locked thulium-holmium doped all-fiber laser *Optics & Laser Technology* **89** 16-20

[5] Keller U, Weingarten K J, Kartner F X, Kopf D, Braun B, Jung I D, Fluck R, Honninger C, Matuschek N and Der Au J A 1996 Semiconductor saturable absorber mirrors (SESAM's) for femtosecond to nanosecond pulse generation in solid-state lasers *IEEE Journal of selected topics in QUANTUM ELECTRONICS* **2** 435-53

[6] Hisyam M B, Rusdi M F M, Latiff A A and Harun S W 2016 Generation of Mode-locked Ytterbium doped fiber ring laser using few-layer black phosphorus as a saturable absorber *IEEE Journal of Selected Topics in Quantum Electronics* **23** 39-43

[7] Chen B, Zhang X, Wu K, Wang H, Wang J and Chen J 2015 Q-switched fiber laser based on transition metal dichalcogenides MoS 2, MoSe 2, WS 2, and WSe 2 *Optics express* **23** 26723-37

[8] Bao Q, Zhang H, Wang Y, Ni Z, Yan Y, Shen Z X, Loh K P and Tang D Y 2009 Atomic-layer graphene as a saturable absorber for ultrafast pulsed lasers *Advanced Functional Materials* **19** 3077-83

[9] Rahman M, Rusdi M, Latiff A, Hisyam M, Dimyati K and Harun S 2019 Passively Q-switched Erbium doped fiber laser by incorporating a segment of Thulium doped fiber saturable absorber. In: *Journal of Physics: Conference Series*: IOP Publishing) p 012010

[10] Rusdi M, Rosol A, Rahman M, Mahyuddin M, Latiff A, Ahmad H, Harun S and Yasin M 2019 Q-switched and mode-locked thulium doped fiber lasers with nickel oxide film saturable absorber *Optics Communications* **447** 6-12

[11] Rahman M, Rusdi M, Harun S and Yasin M 2018 Q-Switched Thulium-Doped Fiber Laser Using Vanadium Oxide Saturable Absorber *Nonlinear Optics, Quantum Optics: Concepts in Modern Optics* **48**

[12] Rahman M, Latiff A, Zaidi U, Rusdi M, Rosol A, Bushroa A, Dimyati K and Harun S 2018 Q-switched and mode-locked thulium-doped fiber laser with pure Antimony film Saturable absorber *Optics Communications* **421** 99-104

[13] Rahman M F A, Latiff A A, Rusdi M F M, Dimyati K and Harun S W 2019 Passively Q-switched fibre laser utilizing erbium-doped fibre saturable absorber for operation in C-band region *Journal of Modern Optics* **66** 235-9

[14] Sobon G, Sotor J, Jagiello J, Kozinski R, Zdrojek M, Holdynski M, Paletko P, Boguslawski J, Lipinska L and Abramski K M 2012 Graphene oxide vs. reduced graphene oxide as saturable absorbers for Er-doped passively mode-locked fiber laser *Optics express* **20** 19463-73

[15] Luo Z, Zhou M, Weng J, Huang G, Xu H, Ye C and Cai Z 2010 Graphene-based passively Q-switched dual-wavelength erbium-doped fiber laser *Optics letters* **35** 3709-11

[16] Rusdi M, Latiff A, Hanafi E, Mahyuddin M, Shamsudin H, Dimyati K and Harun S 2016 Molybdenum disulphide tape saturable absorber for mode-locked double-clad ytterbium-doped all-fiber laser generation *Chinese Physics Letters* **33** 114201

[17] Rahman M, Reddy P, Paul M, Das S, Dhar A, Baharom M, Latiff A, Rusdi M, Wang P and Dimyati K 2019 Titanium dioxide fiber saturable absorber for Q-switched fiber laser generation in the 1-micrometer region *Applied optics* **58** 3495-500

[18] Lee J, Jung M, Koo J, Chi C and Lee J H 2015 Passively Q-Switched 1.89-μm fiber laser using a bulk-structured $B_{i2}Te_3$ topological insulator *IEEE Journal of Selected Topics in Quantum Electronics* **21** 31-6

[19] Baharom M, Rahman M, Latiff A, Wang P and Harun S 2019 Lutetium (III) oxide film as passive mode locker device for erbium-doped fiber laser cavity *Optics Communications*

[20] Baharom M, Rahman M, Latiff A, Wang P, Arof H and Harun S 2019 Lutetium oxide film as a passive saturable absorber for generating Q-switched fiber laser at 1570 nm wavelength *Optical Fiber Technology* **50** 82-6

[21] Rahman M F A, Dhar A, Das S, Dutta D, Paul M C, Rusdi M F M, Latiff A A, Dimyati K and Harun S W 2018 An 8 cm long holmium-doped fiber saturable absorber for Q-switched fiber laser generation at 2-μm region *Optical Fiber Technology* **43** 67-71

Low magnetic field effect of circular conductors on a fiber-optic transmission system

Firas S Mohammed[1]*, Thoalfiqar A Zaker[2], Rana Oday Abdaljalil[1] and Mofaq M B Alsaady[3]

[1]College of sciences, department of physics, Mustansiriyah University, Baghdad, Iraq.
[2]College of education, department of physics Al-hamdaniya University, Iraq.
[3]Faculty of applied sciences & Technology (FAST), department of physics, Universiti Tun Hussein Onn Malaysia, Pagoh, Johor, Malaysia.

fsphd@uomustansiriyah.edu.iq

Abstract. In a communication system that uses fiber optics, light wave transmissions respond to the influence of surrounding magnetic fields, thereby causing undesirable changes in the output signal. Thus, an important task in electrical engineering is analyzing polarization and intensity changes to protect sensitive data transmission systems from unintended sources of magnetic disturbance. In this study, we propose a signal processing method to evaluate fiber-optic transmission systems under the influence of a weak magnetic field. This method is conducted by performing experimental and theoretical measurements on the axis of light propagation using a laboratory setup and simulation software OptiSystem (version 7.0). Results show that the shape, size, and axial positions of the conductor plays a key role in strengthening the magnetic field influence on increasing the bit error rate and decreasing the Q factor of the communication system.

Keywords: Q- Factor, fiber optics, laser, magnetic field effect, bit error rate.

1. Introduction

The challenge of using optical fibers as a waveguide in data transmission through light reaches the end of the fiber only slightly distorted is an attraction in the presence of external perturbations. An optical fiber is a cylindrical dielectric waveguide made of low-loss material that transmits light depending on the process of total internal reflection. However, low power is lost during each reflection [1]. In a communication system that uses fiber optics, light wave transmissions react to changes in their surroundings in various ways, thereby causing undesirable changes in output signal [2]. Several studies have been conducted to analyze the polarization and intensity changes generated by external magnetic fields. Studies have been published on the influence of the magnetic field on optical signal propagation direction [3-5], tunable refractive index and field dependent transmission [6], and degree of overlapping of fibers to control the orientated behavior [7]. An assessment of the magnetic field created by the flow of current through cables or any conductor of various configurations is important in electrical engineering to protect sensitive devices from unintended sources of magnetic disturbance [8]. Biot–Savart law for a volume distribution of current was used to obtain a circulation formula for the magnetic field [9].

In this study, we propose a signal processing method to evaluate fiber-optic transmission systems under the influence of a weak magnetic field. Experimental and theoretical measurements are conducted on the axis of light propagation using a laboratory setup and simulation software OptiSystem7.0, Optiwave Systems Inc.

2. Experimental setup and principle

The experimental setup consists of three main elements: an optical source as a data carrier, an optical fiber as a transmission channel, and a magnetic field source. A magnetic field arises due to the motion of charges on a conductor. When charges move in a conducting wire and produce a current I, the magnetic field can be calculated at any point (P) by adding up the magnetic field contributions, $d\vec{B}$, which arise from small parts of wire $d\vec{s}$ due to the current. According to Biot–Savart law, the infinitesimal current source of wire segments can be written as $Id\vec{s}$ [10, 11].

$$d\vec{B} = \frac{\mu_\circ}{4\pi}\frac{Id\vec{s}}{r^2} \times \frac{r}{r} \quad (1)$$

Where $\mu_\circ$ is permeability of free space

$$\mu_\circ = 4\pi.10^{-7}\,\frac{Vs}{Am}$$

As shown in Figure 1, the position vector from the part of the conductor under consideration to point P is given by r.

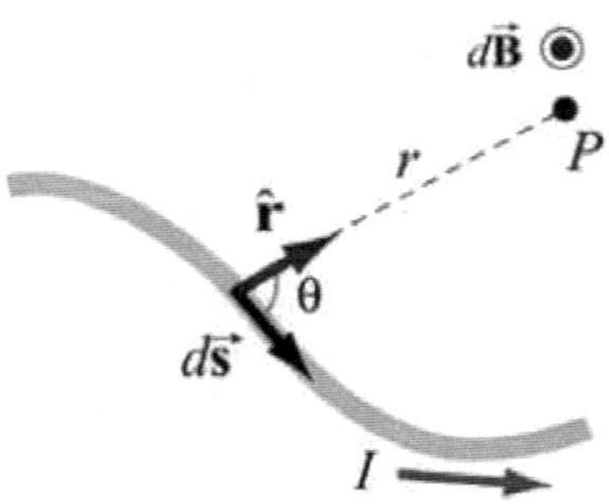

Figure 1. The magnetic field of a current-carrying conductor

The total magnetic field of an infinitely long wire is calculated as follows [10]:

$$\vec{B} = \frac{\mu_\circ I}{4\pi} \times \frac{2}{r} \quad (2)$$

At a distance r from the axis, the field lines are concentric around the cylinder axis, as shown in Figure 2.

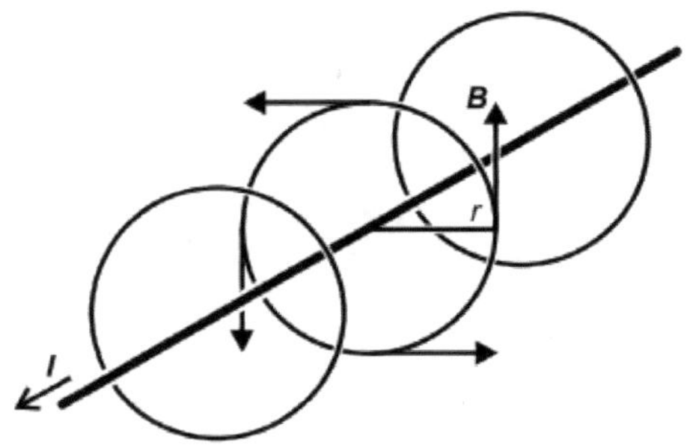

Figure 2. The magnetic field of an infinitely long wire.

The subject of magnetic field influence on fiber-optic transmission systems has not been widely studied. In general, fiber optic loss is expressed in decibels (dB) and calculated [12] with the following equation

$$\text{loss}_{\text{dB}} = 10 \log \frac{\text{P}_{\text{out}}}{\text{P}_{\text{in}}} \quad (3)$$

where P_{in} is the input power to the fiber and P_{out} is the power at the output of the fiber. As shown in Figure 3, the experimental setup consists of the following: a narrow line-width diode laser (650 nm) that is applied as a light source, a transmission optical fiber, a fiber-optic power meter, high-current power supply equipment, a Tesla meter with axial B-probe, a Hall effect sensor, and an adaptor for conductor loops.

The proposed technique is divided into two procedures. The first procedure consists of the following steps:

- The adapter is mounted and attached to a 40 mm conductor loop.
- The optical fiber is aligned with the B-probe and passed through the center of the conductor loop.
- The B-probe is aligned toward the center of the conductor loop and the conductor loop is aligned with the Hall sensor.
- The light from the laser diode (650 nm) is launched and transmitted through the optical fiber and then enter into an optical power meter.
- The magnetic field is controlled by increasing the current (I) from 0 to 20 ampere in intervals of 2 ampere. The measured values of the magnetic field (B) and optical output power are recorded.
- The 40 mm conductor loop is replaced with 80 mm and 120 mm conductor loops. In both cases, the magnetic field is measured. The optical output power and measured values are recorded.

The second procedure is performed by repeating the measurements after placing the optical fiber near but out of the conductor loop.

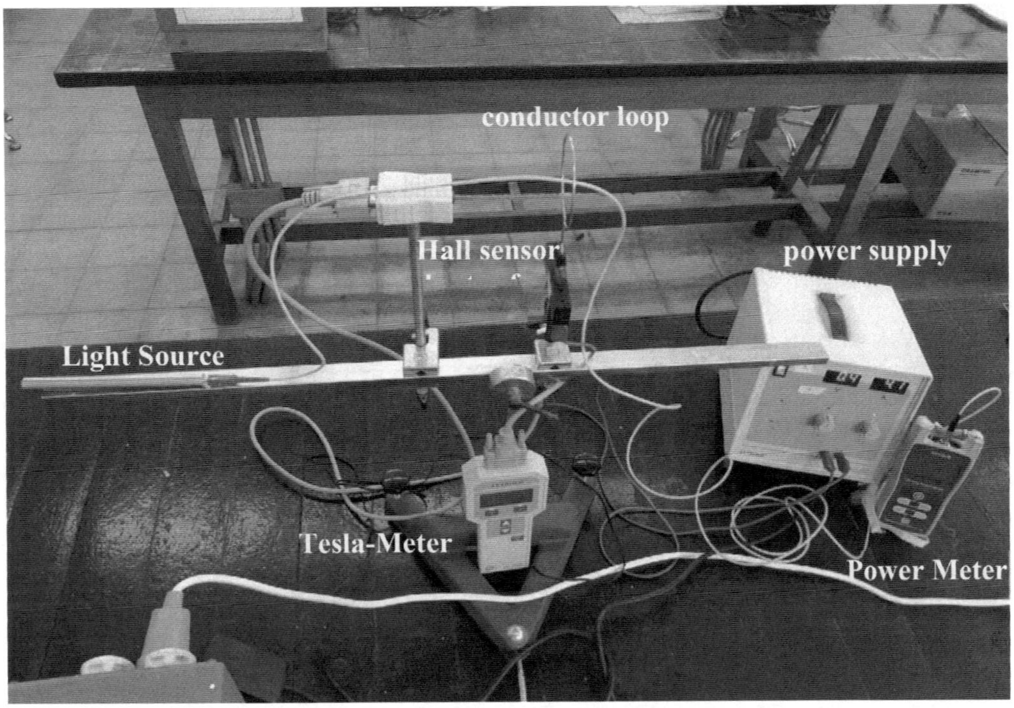

Figure 3. The experimental setup

3. Design and simulations of Propagation setup

Numerical simulations are necessary to investigate the actual performance of data transmission systems under the influence of weak magnetic fields. This task is usually conducted by including the effect of external perturbations on the system performance. The full design of a fiber optic transmission system is presented in Figure 4. The data generated by the pseudo-random bit generator are encoded through a non-return-to-zero pulse generator, and light is modulated by using a Mach–Zehnder modulator. Thus, the CW laser source (1500 nm) acts as the carrier source. The laser passes through the optical fiber channel with a length of 1 km to supply a channel capacity of approximately 30 dBm. The output optical signals are received by an Avalanche photodetector. This simulation uses three visualizers: an optical power meter to measure the power received in dBm and watts, an electrical carrier analyzer, and a BER analyzer to calculate the bit error rate (BER) value and Q factor. For good signal reception, the minimum Q factor is 6 and the BER value is 10^-9 [13]. The design parameters used in the numerical simulation are shown in Table 1.

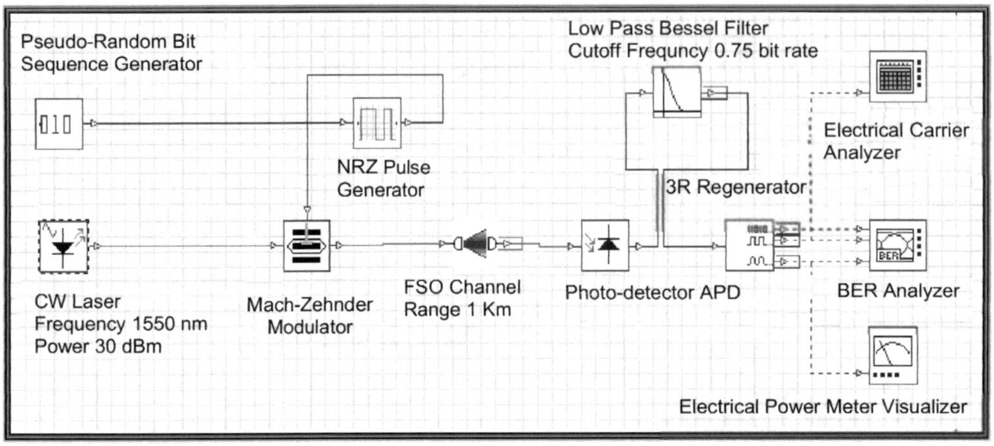

Figure 4. Simulation layout of a fiber optic transmission system

Table 1. Fiber optic system design specifications used in simulation.

Parameter	Value
wavelength	1550 (nm)
Transmitter optical power	30 (dBm)
Transmitter divergence angle	2 (mrad)
Receiver sensitivity	-20 (dBm)
Transmitter aperture	2 (cm)
Receiver aperture	10 (cm)
Transmitter efficiency	0.5
Receiver efficiency	0.5

4. Results and discussion

In the experimental part, the performance of the optical fiber as a waveguide exposed to an external magnetic field was evaluated. Measurement data were collected continuously 10 times for the magnetic field B of a circular conductor loop as a function of the current I. Furthermore, the optical power fluctuation from the fiber as a result of beam propagation through (inside) and near (outside) the circular conductor loop were measured and recorded. Using these data, we determined the magnetic field rate, power fluctuation rate, and power fluctuation percentages of the three circular conductor loops (40, 80, and 120 mm), as shown in Table 2.

Table 2. Experimental measurements

Conductor loop diameter (mm)	Magnetic field rate (mT)		Power fluctuation rate (mW)		Power fluctuation percentage (%)	
	Inside the loop	Outside the loop	Inside the loop	Outside the loop	Inside the loop	Outside the loop
40	0.4	0.388	0.5785	0.570	115.7	114
80	0.239	0.215	0.4938	0.4777	98.8	95.6
120	0.203	0.156	0.4924	0.4757	98.5	95.1

Furthermore, we conducted the simulations and numerical calculations to quantify the efficiency of fiber-optic transmission systems under the influence of a weak magnetic field. Percentages of experimentally measured power fluctuation, as reported in Table 2, were used to predict the power fluctuation for the proposed system, as shown in Table 3.

Table 3. Power fluctuation percentages of 1550 nm fiber optic system.

Conductor loop diameter (mm)	Power fluctuation Percentages (dBm)		BER *E-11	
	Inside the loop	Outside the loop	Inside the loop	Outside the loop
40	34.7	34.2	1.2373	1.2372
80	29.64	28.68	1.2366	1.2364
120	29.55	28.5	1.2363	1.2364

Considering the data in Table 2, we can conclude that the external magnetic field has a significant effect on the stability of the output power in both optical paths (inside and outside) the loop. These optical paths exhibited a linear response to magnetic field increments. The highest sensitivities were obtained at the center of the conductor loops. The optical fiber showed increased losses in receiving power when the loop diameter around the fiber increased linearly, as shown in Tables 2 and 3. Figure 5 shows a reduction in the Q factor when the loop diameter increases. The results also illustrate that all values of the Q factor are above 6; thus, the light reaches the end of the fiber with only a slight distortion in accordance with previous findings [14].

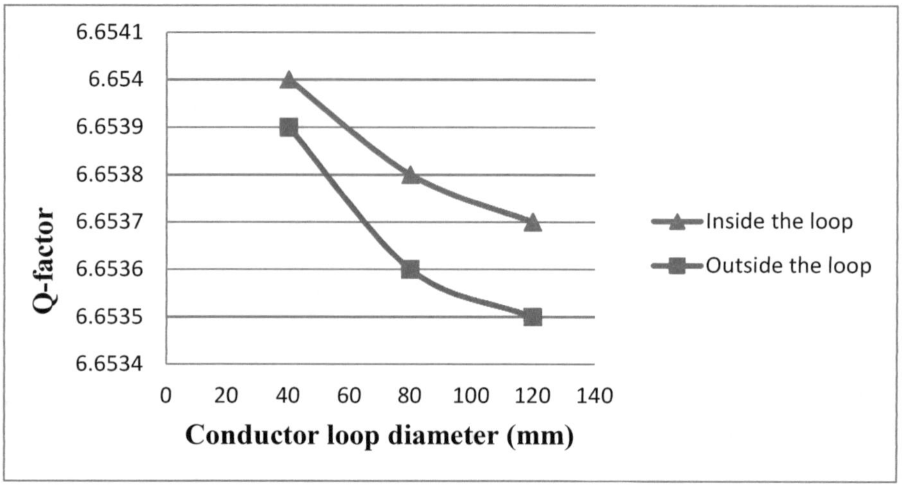

Figure 5. Conductor loop diameter vs Q-factor

5. Conclusion

In this study, the efficiency of the fiber-optic transmission system under the influence of a weak magnetic field was investigated. The results demonstrated that the shape, size, and axial positions of the conductor played a key role in strengthening the magnetic field influence to stabilize the output power, increase the bit error rate, and decrease the Q factor of the communication system. Sensitivities to the magnetic field were determined in both optical paths (inside and outside) the loop. The results also illustrated that all values of the Q factor were above 6, thereby demonstrating that the light reaches the end of the fiber with only slight distortion.

References

[1] Kumar S and Deen M J 2014 Fiber Optic Communications: Fundamentals and Applications *West Sussex, United Kingdom: John Wiley & Sons, Ltd*

[2] Elsarnagawy T 2002 *First annual conf. on Biomedical Engineering ACBME* (Cairo) p 66

[3] Oh K D, Ranade J, Arya V, Wang A and Claus R O 1997 Optical fiber Fabry-Perot interferometric sensor for magnetic field measurement *IEEE Photonics Tech. Lett. Vol 9 pp 797-799.*

[4] Supe A and Porins J 2012 Interaction Between Electromagnetic Field and Optical Signal Transmission in Fiber Optics *Electronics and Electrical Engineering – Kaunas Technologija No. 6(122) pp 83-86.*

[5] Brehler M, Schirwon M, Krummrich P M and Goddeke D 2019 Simulation of Nonlinear Signal Propagation in Multimode Fibers on Multi-GPUSystems *physics.comp-ph, arXiv:1901.01895.*

[6] Alberto N, Domingues M F, Maques C, André P and Antunes P 2018 Optical Fiber Magnetic Field Sensors Based on Magnetic Fluid: A Review *Sensors 18(12),4325, https://doi.org/10.3390/s18124325*

[7] Chen S C, Chung C Y and Tseng Y L 2019 *AIP Conf. Proc.(2065, 040001 (2019); https://doi.org/10.1063/1.5088321*

[8] Ferreira J M and Anacleto J 2018 Magnetic field generated by the flow of AC current through finite length n onmagnetic conductors (cylinders, tubes, coaxial cables) *Electrical Engineering vol 100(3) pp 1379-1391.*

[9] Ferreira J M and Anacleto J 2018 The magnetic field circulation counterpart to Biot-Savart's law *The European Physical Journal Plus 133:234.*

[10] Phillips J A and Sanny J 2008 The Biot-Savart Law: Form Infinitesimal to Infinite *The Physics Teacher vol 46(1) pp 44-47.*

[11] Ferreira J M and Anacleto J 2017 Magnetic field created b y a conducting cylindrical shell of finite length *Electrical Engineering vol 99(3).*

[12] Massa N 2010 Fundamentals of Photonics *SPIE on Fiber Optic and Telcommunication (Bellingham, Wash).*

[13] Majumdar A K and Ricklin J C 2008 Free-Space Laser Communications, Principles and Advances *New York, NY : Springer, ©2008.*

[14] Ismael M N, Al-Dawoodi A, Alshwani S, Ghazi A and Fakhrudeen A M 2019 SDM Over Hybrid Fso Link Under Different Weather Conditions And Ftth Based On Electrical Equalization *Int. Journal of Civil Engineering and Technology vol 10(1) pp 1396-1406.*

Samarium (III) oxide thin film as a saturable absorber for the passively Q-switched Tm-doped fiber laser

K N Rusdi[1], M B Hisyam[2], M F M Rusdi[2], N F Zulkipli[2], A A Latiff [3], S W Harun[2], N Saidin[1]

[1]Department of Electrical and Computer Engineering, International Islamic University Malaysia, Jalan Gombak, 53100 Kuala Lumpur, Malaysia.
[2]Photonics Engineering Laboratory, Department of Electrical Engineering, Faculty of Engineering, University of Malaya, 50603 Kuala Lumpur, Malaysia
[3]Centre for Telecommunication Research & Innovation (CeTRI), Fakulti Kejuruteraan Elektronik & Kejuruteraan Komputer (FKEKK), Universiti Teknikal Malaysia Melaka (UTeM), 76100 Hang Tuah Jaya, Melaka, Malaysia

norazlina@iium.edu.my

Abstract. We have demonstrated a passive Q-switched Thulium-doped fiber laser (TDFL using samarium oxide (Sm_2O_3) nanomaterial as saturable absorber (SA). The Sm_2O_3 based SA was fabricated by simply mixing the Sm_2O_3 powder solution with polyvinyl alcohol (PVA) solution. The homogeneously mixed solution was spread and dry to form a thin film. A piece of 1 mm x 1 mm of the SA thin film is sandwiched between two fiber ferrules and incorporated into a TDFL ring cavity for pulses generation. By controlling the loss and gain in the cavity, stable Q-switching operation was generated. The repetition rate was tunable from 17.62 kHz to 29.20 kHz by varying the pump power from 619 mW to 784 mW. The smallest pulse width of 3.54 μs and the highest pulse energy of 0.20 uJ were obtained at the highest pump power.

Keywords: Fiber laser, Q-switched, saturable absorber

1. Introduction

Fiber lasers operating at 2-micron region have attracted much attention in recent years because of their great potential applications in telecommunication, military, spectroscopy and medical surgery [1-3]. Compared to other lasers operating at 1.0- and 1.55- micron regions, 2-micron fiber laser are advantageous due to its high absorption coefficient in water and gases. For instance, Thulium- and holmium-doped fiber lasers are operated at 1.94 μm and 2.12 μm, which have an absorption coefficient of 114 cm^{-1} and 36 cm^{-1}, respectively. The lasers are also referred to as 'eye-safe lasers'. Due to these characteristics, they are beneficial in many applications including material processing and biomedical diagnostics [3-5]. There are two techniques to generate the Q-switching pulses; active and passive techniques. Passive techniques require saturable absorber (SA) device and they are preferable compared to that of active techniques because of their advantages such as smaller in size, highly efficient, reliable and ease in fabrication. Furthermore, the passive technique requires no external pulse triggering signal while the active technique needs external acousto-optic modulators to actively control and modulate the intra-cavity loss [6].

Graphene, carbon nanotubes (CNTs), graphene, black phosphorus (BP), topological insulators (TIs), transition metal dichalcogenides (TMDs) and quantum dots (QDs) have been widely reported as the Q-switcher. However, these materials have some limitations in a few fields. For instance, graphene has nearly zero bandgaps and thus it has low modulation depth (2.3% per layer) which limits their application in high power laser [7, 8]. Previously, Hisyam et al. reported the mode-locked Ytterbium-doped fiber laser using BP as SA with the repetition rate and pulse energy of 13.5 MHz and 5.93 nJ, respectively [9]. However, BP has a few weakness such as it is hydrophilic and able to change their optical properties when contacted with water and exposed to humid air [10, 11].

To overcome this problem, many new materials have also been proposed [12-16] as alternative SAs including lutetium oxide [17] and cobalt oxide [18] which belongs to the transition metal oxide (TMO) family. Here, we proposed Sm_2O_3 thin film as the new passive SA to generate pulses at the 2-micron region. Sm_2O_3 thin film-based SA has been fabricated through a simple process by embedding the Sm_2O_3 powder into polyvinyl alcohol (PVA) solution. The thin film is incorporated into a ring Thulium-doped fiber laser (TDFL) cavity to initiate and produce Q-switching pulses.

2. Preparation and characterization of Sm_2O_3 based SA

The Sm_2O_3 thin film was fabricated through a simple fabrication process by mixing the Sm_2O_3 powder into PVA solution. At first, the PVA solution was prepared by dissolving 1 gram of PVA powder into 120 ml of distilled (DI) water. After that, the mixing solution was stirred using magnetic stirrer at room temperature until the PVA completely dissolved. Then, 50 ml of the PVA solution was added into the clean beaker with 50 mg of the Sm_2O_3 powder before putting them on the magnetic stirrer for 3 hours. Next, the solution was placed into the ultrasonic bath for 10 minutes. Lastly, the Sm_2O_3 PVA solution was poured into a petri dish and let the solution to dry at room temperature for about 48 hours. The thin film was then formed, and it was peeled off from the petri dish and cut to 1 mm x 1 mm in size, before sandwiching it between two FC/PC fiber ferrules. Figure 1 shows the linear absorption profile of the Sm_2O_3 film SA. As shown, there is some absorption of about 7 dB occurs around the TDFL Q-switched laser operating region.

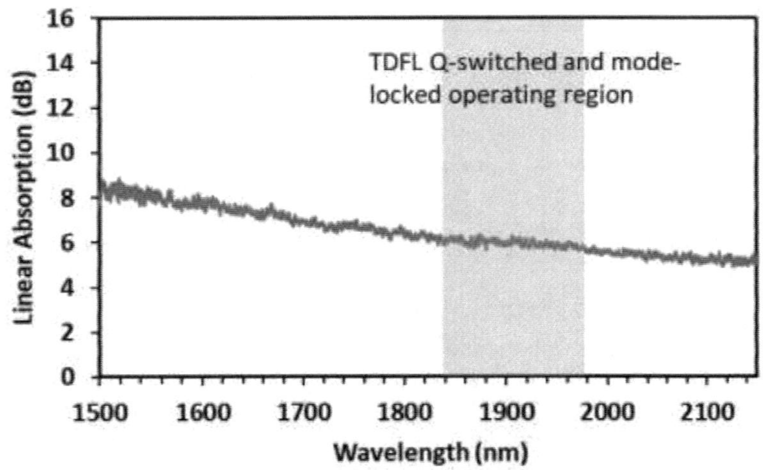

Figure 1. The linear absorption profile of Sm_2O_3.

3. Pulsed fiber laser configuration

Figure 2 shows the experimental configuration of the proposed Q-switched TDFL. As shown in the figure, 5 m long Thulium-doped fiber (TDF) was used as the gain medium and it was pumped by a 1550 nm Erbium-Ytterbium-doped fiber laser (EYDFL) through a 1550/2000 wavelength division multiplexer (WDM). TDF used is Nuferm SM-TSF-9/125 commercial fiber with the numerical aperture (NA) of 0.15

Journal of Physics: Conference Series **1371** (2019) 012026 doi:10.1088/1742-6596/1371/1/012026

and thulium ion absorption of 9.3 dB/m and 27 dB/m at 1180 nm and 793 nm, respectively. The 90/10 optical coupler kept 90% of the light in the ring cavity, while 10% of the light was tapped out for the optical measurement and analysis. All the optical connection in the ring cavity were directly spliced to minimize the optical loss in the cavity except for the SA device. The SA device was constructed by sandwiching 1 mm x 1 mm of the Sm_2O_3 thin film in between two FC/PC fiber ferrules while applying some index matching gel at the surface of fiber ferrules. The equipment used in this experiment for analysis were optical spectrum analyzer (OSA-500 MHz Yokogawa AQ6375), a digital oscilloscope (500 MHz LeCroy Wavejet 352A), a radio frequency spectrum analyzer (RFSA-7.8 GHz Anritsu MS2683A), a digital power meter (THORLABS PMD100D) and a photodetector (PD-EOT ET-5010F InGaAs). The total cavity length in this proposed Q-switched TDFL was 13.1 m.

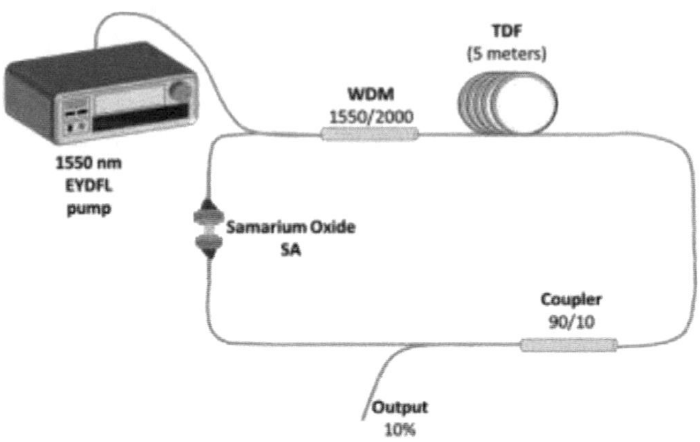

Figure 2. The schematic diagram of the pulsed TDFL setup with Sm_2O_3-SA

4. Result and discussion

With the SA device placed inside the ring cavity, a passively Q-switched TDFL started to appear when the pump power increased above the threshold pump power of 619 mW. Figure 3(a) compares the output spectrum of the Q-switched EDFL with the continuous-wave (CW) laser, which was obtained without the SA at a pump power of 619 mW. As shown in the figure, the Q-switched laser operated at a wavelength of 1935.25 nm, which was at the shorter wavelength compared to the CW laser. The CW laser operated at 1965.77 nm. The operating wavelength shifted to a shorter region with the insertion of the SA device inside the cavity due to the increase in cavity loss of the resonator. The operating laser shifted to a shorter wavelength with a higher gain to compensate for the loss. Figure 3(b) shows the output power and pulse energy versus pump power of the Q-switched TDFL using Sm_2O_3 as SA. As the pump power increased from 619 mW to 784 mW, the output power and the pulse energy increase linearly from 1.3 mW to 5.9 mW and 0.073 μJ to 0.20 μJ, respectively. The calculated slope efficiency is 2.73 % as shown in Figure 3(b). Figure 3(c) shows the typical pulse train of the Q-switched TDFL observed at the maximum pump power of 784 mW. The pulses train has a peak-to-peak separation of 36 μs corresponding to the maximum pulse repetition rate of 29.2 kHz. At a pump power of 784 mW, the output power and pulse energy are measured to be around 5.9 mW and 0.20 μJ, respectively. By optimizing the cavity, the insertion loss of the SA could be reduced thus increasing the pulse energy. The frequency-domain of the pulsed signal at maximum pump power is illustrated in Figure 3(d), showing several frequency harmonics of the Q-switched laser. The pulsed laser has a signal to noise ratio (SNR) of 34.5 dB at the fundamental frequency of 29.2 kHz, which is good matching with the pulse repetition rate obtained in the time domain in Figure 3(c).

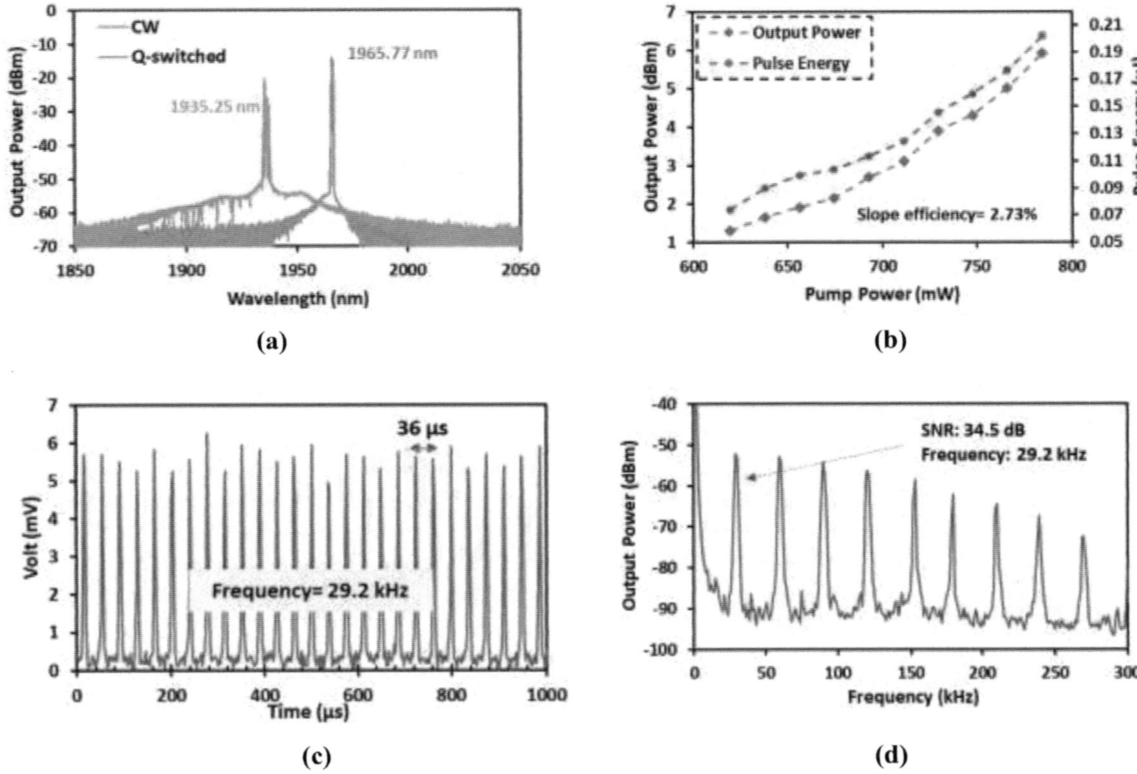

Figure 3. Q-switched laser performances. (a) output spectrum at a pump power of 619 mW. (b) Output power and pulse energy against pump power (c) Typical pulse train (d) RF spectrum.

5. Conclusion

We have successfully demonstrated a stable Q-switched TDFL by integrating Sm_2O_3 thin film as SA in a ring cavity. The Q-switched laser started to appear as the pump power increased to the threshold of 619 mW and remain stable until the pump power up to 784 mW. At the maximum pump power of 784 mW, maximum output power and pulse energy are obtained at 5.9 mW and 0.20 µJ, respectively. The fundamental frequency of 29.2 kHz has a signal to noise ratio (SNR) of 34.5 dB. The results indicate that Sm_2O_3 can become an excellent alternative SA for laser applications.

References

[1] Miller S 2012 *Optical fiber telecommunications*: Elsevier)

[2] Juarez J C, Dwivedi A, Hammons A R, Jones S D, Weerackody V and Nichols R A 2006 Free-space optical communications for next-generation military networks *IEEE Communications Magazine* **44** 46-51

[3] Fried N M and Murray K E 2005 High-power thulium fiber laser ablation of urinary tissues at 1.94 µm *Journal of endourology* **19** 25-31

[4] Geng J, Wang Q, Lee Y and Jiang S 2014 Development of eye-safe fiber lasers near 2 µm *IEEE Journal of Selected Topics in Quantum Electronics* **20** 150-60

[5] Rahman M F A, Dhar A, Das S, Dutta D, Paul M C, Rusdi M F M, Latiff A A, Dimyati K and Harun S W 2018 An 8 cm long holmium-doped fiber saturable absorber for Q-switched fiber laser generation at 2-µm region *Optical Fiber Technology* **43** 67-71

[6] Rusdi M, Rosol A, Rahman M, Mahyuddin M, Latiff A, Ahmad H, Harun S and Yasin M 2019 Q-switched and mode-locked thulium doped fiber lasers with nickel oxide film saturable absorber

Optics Communications **447** 6-12
[7] Luo Z, Zhou M, Weng J, Huang G, Xu H, Ye C and Cai Z 2010 Graphene-based passively Q-switched dual-wavelength erbium-doped fiber laser *Optics letters* **35** 3709-11
[8] Al-Masoodi A, Ahmed M, Latiff A, Arof H and Harun S 2018 Nickel oxide nanoparticles thin film saturable absorber for 1-micron pulsed all-fibre lasers operation *Journal of Modern Optics* **65** 1801-8
[9] Hisyam M B, Rusdi M F M, Latiff A A and Harun S W 2016 Generation of Mode-locked Ytterbium doped fiber ring laser using few-layer black phosphorus as a saturable absorber *IEEE Journal of Selected Topics in Quantum Electronics* **23** 39-43
[10] Rahman M F A, Latiff A A, Zaidi U Z M, Rusdi M F M, Rosol A H A, Bushroa A R, Dimyati K and Harun S W 2018 Q-switched and mode-locked thulium-doped fiber laser with pure Antimony film Saturable absorber *Optics Communications* **421** 99-104
[11] Li J, Luo H, Zhai B, Lu R, Guo Z, Zhang H and Liu Y 2016 Black phosphorus: a two-dimension saturable absorption material for mid-infrared Q-switched and mode-locked fiber lasers *Scientific reports* **6** 30361
[12] Rahman M, Reddy P, Paul M, Das S, Dhar A, Baharom M, Latiff A, Rusdi M, Wang P and Dimyati K 2019 Titanium dioxide fiber saturable absorber for Q-switched fiber laser generation in the 1-micrometer region *Applied optics* **58** 3495-500
[13] Rahman M, Rusdi M, Harun S and Yasin M 2018 Q-Switched Thulium-Doped Fiber Laser Using Vanadium Oxide Saturable Absorber *Nonlinear Optics, Quantum Optics: Concepts in Modern Optics* **48**
[14] Rahman M F, Latiff A A, Rusdi M F, Dimyati K and Harun S W 2018 Q-switched ytterbium-doped fiber laser via a thulium-doped fiber saturable absorber *Applied Optics* **57** 6510-5
[15] Ab Rahman M F, Latiff A A, Rosol A H A, Dimyati K, Wang P and Harun S W 2018 Ultrashort Pulse Soliton Fiber Laser Generation With Integration of Antimony Film Saturable Absorber *Journal of Lightwave Technology* **36** 3522-7
[16] Rahman M, Dhar A, Das S, Dutta D, Paul M, Rusdi M, Latiff A, Dimyati K and Harun S 2018 An 8 cm long holmium-doped fiber saturable absorber for Q-switched fiber laser generation at 2-μm region *Optical Fiber Technology* **43** 67-71
[17] Baharom M, Rahman M, Latiff A, Wang P, Arof H and Harun S 2019 Lutetium oxide film as a passive saturable absorber for generating Q-switched fiber laser at 1570 nm wavelength *Optical Fiber Technology* **50** 82-6
[18] Al-Masoodi A, Alani I, Ahmed M, Al-Masoodi A H, Alani A A, Wang P and Harun S J O F T 2018 Cobalt oxide nanocubes thin film as saturable absorber for generating Q-switched fiber lasers at 1 and 1.5 μm in ring cavity configuration **45** 128-36

A study on relative humidity sensors using PVA and PMMA coating

Zuraidah Harith[2], Huda Adnan Abdullah Zain[1], Malathy Batumalay[3], Sulaiman Wadi Harun[1]

[1]Department of Electrical Engineering, Faculty of Engineering, University of Malaya, 50603 Kuala Lumpur, Malaysia
[2]Faculty of Engineering & Quantity Survey, INTI International University, 71800 Nilai, Negeri Sembilan, Malaysia
[3]Faculty of IT, INTI International University, 71800 Nilai, Negeri Sembilan, Malaysia

zuraidah.harith@newinti.edu.my

Abstract: Simple relative humidity sensors are demonstrated using tapered silica microfibers. The silica microfibers are tapered using flame brushing method to reduce the fiber's waist diameter from 128 micrometers to 2.6 micrometers. The tapered silica microfibers were then coated using polyvinyl acetate (PVA) and polymethyl methacrylate (PMMA) before it was used to sense the relative humidity. The tapered silica microfiber coated with PVA performed better compared to tapered silica microfiber coated using PMMA. The tapered silica microfiber with PVA based sensor has better sensitivity of 0.3023 dB/RH%.

Keywords: Fiber Optic Sensor, Relative Humidity (RH), Sensor, Refractive Index (RI), Tapered Silica Microfibers, Polyvinyl Acetate (PVA) and Polymethyl Methacrylate (PMMA).

1. Introduction

There is a growing interest in using tapered microfiber to build environmental sensors. Tapering is the process of using a chemical, mechanical or thermal method to lower the diameter of optical fiber [1]. The results in increased sensitivity to changes in the environment of the tapered fiber caused by the evanescent mode propagation. When a fiber is tapered, a portion of the light that usually travels through the core radiates into the outside of the fiber creating the evanescent mode and a larger power loss compared to ordinary fiber [2]. The tapered region of single mode fiber behaves like a multimode fiber since light is flowing through the cladding layer. Thus, the surrounding environment of the fiber works as the cladding of the tapered region. This mode coupling in the tapered waist of single mode fiber makes it responsive to changes in the refractive index of its surroundings. The light guided through the cladding of the tapered fiber and affected by the changes in the surrounding environment is called evanescent mode [3].

The evanescent mode in tapered fiber makes it suitable for sensing applications, for example, sensing temperature, humidity, gases, and magnetic fields [3]. The evanescent mode is affected by changes in the fiber's surroundings. Thus, changes in the fiber environments cause changes in the propagation characteristics of the fiber, such as power loss and peak wavelength [4]. This is the working principle of the sensors constructed for this experiment.

The increased evanescent mode with smaller diameters used to measure humidity. Since tapering causes more light to radiate in the evanescent mode, the output power of tapered fiber is highly sensitive to changes in the surrounding environment. When the relative humidity surrounding the tapered fiber increases, the refractive index of the air surrounding the tapered region changes. Additionally, the water particles in the air increase the scattering near the tapered region. Consequently, the power loss of the tapered fiber influenced by this change in humidity [5]. The changes of relative humidity in the fiber environment cause the power loss of the fiber to also increase and affect the output power of the fiber [6].

Tapering increases fiber sensors ability to measure relative humidity, as shown in the previous chapter. This sensitivity can be improved by coating the tapered region with materials that react to changes in relative humidity. The coating layer act as a new cladding layer for the tapered region since its original cladding severely diminished after flame brushing. Thus, any changes in the coatings refractive index affect the evanescent mode propagation and the power loss of the fiber [4]. In our work, we choose Polyvinyl acetate (PVA) and Polymethyl methacrylate (PMMA) due to their high sensitivity to relative humidity. The refractive index of both of PVA and PMMA changes when relative humidity increases. As a result, the propagation characteristics of the coated fiber vary with changing relative humidity [7].

In this experiment, a comparison between the performance of PVA and PMMA coated fiber was proposed. The output power of PVA coated microfiber was measured while the relative humidity was varied. These steps were repeated for PMMA coated microfiber as well. Finally, a comparison obtained based on the sensitivity with changes in relative humidity for both of the coated sample as well as bare fiber sample.

2. Experiment

2.1. Fabrication of Tapered Fiber

The setup for the process consisted of two holders for the fiber, a slider stage, two stepper motors, a microcontroller board as shown in Figure 1 [8]. Flame brushing method used to taper the fiber. The flame is using butane gas and shaped by oxygen. To prepare the tapered fiber, single-mode fiber with 128 μm diameters (SMF, Corning, 128) was used. A microscope is employed to ensure the required diameter for the tapered fiber [9].

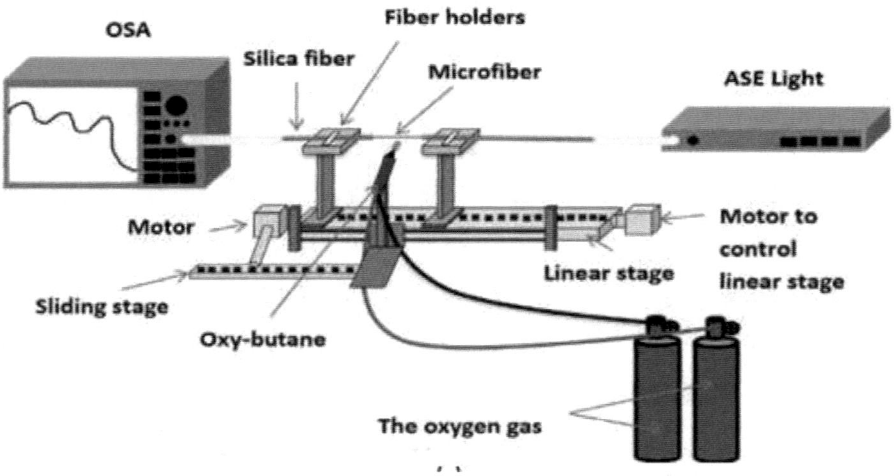

Figure 1. Setup for the fabrication of tapered fiber

2.2. Experimental setup

To measure the tapering effect on relative humidity sensitivity, the following setup as per Figure 2 was proposed. Firstly, the tapered fiber was spliced into a patch cord and connected to an optical power meter on one end and an ASE (amplified spontaneous emission) light source, with a center wavelength of $1550\mu m$, on the other end. Secondly, the tapered region of the sample was enclosed in a locked chamber where humidity was steadily increased using saturated salts (Sodium Hydroxide). Additionally, the relative humidity in the chamber was measured using an electronic humidity sensor for reference.

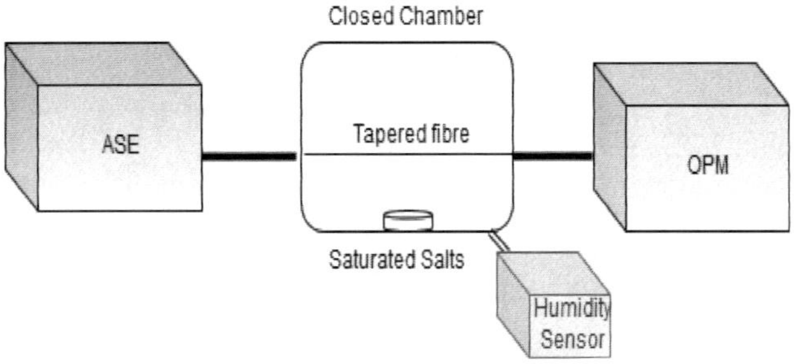

Figure 2. The experiment setup for bare fiber testing

2.3 Coating of tapered fiber with PMMA

Microfiber with a waist diameter of 2.6 μm is fixed since it had the highest sensitivity to relative humidity and coated with PMMA. The coating layer consists of 1 mg of crystal PMMA with 10 ml of isopropanol. These materials were mixed and heated at 100 C° with 700 rpm for an entire hour on a hot plate. The resulting solution was carefully dripped onto the microfiber sample. The coated sample was left to dry for 24 hours [7, 8, 10].

2.4 Coating of tapered fiber with PVA

To fabricate PVA, 5 mg of PVA crystals were mixed with 12 ml of distilled water. Next, the mixture was cleaned in an ultrasonic bath for 6 hours at 20° C. The resultant solution was carefully dripped onto the selected microfiber sample's tapered region and left to dry for an hour.

2.5 Humidity sensing experiment

The prepared sample was connected to a patch cord on both ends using a splicer in a closed chamber, and saturated salts are used to increase the relative humidity. The patch cord had the light of peak wavelength of 1550 μm and the output power were recorded using the OPM. The relative humidity inside the chamber was monitored using an electronic humidity sensor for reference. The experiment setup is shown in Figure 3.

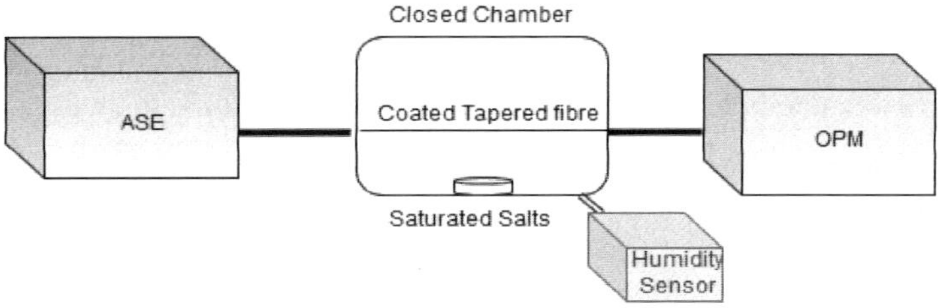

Figure 3. Experiment setup for testing the PVA coated and PMMA coated tapered fiber.

3. Results and Discussion

The results of several samples of tapered fiber are presented in Figure 4. Based on this data, a summary of the analysis is shown in Table 1. Apparently, with an increase of relative humidity, power loss increases as well, due to the additional scattering. The water particles in the chamber change the refractive index of the surroundings of the tapered region. Thus, more light travels in the evanescent mode and the output power of the sensor are decreased, causing the changes to the output power of the tapered fiber [11]. In Figure 4, the output power of each tapered sample was recorded while varying the relative humidity from 35 % to 85% in the sealed chamber. The slope of the output power of each sample against relative humidity provides the sensitivity of the proposed sensor. Figure 4 also shows that the tapered fibers with diameter below 4 μm have greater sensitivity with peak sensitivity at 2.6 μm. The tapered fiber with diameters below 4 μm has larger sensitivity due to the larger effect of the evanescent mode that makes the sensor more susceptible to changes in its environment [12]. As the diameter of the fiber gets smaller with tapering, more light propagates in the evanescent mode outside the core. Consequently, smaller diameters are more susceptible to changes in the environment.

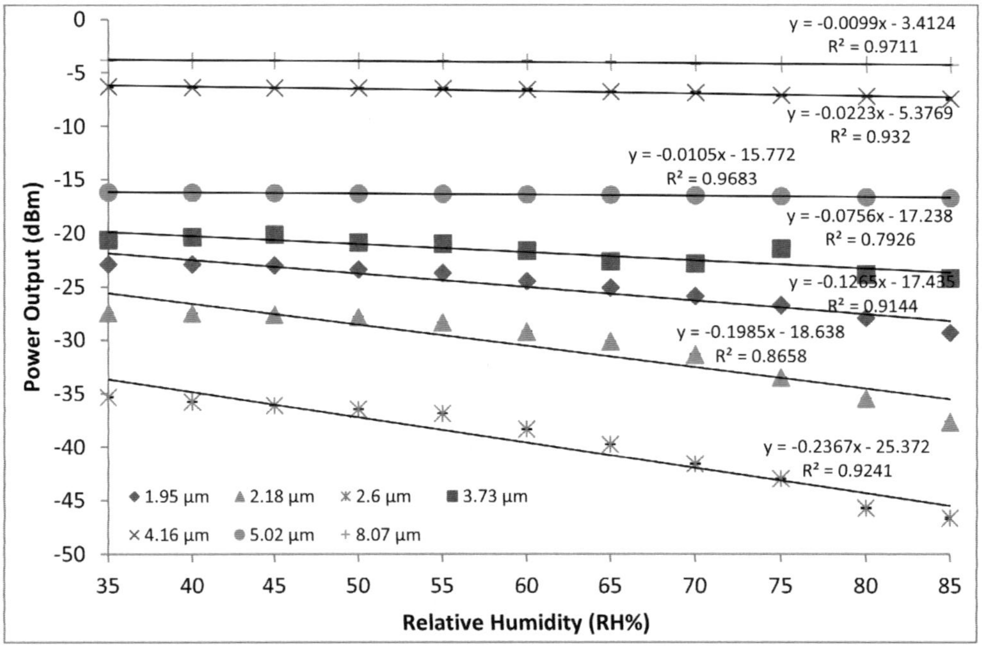

Figure 4. The output power of various tapering diameters with varying humidity.

The bar chart in Figure 5 depicts that the sensitivity increases with diameters below 4 μm. The highest sensitivity of 0.2367 dB/ %RH was achieved at 2.6 μm in diameter. Additionally, fibers with diameters below 4 μm performed much better than samples with larger diameters. This shows that the sensitivity of tapered fiber increases and it able to detect changes in the relative humidity.

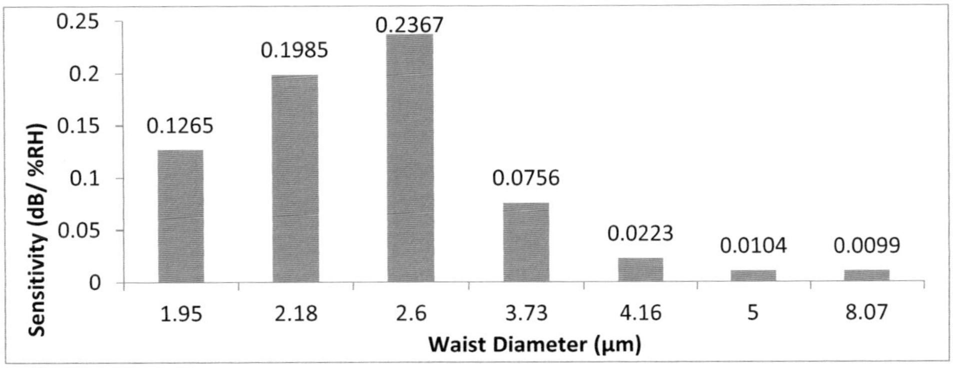

Figure 5. The sensitivity to relative humidity variation of various tapering diameters.

In Figure 6, 7, and 8, the output power of the samples for 2.6 μm tapered fibers before and after coating with PVA and PMMA is presented. For each figure, it can be observed that the output power of the sensor against the relative humidity was increasing gradually. The samples were tested three times in humidity ranges from 35-85%RH.

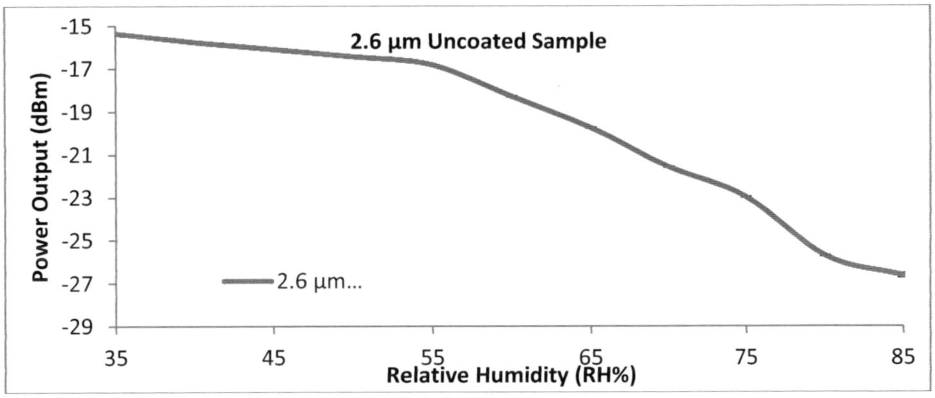

Figure 6. The output power of uncoated tapered fiber with 2.6 μm diameter with varying humidity.

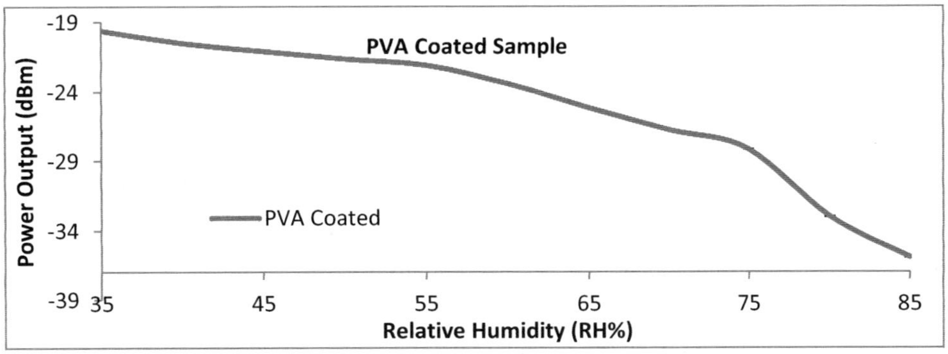

Figure 7. The output power of PVA coated tapered fiber with 2.6 μm diameter with varying humidity.

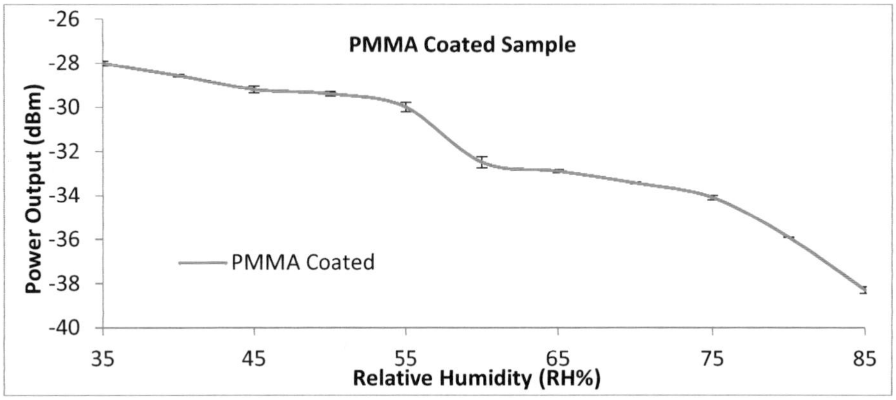

Figure 8. The output power of PMMA coated tapered fiber with 2.6 μm diameter with varying humidity.

From the graphs above, it can be concluded that the output power decreases with the change of humidity. The PVA coating increases the total power loss by more than 3dBm and PMMA coating introduces even a larger loss of more than 12 dBm.

The analysis of the three samples is presented in Figure 9. Firstly, using coating lowers the overall output power regardless of the relative humidity. This is due to the tension, weight and insertion loss of the coated fiber. Secondly, the PVA coated sample shows an increased sensitivity of 0.3023 dB/RH% compared to the bare sample's 0.2367 dB/RH%. This shows the effectiveness of PVA coating tapered fiber in sensing relative humidity. The PVA coated on the tapered fiber acts as a new cladding layer for the tapered region. The increasing humidity in the chamber changes the refractive index of the cladding. Thus, the loss of the tapered fiber increases with the changes in humidity. The PVA coating reacts more strongly to changes in the relative humidity than air so the sensitivity of PVA coated fiber is higher [13].

On the other hand, PMMA coating shows a lower sensitivity in detecting changes in relative humidity than both bare and PVA coated fiber, with a sensitivity of 0.1936dB/RH%. Therefore, it can be observed that PVA that is coated on the tapered fiber is the most effective approach to build relative humidity sensors compared to PMMA or bare fiber.

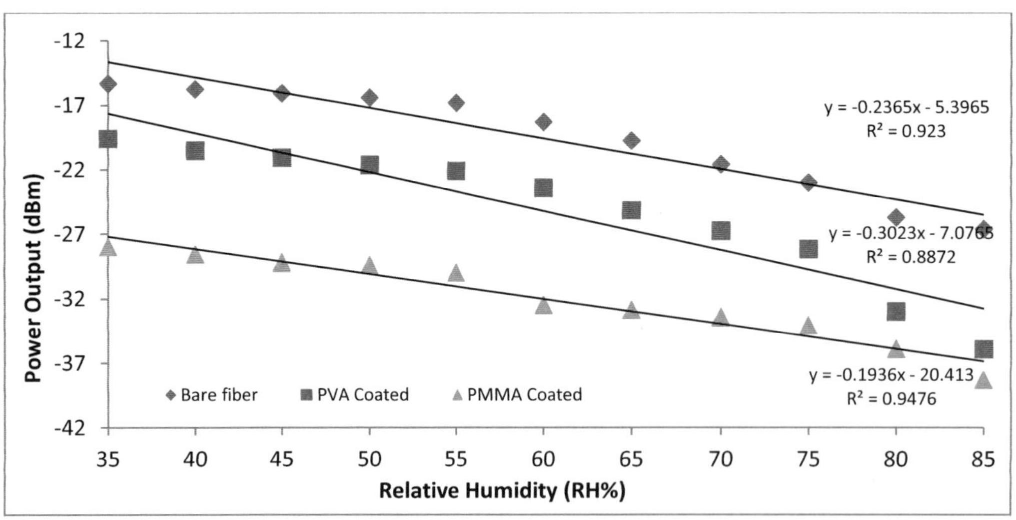

Figure 9. The transmitted power levels for tapered fiber with varying humidity.

Table 1 compares the performance of the three samples. The uncoated fiber has slightly higher linearity than the PVA and PMMA coated samples. All the methods have linearity larger than 90%, and the experiment was conducted with relative humidity between 35-85 %.

Resolution of the PVA coated sample shows the best response to relative humidity changes. The bare fiber, PVA coated, and PMMA coated have fine resolution below 1 %RH. The PMMA coated sample shows the least promising performance with a resolution of 0.5 %RH. PVA coating improves the resolution of the sensor by a factor of two. Finally, this shows that coating bare tapered fiber by PVA able to increase its sensitivity and resolution as a relative humidity sensor.

Table 1. Performance analysis of tapered fiber performance as a relative humidity sensor.

Diameter	2.6 µm	2.6 µm PVA	2.6 µm PMMA
Sensitivity (dBm/ %RH)	0.2367	0.3023	0.1936
Linearity (100%)	96.1300	94.19	97.34
Standard deviation (dBm)	0.0223	0.01734	0.1142
Resolution (%RH)	0.0942	0.0574	0.5899
Range (% RH)	35-85	35-85	35-85

The experiment was repeated three times for all the samples. From graphs 10, 11 and 12, the PVA coated sample shows good repeatability compared to the uncoated fiber and the PMMA coated fiber. This is due to the fact that the PVA coating is solidified around the tapered region providing additional support to its fragile structure. The fragile nature of uncoated tapered fiber more prone to bending and behaving unexpectedly.

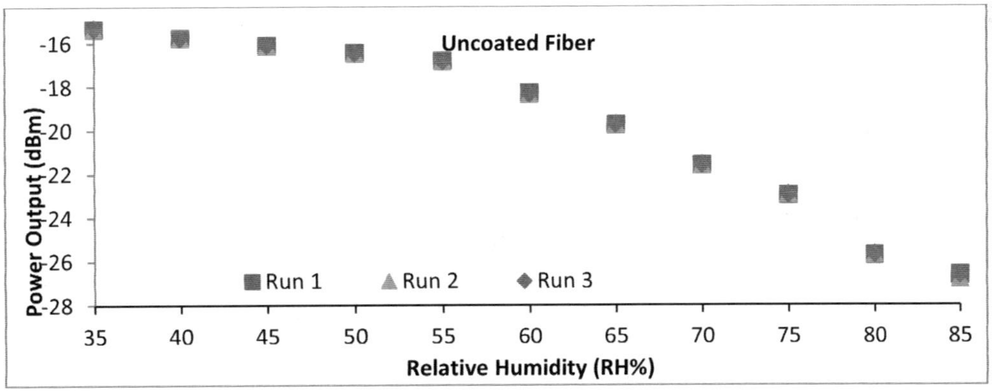

Figure 10. The output power of uncoated tapered fiber in three runs.

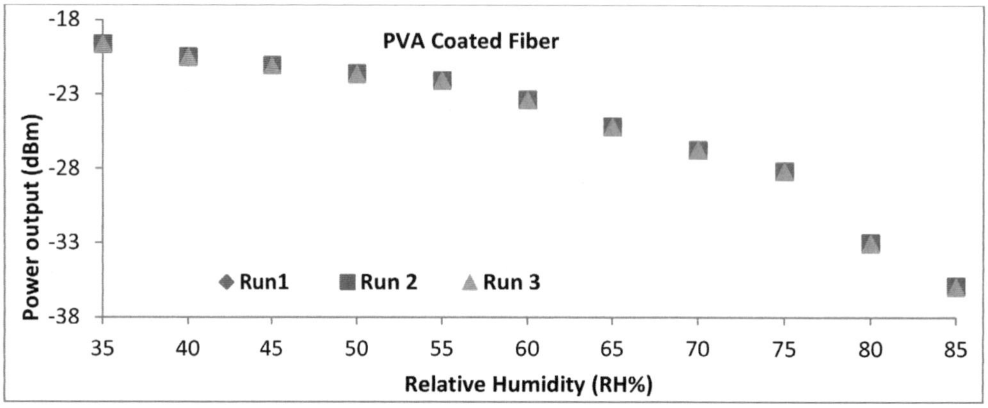

Figure 11. The output power of PVA coated tapered fiber in three runs.

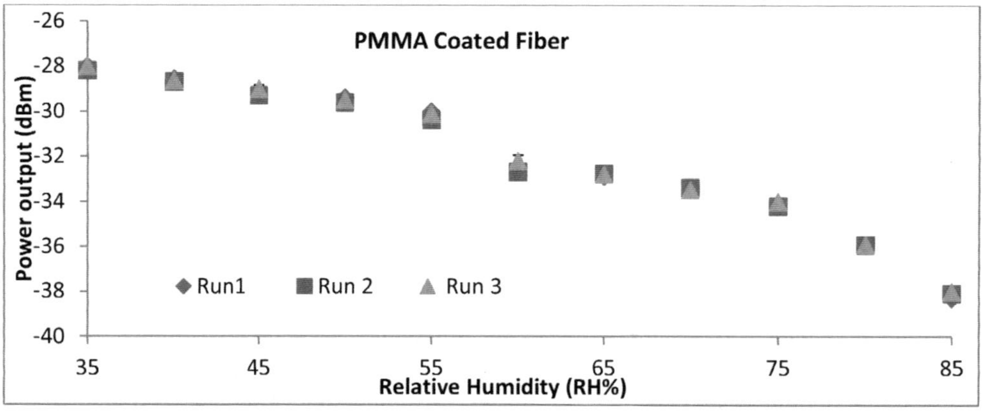

Figure 12. The output power of PMMA coated tapered fiber in three runs.

4. Conclusion

The accuracy of relative humidity measurements is vital to many industries. Current electronic relative humidity sensors are susceptible to electromagnetic interference and do not work well in volatile environments. Fiber optics based sensors provide a solution to these challenges. Due to their stability, small size, cost, and immunity to electromagnetic interference fiber optics cable can work as relative humidity sensors. In this proposed work, relative humidity sensors were fabricated using tapered fiber optics cable. The performance of several tapered samples of diameters ranging from 10 to 2 μm was tested where samples with diameter below 5 μm had a larger sensitivity in detecting changes in relative humidity and the sample of 2.6 μm is having the highest sensitivity of 0.2367 db/RH%. This sample size was fixed for testing the effect of PVA and PMMA coating. The results showed that PVA coating increases the sensitivity of the tapered sensors to 0.3023 dB/RH% from the bare fiber's sensitivity of 0.2367 db/RH%. However, the sample coated with PMMA showed the lowest sensitivity of 0.192 db/RH%. The experiment was repeated for another three times for all the samples. The PVA coated sample shows good repeatability compared to the uncoated fiber and the PMMA coated fiber. This is due to the fact that the PVA coating is solidified around the tapered region providing additional support to its fragile structure. The fragile nature of uncoated tapered fiber is more prone to bending and behaving unexpectedly.

Acknowledgement
The authors would like to acknowledge the University of Malaya and INTI International University for financial support.

References

[1] Chen J-H, Li D-R and Xu F 2019 Optical Microfiber Sensors: Sensing Mechanisms, and Recent Advances *Journal of Lightwave Technology* **37** 2577-89
[2] Yeo T, Sun T and Grattan K 2008 Fibre-optic sensor technologies for humidity and moisture measurement *Sensors and Actuators A: Physical* **144** 280-95
[3] Matias I R, Ikezawa S and Corres J 2016 *Fiber Optic Sensors: current status and future possibilities* vol 21: Springer)
[4] Ascorbe J, Corres J, Arregui F and Matias I 2017 Recent developments in fiber optics humidity sensors *Sensors* **17** 893
[5] Batumalay M, Harun S, Ahmad F, Nor R, Zulkepely N and Ahmad H 2014 Study of a fiber optic humidity sensor based on agarose gel *Journal of Modern Optics* **61** 244-8
[6] Lokman A, Nodehi S, Batumalay M, Arof H, Ahmad H and Harun S W 2014 Optical fiber humidity sensor based on a tapered fiber with hydroxyethylcellulose/polyvinylidenefluoride composite *Microwave and Optical Technology Letters* **56** 380-2
[7] Irawati N, Rahman H, Ahmad H and Harun S 2017 A PMMA microfiber loop resonator based humidity sensor with ZnO nanorods coating *Measurement* **99** 128-33
[8] Harun S W, Lim K, Tio C, Dimyati K and Ahmad H 2013 Theoretical analysis and fabrication of tapered fiber *Optik-International Journal for Light and Electron Optics* **124** 538-43
[9] Lim K, Ahmad H, Arof H and Harun S 2012 *Fabrication and applications of microfiber*: INTECH Open Access Publisher)
[10] Isa N M, Irawati N, Rahman H A, Yusoff M H M and Harun S W 2018 Polyaniline-Doped Poly (Methyl Methacrylate) Microfiber for Methanol Sensing *IEEE Sensors Journal* **18** 2801-6
[11] Peng Y, Zhao Y, Chen M Q and Xia F 2018 Research advances in microfiber humidity sensors *Small* **14** 1800524
[12] Jali M H, Rahim H R A, Johari M A M, Hamid S S, Yusof H H M, Thokchom S, Wang P and Harun S W 2019 Optical characterization of different waist diameter on microfiber loop resonator humidity sensor *Sensors and Actuators A: Physical* **285** 200-9
[13] Jiao S-x, Zhao Y and Gu J-j 2018 Simultaneous measurement of humidity and temperature using a polyvinyl alcohol tapered fiber bragg grating *Instrumentation Science & Technology* **46** 463-74

Q-switched erbium-doped fiber lasers based on copper nanoparticles saturable absorber

E I Ismail[1], F Ahmad[1], S Ambran[1], A A Latiff[2] and S W Harun[3]

[1]Malaysia-Japan International Institute of Technology, Universiti Teknologi Malaysia, Jalan Sultan Yahya Petra, 54100 Kuala Lumpur, Malaysia
[2]Centre for Telecommunication Research & Innovation (CeTRI), Fakulti Kejuruteraan Elektronik & Kejuruteraan Komputer (FKEKK), Universiti Teknikal Malaysia Melaka (UTeM),76100 Durian Tunggal, Melaka, Malaysia
[3]Photonics Engineering Laboratory, Department of Electrical Engineering, University of Malaya, 50603 Kuala Lumpur, Malaysia

ezzatulirradah5@gmail.com; fauzan.kl@utm.my

Abstract. We reported a generation of passively Q-switching pulses in an erbium-doped fiber laser by exploiting copper nanoparticles polyvinyl alcohol polymer composite (CuNP-PVA) based saturable absorber (SA). The modulation depth and the saturation power intensity of the CuNP-PVA film are calculated as 18 % and 0.021 kW/cm^2, respectively. A maximum repetition rate of 104.2 kHz, the shortest pulse width of 5.1 μs and pulse energy up to 20.4 nJ was observed at 1558 nm wavelength.

1. Introduction

The evolution of 1-, 2- and 3-Dimensional material in the field of material sciences has allowed the realization of a multitude of new devices, called saturable absorber (SAs). These SAs are the key to generate passively Q-switched fiber lasers. Passively Q-switched fiber lasers have been extensively used in optical communications, range finding and material processing due to its flexible and compact advantages [1,2]. Essentially, Q-switched laser has its convenience especially in terms of relatively high pulse energy and tunable repetition rate. Originally, saturable absorber begins with semiconductor saturable absorber mirrors (SESAMs) to generate pulsed outputs in laser cavities [3]. However, due to their complex procedure, limited bandwidth and expensive costs lead the researchers to explore material which acquires high laser damage threshold, good fiber compatibility and low optical saturable intensity [4-8]. Two-dimensional (2D) layered materials including graphene and transition metal dichalcogenides (TMDCs) have been employed as SA considering their simplicity of manufacture and inexpensive cost contrasting to SESAMs [9-12]. Graphene normally has a low optical saturation which is 2.3% per layer and TMDCs are progressively available in the visible region because of their substantial bandgaps. Still, 2D materials in specific, are acknowledged as a strong prospect for the cutting-edge photonics innovation in light of their ultrafast carrier elements and wideband responses [13,14]. Next, black phosphorus (BP) has been regained as one of the next SA for optoelectronic applications [15]. However, they are also vulnerable to high laser optical damage threshold and easily oxidized to oxygen and moisture environment.

Quite recently, metal nanomaterial such as silver and copper attract many researchers as the next generation of SAs as it has high potential in the optic application. Copper especially has broad saturable absorption, large third-order nonlinearity and ultrafast response time [16,17]. Muhammad et al. proposed a Q-switched fiber laser with copper-PVA at the 1.5 wavelength region. The laser operated at a repetition rate and pulse width of 101.2 kHz and 4.28 μs respectively, while the maximum average output power was obtained at 1.86 mW [18]. Sinan et al. reported Q-switch pulse using copper oxide-

PVA at 1560 nm with a repetition rate of 83 kHz, pulse width 2.6 μs of and the maximum pulse energy of 66 nJ [19].

In the present work, a passively Q-switched EDFL was successfully realized at C-band region by incorporating a copper nanoparticles polymer composite based SA into a ring fiber laser cavity. The modulation depth and the saturation power intensity of the CuNP-PVA film are 18% and 0.021 kW/cm^2 respectively. The laser generates a Q-switching pulse train with the shortest pulse width of 5.1 μs, a repetition rate of 104.2 kHz and an output power of 2.1 mW at the maximum input pump power of 198.0 mW.

2. Preparation and optical characterization of the CuNP-PVA based SA.

The copper nanoparticles PVA film fabricated in this work is utilized as SA to generate laser pulses in erbium-doped fiber laser (EDFL). Commercially available copper nanoparticle powder from Sigma Aldrich are used to fabricate the SA. The copper nanoparticles are obtained in a powder consist of an average of 25 nm particles size. The film based SA is prepared by first dissolving 25 mg of copper nanoparticles powder in tetrahydrofuran (THF) and undergone ultrasonic bath for 6h to ensure the dispersion of the copper nanoparticles. Apart from that, the polymer host is prepared by dissolving 1g of PVA powder into 120 ml DI water. The mixture is stirred for 6 hours to ensure the homogeneous solution at 500 rpm at 145 °C. Then, 10 ml of the dispersed copper nanoparticles in THF solution is added into approximately 10 ml of the dissolved PVA solution to form a CuNP-PVA composite solution with a composition ratio of 1:1. The prepared CuNP-PVA composite solution is stirred for another 1 hour at room temperature to ensure the good binding of the composite as well as to allow any excess solvent to evaporate. Finally, about 20% of the resulting CuNP-PVA composite solution is taken for analysis while the remainder is poured into a petri dish and dried in room temperature for 3 days to produce the desired free-standing film of CuNP-PVA film.

Figure 1(a) shows the transmission electron microscope (TEM) of copper nanoparticles a 50 nm scale bar. Figure 1(b) demonstrates the nonlinear transmission profile of twin detector measurement. According to data fitting as shown in the figure below, the modulation depth and the saturation power intensity of the CuNP-PVA film calculated as 18% and 0.021 kW/cm^2, respectively. The measured modulation depth of CuNP-PVA thin film is comparable to that of other materials such as Cu-PVA at 36 % [18], CuO-PVA at 3.5 % [19] and silver nanoparticles-PVA at 19% [20].

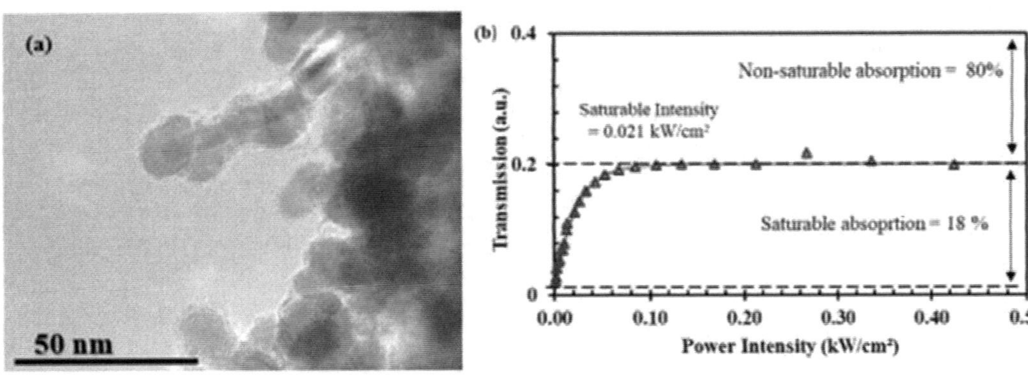

Figure 1. Characteristics of CuNP-PVA film (a) TEM image (b) nonlinear transmission

3. Experimental setup

Figure 2 illustrates the experimental set-up of the proposed passively Q-switched EDFL with the CuNP-PVA thin film. The cavity is composed of a 975 nm laser diode (LD) as the pump sources and is

Journal of Physics: Conference Series **1371** (2019) 012028 doi:10.1088/1742-6596/1371/1/012028

associated with a 980/1550 nm wavelength division multiplexer (WDM) port. The normal port of the WDM is currently connected to a 3.0 m long erbium-doped fiber (EDF). The EDF has a numerical aperture (NA) of 0.16 and erbium ion absorption of 23 dB/m at 980 nm with a core and cladding diameters of 4 μm and 125 μm respectively. To secure unidirectional propagation of light in the ring cavity, a polarization-independent isolator was used. Then, the end of the isolator is associated with a fiber-based 90:10 optical coupler, which is utilized to remove roughly 90% of the proliferating signal for examination. The 90% port of the coupler then again is associated with an integration of the SA between the two fiber ferrules. At last, the end of the SA is connected back with the 1550 nm port of the WDM. The end coupler of 10% port is associated with another 3 dB optical coupler to monitor simultaneously the results of the Q-switching performance. With one end of 50%, the coupler is associated with a Yokogawa AQ6370B optical range analyzer (OSA) for optical spectrum range. The opposite end of the coupler is associated with an InGaAs photodetector (1.3 GHz Thorlabs DET10D/M) coupled with 350 MHz oscilloscope (OSC, GW Instek GDS-3352), and 7.8 GHz Radio Frequency spectrum analyzer (RFSA, Anritsu MS2683A) and a Thorlabs control power meter for the estimation of the output power.

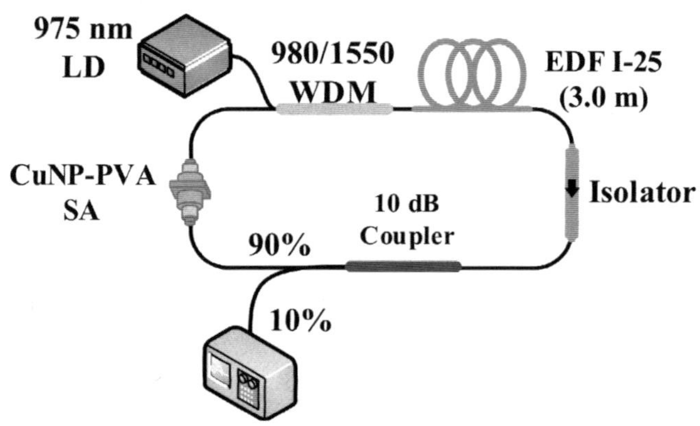

Figure 2. Schematic of the Q-switched fiber laser utilizing CuNP-PVA SA.

4. Result and discussion

The laser operates within the continuous wave laser (CW) at 32.76 mW input pump power. After the integration of the SA in the cavity, Q-switched operation is realized at the increasing input pump power of 83.61 mW up to the maximum input pump power of 198.0 mW. The relatively high threshold for the CW operation indicated too high overall cavity loss and high insertion loss of the SA device. The SA presented intracavity loss modulation and transformed the working from CW to Q-switching operation. When the pump power is increasing to 83.61 mW, a self-starting stable Q-switch pulse is observed with a repetition rate of 52.74 kHz and a pulse-width of 9.44 μs. It remains noticeable until the pump power extent to 198.0 mW. The output spectrum of the Q-switched laser operates at a wavelength of 1558 nm is shown in Figure 3.

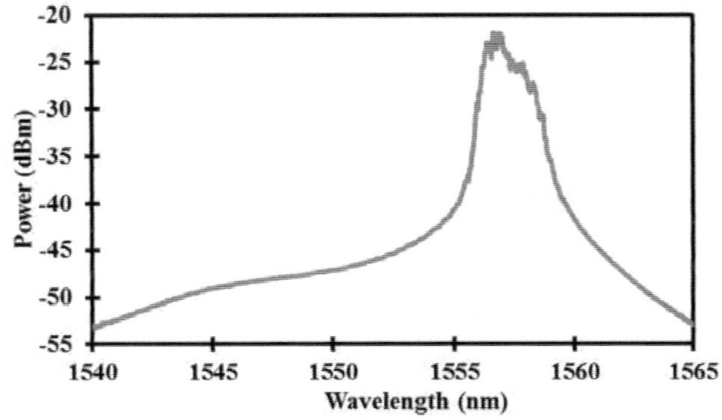

Figure 3. Output Spectrum of the Q-switched pulse at a maximum pump power of 198.0 mW.

Figure 4(a) indicates an oscilloscope trace of the Q-switched pulse at the maximum LD pump power. It can be seen that the pulses trains become narrower as the pump power increases. The time interval between the adjacent pulses was about 14.3 µs, which corresponds to 104.2 kHz pulse repetition rate. The pulse envelope full width at half-maximum (FWHM) value of 5.1 µs. It is also observed there is no trace of mode-locking operation during the Q-switching process.

The stability of the generated Q-switched output is confirmed by monitoring the pulse in the radio frequency spectrum analyser at the maximum LD pump power. Figure 4(b) shows the stability of the Q-switched pulse recorded in RFSA. The highest repetition rate of 104.2 kHz is measured with a high signal to noise ratio (SNR) of about 54 dB at the maximum input pump power of 198.0 mW. The RF spectrum also showed no frequency components other than the multiples of the fundamental frequency which indicates excellent stability of the oscillating Q-switched pulses in the laser cavity.

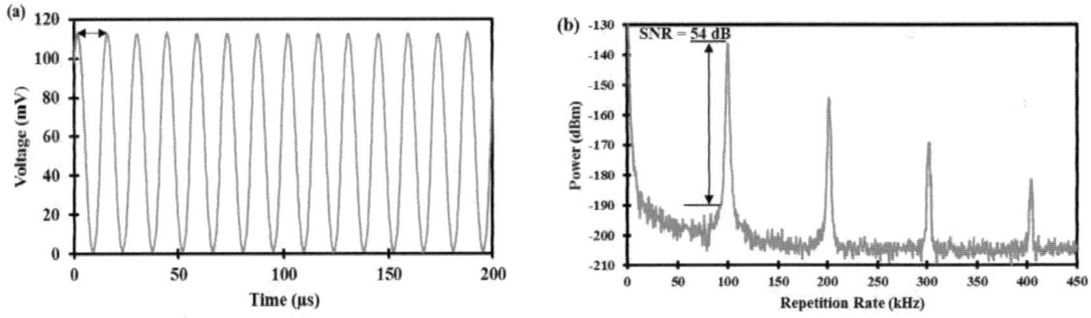

(a) (b)

Figure 4. Typical output pulse characteristics (a) oscilloscope trace of the pulse train (b) radio frequency spectrum of the proposed Q-switched fiber laser at the maximum input pump power of 198.0 mW.

Figure 5(a) illustrates the Q-switching performance of calculated pulse energy and output power of the pulse against the input pump power. As expected, both pulse energy and output power show an increasing trend against the input pump power. Over the input pump power range of 83.61–198.0 mW, the output power of the Q-switched laser is seen to increase from 0.74 mW to 2.13 mW. Similarly, the pulse energy also increases from 14.12 nJ to 20.44 nJ over the same power range, although at a pump power above 153.53 mW, the pulse energy is seen to decrease slightly. This is attributed to insufficient time for the CuNP based SA to fully recover from saturation at the high pump power. The highest pulse energy calculated was 20.44 nJ obtained at a maximum input pump power of 198.0 mW.

The relationship between the repetition rate and pulse width is shown in Figure 5(b). It can be seen the detailed evolution of the repetition rate increased steadily from 52.74 kHz to 104.2 kHz from input pump power of 83.61 mW to 198.0 mW, while the pulse width, on the other hand, decreases from 9.44 to 5.08 μs. This is the trendline behaviour for passively Q-switching operation where when the increasing input pump power is almost linearly with increasing repetition rate and decreasing the pulse width.

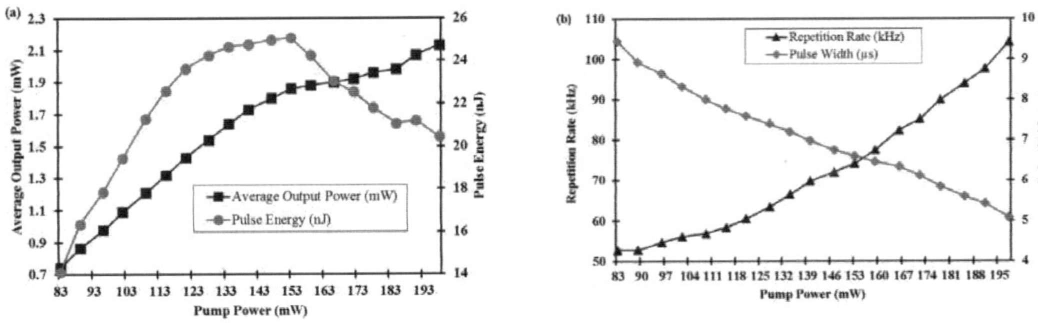

Figure 5. Q-switching performances (a) Average output power and output pulse energy against power (b) Repetition rate and pulse width against pump power

5. Conclusion

In this work, a passively Q-switched EDFL is proposed and demonstrated. The system employs a CuNP-PVA SA in the ring cavity and provides the desired pulse output. The system is capable of generating pulsed outputs with a maximum repetition rate of 104.2 kHz, the narrowest pulse width of 5.1μs and pulse energy of 20.44 nJ at the maximum input pump power of 198.0 mW. The RF spectrum showed the signal noise to a ratio of about 54 dB.

References

[1] Fermann, Martin E., and Ingmar Hartl. "Ultrafast fibre lasers." Nature photonics 7.11 (2013): 868.

[2] Lee, Junsu, Minwan Jung, Joonhoi Koo, Cheolhwan Chi, and Ju Han Lee. "Passively Q-Switched 1.89-μm fiber laser using a bulk-structured Bi 2 Te 3 topological insulator." IEEE Journal of Selected Topics in Quantum Electronics 21, no. 1 (2014): 31-36.

[3] Paschotta R, Häring R, Gini E, Melchior H, Keller U, Offerhaus HL, Richardson DJ. Passively Q-switched 0.1-mJ fiber laser system at 1.53? m. Optics letters. 1999 Mar 15;24(6):388-90.

[4] Keller, Ursula. "Recent developments in compact ultrafast lasers." nature 424.6950 (2003): 831.

[5] Dvoyrin, Vladislav V., V. M. Mashinsky, and E. M. Dianov. "Yb-Bi pulsed fiber lasers." Optics letters 32.5 (2007): 451-453.

[6] Kurkov, A. S., E. M. Sholokhov, and O. I. Medvedkov. "All fiber Yb-Ho pulsed laser." Laser Physics Letters 6, no. 2 (2008): 135.

[7] Kurkov, A. S., Ya E. Sadovnikova, A. V. Marakulin, and E. M. Sholokhov. "All fiber Er-Tm Q-switched laser." Laser Physics Letters 7, no. 11 (2010): 795.

[8] Sholokhov, E. M., A. V. Marakulin, A. S. Kurkov, and V. B. Tsvetkov. "All-fiber Q-switched holmium laser." Laser Physics Letters 8, no. 5 (2011): 382.

[9] Woodward, Robert, and Edmund Kelleher. "2D saturable absorbers for fibre lasers." Applied Sciences 5.4 (2015): 1440-1456.

[10] Set, Sze Y., Hiroshi Yaguchi, Yuichi Tanaka, and Mark Jablonski. "Ultrafast fiber pulsed lasers

incorporating carbon nanotubes." IEEE Journal of selected topics in quantum electronics 10, no. 1 (2004): 137-146.

[11] Novoselov, Kostya S., Andre K. Geim, Sergei V. Morozov, D. Jiang, Y_ Zhang, Sergey V. Dubonos, Irina V. Grigorieva, and Alexandr A. Firsov. "Electric field effect in atomically thin carbon films." science 306, no. 5696 (2004): 666-669.

[12] Geim, Andre Konstantin. "Graphene: status and prospects." science 324.5934 (2009): 1530-1534.

[13] Ayari, Anthony, Enrique Cobas, Ololade Ogundadegbe, and Michael S. Fuhrer. "Realization and electrical characterization of ultrathin crystals of layered transition-metal dichalcogenides." Journal of applied physics 101, no. 1 (2007): 014507.

[14] Wang, Kangpeng, Jun Wang, Jintai Fan, Mustafa Lotya, Arlene O'Neill, Daniel Fox, Yanyan Feng et al. "Ultrafast saturable absorption of two-dimensional MoS2 nanosheets." ACS nano 7, no. 10 (2013): 9260-9267. Phys. A 71,593–596 (2000).

[15] Lu, S. B., L. L. Miao, Z. N. Guo, X. Qi, C. J. Zhao, H. Zhang, S. C. Wen, D. Y. Tang, and D. Y. Fan. "Broadband nonlinear optical response in multi-layer black phosphorus: an emerging infrared and mid-infrared optical material." Optics express 23, no. 9 (2015): 11183-11194.

[16] Guo, Hao, Ming Feng, Feng Song, Haoyu Li, Aibing Ren, Xukang Wei, Yigang Li, Xiaoxuan Xu, and Jianguo Tian. "$ Q $-Switched Erbium-Doped Fiber Laser Based on Silver Nanoparticles as a Saturable Absorber." IEEE Photonics Technology Letters 28, no. 2 (2015): 135-138.

[17] Glubokov, Dmitrii Anatol'evich, Vladimir Vasil'evich Sychev, Aleksandr Sergeevich Mikhailov, Andrei Evgen'evich Korolkov, Dmitrii Anatol'evich Chubich, Boris Isaakovich Shapiro, and Alexey Grigor'evich Vitukhnovskii. "Saturable absorber based on silver nanoparticles for passively mode-locked lasers." Quantum Electronics 44, no. 4 (2014): 314.

[18] Muhammad, A. R., M. T. Ahmad, R. Zakaria, H. R. A. Rahim, S. F. A. Z. Yusoff, K. S. Hamdan, H. H. M. Yusof, H. Arof, and S. W. Harun. "Q-switching pulse operation in 1.5-μm region using copper nanoparticles as saturable absorber." Chinese Physics Letters 34, no. 3 (2017): 034205.

[19] Sadeq, Sinan Abdulhameed, Sarah Kadhim Al-Hayali, Sulaiman Wadi Harun, and Abdulhadi Al-Janabi. "Copper oxide nanomaterial saturable absorber as a new passive Q-switcher in erbium-doped fiber laser ring cavity configuration." Results in Physics 10 (2018): 264-269.

[20] Lokman, Muhammad Quisar, Siti Fatimah Az Zahra Yusoff, Fauzan Ahmad, Rozalina Zakaria, Hafizal Yahaya, Suhaidi Shafie, Rizuan Mohd Rosnan, and Sulaiman Wadi Harun. "Deposition of silver nanoparticles on polyvinyl alcohol film using electron beam evaporation and its application as a passive saturable absorber." Results in Physics 11 (2018): 232-236.

Microfibre deposited NiS_2 for all-fibre optical thresholding

Shi Li[1], Yu Yin[1], and Pengfei Wang[1,2]

[1]Key Laboratory of In-fibre Integrated Optics of Ministry of Education, College of Physics and Optoelectronic Engineering, Harbin Engineering University, Harbin 150001, China
[2]Key Laboratory of Optoelectronic Devices and Systems of Ministry of Education and Guangdong Province, College of Optoelectronic Engineering, Shenzhen University, Shenzhen, 518060, China

m18204313488@163.com; pengfei.wang@dit.ie

Abstract. We demonstrate a microfibre deposited NiS_2 for optical thresholding. By employing the fabricated NiS_2 device, an all-optical thresholding function based on the saturable absorption effect has been successfully realized. The signal-to-noise ratio of the input pulse signal can be improved from 2.7 dB to 14.5 dB. The results achieved indicate that this NiS_2-based microfibre device could be operated as an excellent nonlinear photonics device for all-optical signal processing.

1. Introduction

Saturable absorption (SA) effect has been widely used in nonlinear photonics and ultrafast optics. The excellent performance of SA effect results in that "high-intensity light" passes through and "low-intensity light" is efficiently absorbed [1]. The unique performance of the SA effect promotes numerous SA device found and investigated. The mostly investigated method to achieve SA effect is using real saturable absorber materials. Recently, numerous materials with saturable absorption capacity have been developed including graphene [2], black phosphorus [3] and transition metal dichalcogenides (TMDs) [4] etc.

Graphene is the most investigated materials for ultrafast photonics and optical modulators. In addition, due to its saturable absorption capacity and optical nonlinearity, the graphene-based device has been developed as loss and phase modulator, respectively [2]. However, graphene has drawbacks of weak absorption, complex fabrication and environmental instability, which weaken its modulation ability a lot [2]. Compared to graphene, TMDs has the non-zero bandgap and are more flexible. Furthermore, TMDs have been investigated and confirmed with nonlinear optical absorption, which indicates that TMDs can be used for ultrafast pulse operation and optical switching. WS_2 [4], SnS_2 [5], TiS_2 [1], three types of TMDs have been widely investigated in the fibre laser applications and optical thresholding. Research on the application of TMDs to achieve mode-locked pulse operation and optical modulation remains in progress. Hence, it is of great interest to explore new TMDs for ultrafast photonics and optical modulation.

Nickel disulfide (NiS_2), another new type of TMD, has attracted significant attention in recent years owing to its superior electrical and optoelectronic properties [6-10]. NiS_2 is a semiconductor, similar to WS_2 and MoS_2. Compared with typical large bandgap TMDs (above 1 eV), the bandgap of NiS_2 is 0.3 eV [11]. However, the nonlinear optical properties of NiS_2 and its application in fibre laser or optical thresholding have seldom been reported. Thus, it is a very valuable work to investigate the applications of NiS_2.

In this paper, a microfibre deposited NiS_2 has been investigated to achieve all-optical thresholding. The signal-to-noise ratio (SNR) of the input pulse signal can be improved from 2.7 dB to dB 14.5 with long-term stability, which indicates the NiS_2 device can be used as a high-quality all-optical thresholding device to suppress the noise and improve the SNR. Our results suggest that NiS_2-deposited SA could be used as an effective highly nonlinear photonic device for optical communication.

2. Experiment and results

In order to measure whether the microfibre deposited NiS_2 could be used for all-optical thresholding, the experimental setup was proposed and demonstrated as shown in Fig. 1(a). The pulse laser was a 10 MHz mode-locked fibre laser with 800 fs pulse duration at 1550.3 nm centre wavelength. The noise source is centred at 1550 nm with a 30 MHz radio frequency modulated low-power. The lights came from the noise and pulse sources were coupled into an optical coupler (50:50). The output of the three points marked as N, P and I in this figure corresponded to the three inset oscillograms. The variable optical attenuator (VOA) was adopted to modulate the input signal incident to the sample. Another optical coupler (10:90) was used to separate the light into two different segments: 10% port connected with an optical power meter to measure the output power, which can confirm the total output power; the 90% port connected with the NiS_2-based microfibre SA device. The microfiber deposited NiS_2 was shown in the inset of Fig. 1(a). The diameter of the microfibre is about 3 μm.

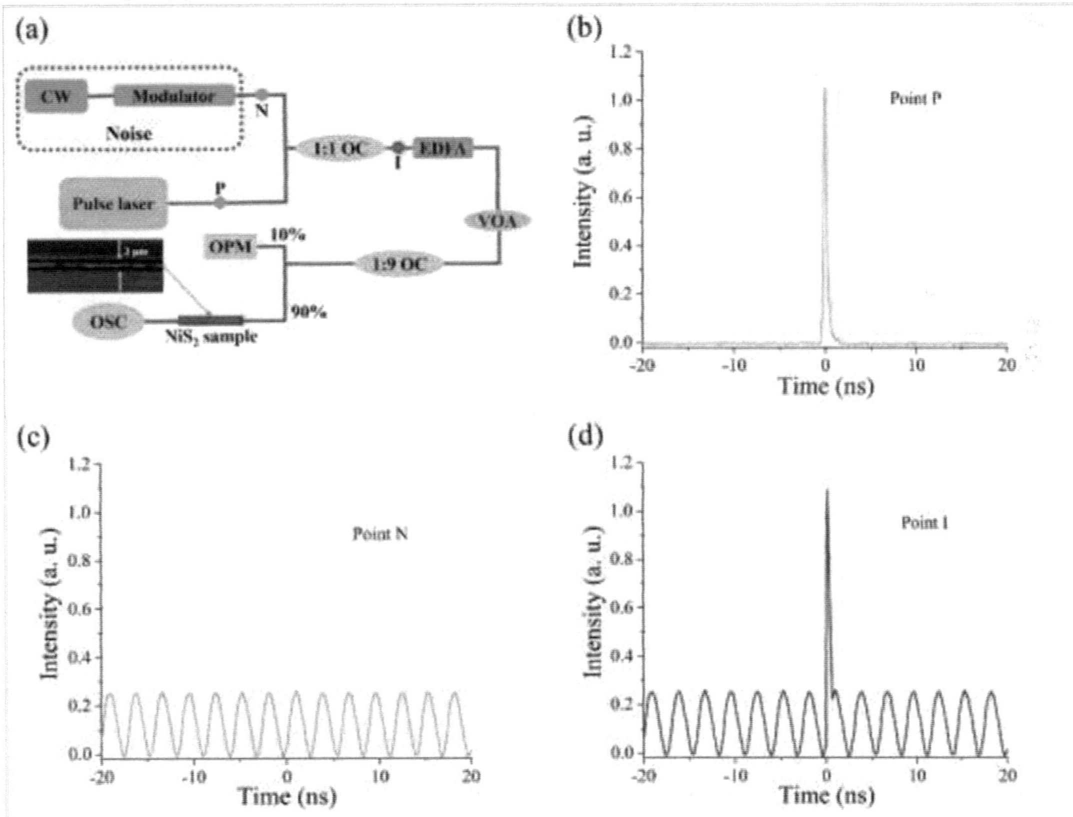

Fig. 1. (a) Schematic diagram of the experiment setup for all-optical thresholding with NiS_2-coated microfibre. (b) The waveform corresponding to the point marked with P. (c) The waveform corresponding to the point marked with N. (d) The waveform corresponding to the point marked with I.

By adjusting the VOA, a pulse with notably suppressed noise was successfully obtained and shown in Fig. 2. Fig. 2(a) and 2(b) show the pulse waveform before and after the introduction of this nonlinear photonics device when the VOA value set as 17.5 dB. The SNR of the oscillograph was 2.7 dB before the sample and 14.5 dB after the sample. The SNR after the sample is significantly much higher than before. Meanwhile, the noise is significantly suppressed. The long-term stability of the NiS_2 as an optical thresholding device was also verified, as shown in Fig. 2(c). The pulse SNR remains unchanged for 8 h, which strongly suggests the excellent stability of this device. Fig. 2(d) recorded the corresponding SNR with the increasing value of VOA. The SNR increased from 4 dB to 14.5 dB with the VOA value from 6 dB to 17.5 dB. It is also clear that before the NiS_2 sample, the SNR has no change with the increasing VOA value. Our experiments indicate that NiS_2 is suitable for all-optical thresholding devices.

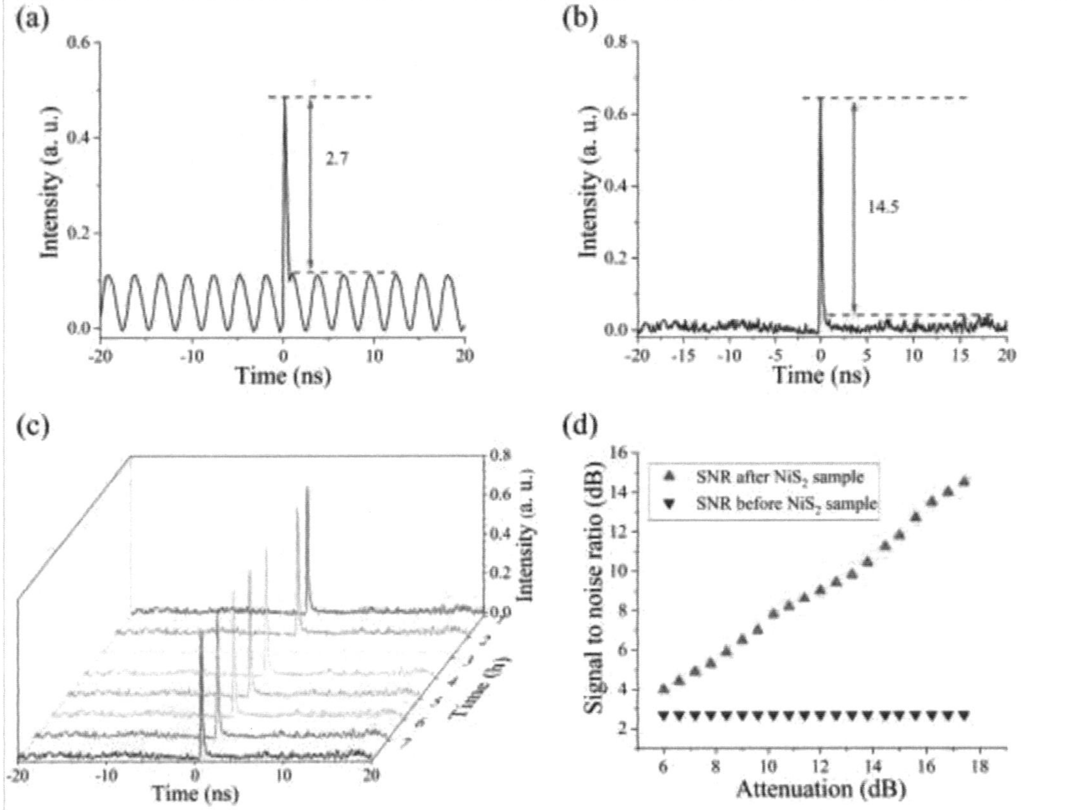

Fig. 2. Experimental results of the all-optical thresholding. a) The optical pulse waveform before the sample. b) The optical pulse waveform after the sample. c) The long-term stability while the output of EDFA is kept at 17.5 dBm for 8 h. d) The corresponding SNR against the value of VOA.

3. Conclusion

In conclusion, a microfibre deposited NiS_2 was successfully demonstrated for all-optical thresholding. The signal-to-noise ratio of the input pulse signal can be improved from 2.7 dB to 14.5 dB. The achieved results indicate that this NiS_2-based microfibre device could be operated as an excellent nonlinear photonics device for all-optical signal processing.

Acknowledgement

National Key R&D Program of China (2016YFE0126500); National Natural Science Foundation of China (NSFC) (61575050); The fundamental research funds for the central universities (HEUCFG201841); Key Program for Natural Science Foundation of Heilongjiang Province of China

(ZD2016012); Open Fund of the State Key Laboratory on Integrated Optoelectronics (IOSKL2016KF03); the 111 project to the Harbin Engineering University (B13015).

References

1. Y. Ge, Z. Zhu, Y. Xu, Y. Chen, S. Chen, Z. Liang, Y. Song, Y. Zou, H. Zeng, and S. Xu 2018 Adv. Opt. Mater. **6** 1701166.
2. Z. Sun, T. Hasan, F. Torrisi, D. Popa, G. Privitera, F. Wang, F. Bonaccorso, D. M. Basko, and A. C. Ferrari 2010 ACS nano **4** 803-810.
3. S. Li, Y. Yin, E. Lewis, G. Garrell, and P. Wang 2019 Opt. Commun. **437** 342-345.
4. S. Li, Y. Yi, Y. Yin, Y. Jiang, H. Zhao, Y. Du, Y. Chen, E. Lewis, G. Farrell, S. W. Harun and P. Wang 2018 J. Lightwave Technol. **36** 5633-5639.
5. K. Niu, R. Sun, Q. Chen, B. Man, and H. Zhang 2018 Photonics Res. **6** 72-76.
6. J. Honig, and J. Spałek 1998 Chem. Mater. **10** 2910-2929.
7. G. Krill, M. Lapierre, C. Robert, F. Gautier, G. Czjzek, J. Fink, and H. Schmidt 1976 J. Phys. C Solid State Phys. **9** 761.
8. J. Osuwa, and P. Uwaezi 2011 Chalcog. Lett. **8** 587-594.
9. K. Sato 1984 J. Phys. Soc. Japan **53** 1617-1620.
10. R. Kautz, M. Dresselhaus, D. Adler, and A. Linz 1972 Phys. Rev. B **6** 2078.
11. T. Anand, K. Rajes, and M. Zaidan 2013 Rep. Electrochem **3** 25-29.

AUTHOR INDEX